Ursula Weber, Herbert Thiele

NMR Spectroscopy: Modern Spectral Analysis

WILEY-VCH

Spectroscopic Techniques:
An Interactive Course

Pretsch/Clerc
Spectra Interpretation of Organic Compounds
1997. ISBN 3-527-28826-0

Bigler
NMR Spectroscopy: Processing Strategies
1997. ISBN 3-527-28812-0

Weber/Thiele
NMR Spectroscopy: Modern Spectral Analysis
1998. ISBN 3-527-28828-7

In Preparation:

Jenny
NMR Spectroscopy: **Data Acquisition**

Fröhlich/Thiele
NMR Spectroscopy: **Intelligent Data Management**

Ursula Weber,
Herbert Thiele

NMR Spectroscopy: Modern Spectral Analysis

WILEY-VCH

Weinheim · New York · Chichester · Brisbane · Singapore · Toronto

Dr. Ursula Weber,
Dr. Herbert Thiele
Bruker-Franzen Analytik GmbH
Fahrenheitstr. 4
D-28359 Bremen
Germany

A CD-ROM containing teaching versions of several WIN-NMR programs (© Bruker Analytik GmbH) is included with this book. Readers can obtain further information on this software by contacting:
Bruker Analytik GmbH, Silberstreifen, D-76287 Rheinstetten, Germany.

> This book and CD-ROM was carefully produced. Nevertheless, author and publisher do not warrant the information contained therein to be free of errors. Readers are advised to keep mind that statements, data, illustrations, procedural details or other items may inadvertently be inaccurate.

Library of Congress Card No. applied for
A catalogue record for this book is available from the British Library

Die Deutsche Bibliothek – CIP-Einheitsaufnahme
NMR spectroscopy: modern spectral analysis / Ursula Weber ; Herbert Thiele.
- Weinheim ; New York ; Chichester ; Brisbane ; Singapore ; Toronto : WILEY-VCH
  (Spectroscopic techniques)
  ISBN 3-527-28828-7

Buch. 1998
Gb.

CD-ROM. 1998

© WILEY-VCH Verlag GmbH, D-69469 Weinheim (Federal Republic of Germany), 1998,
ISBN 3-527-28828-7

Printed on acid-free and low chlorine paper

All rights reserved (including those of translation into other languages). No part of this book may be reproduced in anyform – by photoprinting, microfilm, or any other means – nor transmitted or translated into a machine language without written permission from the publishers. Registered names, trademarks, etc. used in this book, even when not specifically marked as such, are not to be considered unprotected by law.
Composition: Kühn & Weyh, D-79111 Freiburg
Printing: Betzdruck GmbH, D-64291 Darmstadt
Bookbinding: Wilhelm Osswald & Co., D-67933 Neustadt

Printed in the Federal Republic of Germany

Preface

In recent years Nuclear Magnetic Resonance spectroscopy has become one of the most popular and elegant methods for the elucidation of chemical structures. The rapid development in FOURIER transform techniques such as n-dimensional and imaging experiments has established NMR spectroscopy as the primary analytical method, not only in chemistry but also in the realms of biology, physics, medicine and related sciences.

The masses of raw data *acquired* with the NMR spectrometer need to be *processed, analyzed, interpreted* and *archived* for use in future structural problems.

Action	Tool	Result	Volume
Acquisition	⇨	FID	2
⬇		⬍	
Processing	⇨	Spectrum	1
⬇		⬍	
Analysis	⇨	Parameter	3
⬇		⬍	
Interpretation	⇨	Structure	1,3,4
⬇		⬍	
Archiving	⇨	Knowledge	4

Fig. 1: Flow scheme of data preparation in NMR spectroscopy.

In Fig. 1, the vertical black arrows (⬇) on the left hand side symbolize the steps required in the analysis to obtain the information listed on the same row in the *Result* column, e.g. to obtain a spectrum (*Result*), it is first necessary to acquire and process (*Action*) the raw data. The horizontal white arrows (⇨) symbolize the software tools (*Tool*) required to obtain the desired *Result*. The strategy necessary to successfully perform the appropriate *Action* is described in the corresponding *Volume* in the series **Spectroscopic Techniques: An Interactive Course**. Thus the present volume, **NMR Spectroscopy: Modern Spectral Analysis** will discuss the various software tools and techniques required to perform the spectrum *Analysis* to obtain the correct NMR *Parameters*. Finally, the bi-directional vertical black arrows in the *Result* column symbolize the strong relationships between the different types of data that can be obtained which are discussed in detail in the various books in this series.

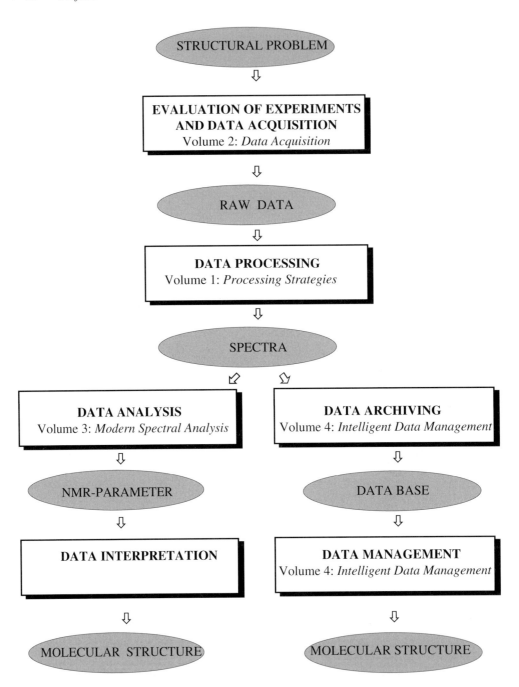

Fig. 2: Concept in the series of the interactive course: NMR Spectroscopy.

- Volume 1 – Processing Strategies
- Volume 2 – Data Acquisition
- Volume 3 – Modern Spectral Analysis
- Volume 4 – Intelligent Data Management

The concept of the series is visualized in Fig. 2, and the contents of volumes 1–4 may be summarized as follows:

Volume 1: ***Processing Strategies***

Processing NMR data transforms the acquired time domain signal(s) – depending on the experiment – into 1D or 2D spectra. This is certainly the most central and important step in the whole NMR analysis and is probably the part, which is of interest to the vast majority of NMR users. Not everyone has direct access to an NMR spectrometer, but most have access to some remote computer and would prefer to process their own data according to their special needs with respect to their spectroscopic or structural problem and their ideas concerning the graphical layout *i.e.* for presentation of reports, papers or thesis. It is essential for the reliability of the extracted information and subsequent conclusions with respect to molecular structure, that a few general rules are followed when processing NMR data. It is of great advantage that the user is informed about the many possibilities for data manipulation so they can make the best use of their NMR data. This is especially true in more demanding situations when dealing with subtle, but nevertheless important spectral effects. Modern NMR data processing is not simply a FOURIER transformation in one or two dimensions, it consists of a series of additional steps in both the time and the frequency domain designed to improve and enhance the quality of the spectra.

Processing Strategies gives the theoretical background for all these individual processing steps and demonstrates the effects of the various manipulations on suitable examples. The powerful BRUKER 1D WIN-NMR, 2D WIN-NMR and GETFILE software tools, together with a set of experimental data for two carbohydrate compounds allow you to carry out the processing steps on your own remote computer, which behaves in some sense as a personal "NMR processing station". You will learn how the quality of NMR spectra may be improved, experience the advantages and limitations of the various processing possibilities and most important, as you work through the text, become an expert in this field. The unknown structure of one of the carbohydrate compounds should stimulate you to exercise and apply what you have learnt. The elucidation of this unknown structure should demonstrate, how powerful the combined application of several modern NMR experiments can be and what an enormous and unexpected amount of structural information can thereby be obtained and extracted by appropriate data processing. It is this unknown structure which should remind you throughout this whole educational series that NMR data processing is neither just "playing around" on a computer nor some kind of scientific "l'art pour l' art". The main goal for measuring and processing NMR data and for extracting the structural information contained in it, is to get an insight into how molecules behave.

Furthermore, working through ***Processing Strategies*** should encourage you to study other topics covered by related volumes in this series. This is particularly important if you intend to operate a NMR spectrometer yourself, or want to become familiar with additional powerful software tools to make the best of your NMR data.

Volume 2: ***Data Acquisition***

Any NMR analysis of a structural problem usually starts with the selection of the most appropriate pulse experiment(s). Understanding the basic principles of the most common experiments and being aware of the dependence of spectral quality on the various experimental parameters are the main prerequisites for the successful application of any NMR experiment. Spectral quality on the other hand strongly determines the reliability of the structural information extracted in subsequent steps of the NMR analysis. Even if you do not intend to operate a spectrometer yourself, it would be beneficial to acquire some familiarity with the interdependence of various experimental parameters e.g. acquisition time and resolution, repetition rate, relaxation times and signal intensities. Many mistakes made with the application of modern NMR spectroscopy arise because of a lack of understanding of these basic principles. ***Data Acquisition*** covers these various aspects and exploits them in an interactive way using the Bruker software package NMRSIM. Together with 1D WIN-NMR and 2D WIN-NMR, NMRSIM allows you to simulate routine NMR experiments and to study the interdependence of a number of NMR parameters and to get an insight into how modern multiple pulse NMR experiments work.

Volume 3: ***Modern Spectral Analysis***

Following the strategy of spectral analysis, the evaluation of a whole unknown structure, of the local stereochemistry in a molecular fragment or of a molecules dynamic properties, depends on NMR parameters. Structural informations are obtained in subsequent steps from chemical shifts, homo- and heteronuclear spin-spin connectivities and corresponding coupling constants and from relaxation data such as NOEs, ROEs, T_1s or T_2s and assumes that the user is aware of the typical ranges of these NMR parameters and of the numerous correlations between NMR and structural parameters, i.e. between coupling constants, NOE enhancements or line widths and dihedral angles, internuclear distances and exchange rates respectively. However, the extraction of these NMR parameters from the corresponding spectra is not always straightforward,
- The spectrum may exhibit extensive signal overlap, a problem common with biomolecules.
- The spectrum may contain strongly coupled spin systems.
- The molecule under investigation may be undergoing dynamic or chemical exchange.

Modern Spectral Analysis discusses the strategies needed to efficiently and competently extract NMR parameters from the corresponding spectra. You will be shown how to use the spectrum simulation package WIN-DAISY to extract chemical shifts, coupling constants and individual line widths from even highly complex NMR spectra. In addition, the determination of T_1s, T_2s or NOEs using the special analysis tools of 1D WIN-NMR will be explained. The simulation of double resonance spectra are shown using the program WIN-DR and the calculation of dynamic NMR spectra with WIN-DYNAMICS is trained. Sets of spectral data for a series of representative compounds, including the two carbohydrates mentioned in volume 1 are used as instructive examples and for problem solving. NMR analysis often stops with the plotting of the spectrum thereby renouncing a wealth of structural data. This part of the series should encourage you to go further and fully exploit the valuable information "hidden" in the carefully determined NMR parameters of your molecule.

Volume 4: ***Intelligent Data Management***

The evaluation and interpretation of NMR parameters to establish molecular structures is usually a tedious task. An alternative way to elucidate a molecular structure is to directly compare its measured NMR spectrum – serving here as a fingerprint of the investigated molecule – with the corresponding spectra of known compounds. An expert system combining a comprehensive data base of NMR spectra with associated structures, NMR spectra prediction and structure generators not only facilitates this part of the NMR analysis but makes structure elucidation more reliable and efficient.

In ***Intelligent Data Management***, an introduction to the computer-assisted interpretation of molecular spectra of organic compounds using the Bruker WIN-SPECEDIT software package is given. This expert system together with the Bruker STRUKED software tool is designed to follow up the traditional processing of NMR spectra using 1D- and 2D WIN-NMR in terms of structure-oriented spectral interpretation and signal assignments. WIN-SPECEDIT offers not only various tools for automatic interpretation of spectra and for structure elucidation, including the prediction of spectra, but also a number of functions for so-called "authentic" archiving of spectra in a database, which links molecular structures, shift information and assignments with original spectroscopic data. You will learn to exploit several interactive functions such as the simple assignment of individual resonances to specific atoms in a structure and about a number of automated functions such as the recognition of signal groups (multiplets) in ^{1}H NMR spectra. In addition, you will also learn how to calculate and predict chemical shifts and how to generate a local database dedicated to your own purposes. Several examples and exercises, including the two carbohydrate compounds, serve to apply all these tools and to give you the necessary practice for your daily spectroscopic work.

The concept of the series is the complementary self-teaching method of textbook and interactive data treatment using the *Teaching Versions* of the WIN-NMR family of programs stored on your own personal computer (PC). Each book is delivered with a CD-ROM which includes all the appropriate software tools and the corresponding data base to enable you to perform all the "Check it" exercises in the textbook.

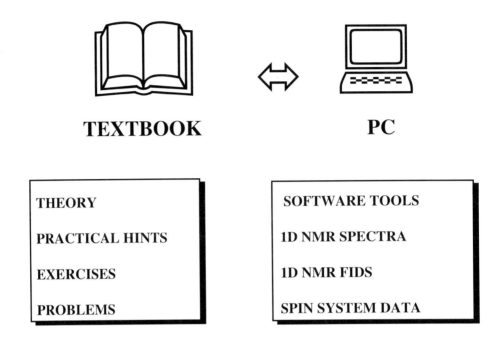

Fig. 3: Mutual interaction between textbook and PC.

The textbook contains a brief theoretical background description stressing the practical aspects of *why* things are done in a particular manner. The "Check it" exercises in the textbook describe the treatment of the NMR data using the software tools and initiate the switching from the textbook to the software tools loaded on your PC (Fig. 3). The way *how* to solve the problems with respect to the theoretical descriptions is explained in the "Check it" text. The results may be verified by comparing your own results either with the comments made in the text or by using the solutions delivered with the data base. Using this "learning-by-doing" method with its complementary *why*-and-*how* approach you will be shown the most effective way to solve problems and at the same time you will also become familiar with the software tools that make up the WIN-NMR family of programs.

The Teaching Versions are identical with the full program versions, but they can only process the special NMR data sets delivered on the CD-ROM. To order the full program version, contact your local BRUKER office, or send your order to:

BRUKER Analytik GmbH   Tel.: +49 (0)721 5161 0
Silberstreifen   Fax: +49 (0)721 5161 297
D-76287 Rheinstetten/Karlsruhe
FRG

When you begin this text, it is assumed that you are already skilled in the processing of NMR data, which is the subject of the first volume of this series, **NMR Spectroscopy: Processing Strategies**. The present volume, **NMR Spectroscopy: Modern Spectral Analysis**, deals with the analysis of one-dimensional NMR spectra. Methods to determine NMR parameters such as chemical shifts, scalar coupling constants, half-height line widths, relaxation times and rate constants are described. Special emphasis is laid on the understanding of coupled spin systems and on the abstract concept of spin systems but the aim is always to use this information in a *practical* way to obtain structural information. The book contains many examples and discusses in detail the interpretation of NMR parameters in order either to verify a model structure or to determine an unknown structure and simultaneously teaches you how to analyze your own 1D NMR data in the most effective manner.

It is the primary aim of the series to teach the user how NMR spectra may be obtained from the data acquired on a spectrometer and how these spectra may be used to establish a molecular structure following one of the two strategies outlined before. The series of volumes therefore emphasizes the methodical aspect of NMR spectroscopy, rather than the more usual analytical aspects *i.e.* the description of the various NMR parameters and of how they depend on structural features, presented in numerous text books.

Acknowledgments

We are greatly obliged to Dr. B. F. Taylor, University of Sheffield (UK) for many instructive discussions, suggestions and comments regarding the data, the program and the text. His assistance in checking the interactive parts and proofreading the complete volume was a great and inestimable help in preparing this manuscript. We would also like to thank Dr. T. Mitchell, University of Dortmund, (FRG) for proofreading and Prof. Dr. G. Hägele, University of Düsseldorf, (FRG) for his support, his encouraging discussions and contributions in the progress of the present work. We are also very grateful to Dr. A. Germanus (Bruker-Franzen Analytik, Bremen) for his helpful advice and excellent collaboration.

The wide range of example spectra included in this volume was made available by cooperation with a number of groups. Therefore we wish to express our thanks, in alphabetical order, to the following people and their co-workers: Dr. P. Bigler, University of Bern (CH), Dr. B. Diehl, Spectral Service Cologne (FRG), Prof. Dr. G. Hägele, University of Düsseldorf (FRG), Prof. Dr. G. Heckmann, University of Stuttgart (FRG), Dr. K. Karaghiosoff, University of Munich (FRG), Prof. Dr. D. Leibfritz, University of Bremen (FRG), Dr. H. Mayer, University of Tübingen (FRG), Prof. Dr. E. Pretsch, ETH Zürich (CH), Prof. Dr. H.-U. Siehl, University of Ulm (FRG) and Dr. B. F. Taylor, University of Sheffield (UK).

Finally, we would like to thank BRUKER Analytik, Wiley-VCH, all our colleagues at BRUKER-Franzen Analytik and all our friends who have shown a keen interest in this project for their fruitful discussions, encouragement, assistance, support and patience.

Bremen, September 1997

U. Weber
H. Thiele

Table of Contents

1	**Introduction**	**1**
1.1	Scope and Audience	1
1.2	Organisation	2
1.3	Personal Qualifications	4
1.4	Contents	5
1.5	Technical Requirements	7
1.6	Software	8
1.6.1	General Remarks	8
1.6.2	Installation	8
1.6.3	Basic Program Knowledge	10
1.6.4	Book Formatting	13
1.7	Recommended Reading and References	16
2	**General Characteristics of Spin Systems**	**19**
2.1	The Chemical Shift	20
2.1.1	Spin Systems Without Coupling	20
2.1.2	The NMR Resonance Frequency	22
2.1.2.1	The NMR Transition Frequency	24
2.1.2.2	The NMR Transition Intensity	27
2.1.2.3	The NMR Lineshape	27
2.1.3	Two Non-Coupled Spins	29
2.1.3.1	Two Isochronous Spins	30
2.1.3.2	Two Non-Isochronous Spins Without Coupling	32
2.1.4	Basic Notation of Spin Systems	33
2.1.5	Summary and Examples	34
2.2	First Order Spin Systems	36
2.2.1	Introduction to Spin-Spin Coupling	36
2.2.2	The AX Spin System	38
2.2.2.1	Heteronuclear AX Spin Systems	41
2.2.2.2	Homonuclear AX Spin Systems	41
2.2.2.3	Summary	46
2.2.3	The AMX Spin System	46
2.2.4	Extended First Order Spin Systems	57
2.2.4.1	AMRX Spin Systems	57

2.2.4.2	Fragmentation of Spin Systems	58
2.3	Second Order Spin Systems	63
2.3.1	Notational Aspects	63
2.3.2	The AB Spin System	64
2.3.2.1	The Graphical Spectrum Analysis	69
2.3.2.2	Suppression of Particular Transitions	74
2.3.3	The ABX Spin System	77
2.3.3.1	Analytical Solution for the ABX Spin System	81
2.3.3.2	First Order Based ABX Solution	88
2.3.4	Extended Spin Systems	93
2.3.4.1	Taking Advantage of Field Effects	95
2.4	Magnetic Equivalence	98
2.4.1	Notational Aspects	99
2.4.2	The Coupled A_2 Spin System	99
2.4.3	The Composite Particle Approach	102
2.4.4	The A_n Spin Systems	105
2.4.5	First Order Spin Systems	108
2.4.5.1	The A_2X Spin System	108
2.4.5.2	The AX_3 Spin System	113
2.4.5.3	A_nX_m Spin Systems	116
2.4.5.4	First Order Spin Systems with more than one Coupling	119
2.4.6	Second Order Spin Systems	121
2.4.6.1	The A_2B Spin System	121
2.5	References	129

3 Structure and Spin System Parameters — 131

3.1	Symmetry Effects	131
3.1.1	Structural and Notational Aspects	132
3.1.2	The $[AX]_2$ Spin System	134
3.1.3	The $[AB]_2$ spin system	149
3.2	Configuration Isomers in NMR	160
3.2.1	Cis/trans Isomerism	160
3.2.1.1	Determination of the Signs of Coupling Constants	161
3.2.1.2	Extended Spin Systems	167
3.2.2	Enantiomers and Diastereomers	173
3.3	Conformation and NMR	175
3.3.1	Taking Advantage of Field and Solvent Effects	180
3.3.1.1	Linewidths	186
3.3.2	Symmetry of Rigid and Non-Rigid Compounds	189
3.3.3	Advanced Examples	195
3.3.3.1	Glucose	202
3.3.3.2	Piperine and Nicotine	219
3.4	References	228

4	**Spin Systems and the Periodic Table**	**229**
4.1	Spin-½ Pure Elements	229
4.1.1	Spin Systems Containing ^{19}F	230
4.1.1.1	All-Fluorine Spin Systems	231
4.1.1.2	Heteronuclear Spin Systems	235
4.1.1.3	Symmetry	239
4.1.1.4	Isotope Effect	242
4.1.2	Spin Systems Containing ^{31}P	244
4.1.3	Spin Systems Containing ^{103}Rh	258
4.2	Spin-½ Rare Isotopes	259
4.2.1	^{13}C Spectra and Spin Systems with ^{13}C	259
4.2.1.1	Double Resonance Methods	260
4.2.1.2	Satellite Spectra	263
4.2.2	Spin Systems Containing ^{15}N	273
4.2.3	Spin Systems Containing ^{29}Si	277
4.2.4	Spin Systems Containing ^{195}Pt	279
4.2.5	Spin Systems Containing ^{77}Se	281
4.2.6	Spin Systems Containing ^{115}Sn / ^{117}Sn / ^{119}Sn	286
4.3	Spin Systems with Quadrupolar Nuclei	293
4.3.1	Basic Features of Quadrupolar Nuclei	293
4.3.2	Spin Systems Including ^{2}H	296
4.3.2.1	Chloroform	296
4.3.2.2	Deuterium Substitution	299
4.3.2.3	Methylenechloride	300
4.3.2.4	Dimethylsulfoxide	302
4.3.2.5	Ammonia Hydrogen Isotopomers	304
4.3.3	Spin Systems Containing ^{14}N	305
4.3.4	Spin Systems Containing ^{10}B / ^{11}B	306
4.3.5	Spin Systems Containing ^{17}O	310
4.3.6	Spin Systems Containing ^{9}Be	312
4.4	References	314
5	**Time Dependent Phenomena**	**317**
5.1	Relaxation Times	317
5.1.1	Spin-Lattice Relaxation Time	318
5.1.1.1	Quadrupolar Nuclei	318
5.1.1.2	Spin-½ Nuclei	319
5.1.2	Spin-Spin Relaxation Time	326
5.2	Nuclear Overhauser Effect	327
5.3	Magnetic Site Exchange	339
5.3.1	Non-Coupled Spin Systems	340
5.3.2	Systems Involving Coupling	349
5.4	References	361

6	**Appendix**	**363**
6.1	Quantum Mechanical Theory	363
6.1.1	The AB Spin System and the General Case	363
6.1.2	Composite Particle Theory	370
6.1.2.1	The A_2 Spin System	370
6.1.2.2	The A_2B Spin System	373
6.1.2.3	Total Intensity Theorem	375
6.2	Spin System Probabilities	376
6.3	The Lineshape Fitting	378
6.4	Glossary	382
6.5	References	384
	Index	385

1 Introduction

1.1 Scope and Audience

From the discovery of the magnetic resonance effect in 1945 [1.1], [1.2] up to the present day Nuclear Magnetic Resonance (NMR) spectroscopy has achieved an enormous importance and acceptance in chemical research for the elucidation of molecular structure. Experimental techniques underwent a revolution with the introduction of pulsed FOURIER transform spectroscopy [1.3] and highly sophisticated pulse sequences have been developed for the extraction of the desired information from one, two and even three-dimensional NMR spectra [1.4].

It was soon discovered that even 'simple' one-dimensional NMR spectra contained a wealth of information about coupled spin systems and because the extraction of NMR parameters was not always straightforward, methods for analyzing spectra were developed [1.5]–[1.8]. To understand the complexity of the measured spectra, computer programs evolved to perform the quantum mechanical calculation of NMR spectra. Within a few years a variety of iterative programs (operating on main frame computers) were written, optimizing parameters on the basis of experimental data to extract the true chemical shifts and coupling constants [1.9]–[1.11]. A review of the available programs is given in the appendix of reference [1.12]. Although numerous books and review articles have dealt with the analysis of high-resolution NMR spectra, the combination of analytical formulae, quantum mechanics and, for computer aided analysis, "user-unfriendly" software has meant that spin system analysis has not been very popular. Indeed, in many papers in the literature little or no attempt is made to analyze spin systems, signals often being reported as a "multiplet".

The introduction of new experimental techniques, particular two-dimensional experiments, has tended to push both the use and the teaching of one-dimensional spectral analysis further into the background. Consequently there is now a lack of understanding of NMR spin systems which can have serious repercussions on both the experimental and theoretical analysis of NMR data.
- In cases where the 'simple' single-pulse one-dimensional spectrum is not fully understood, the interpretation of multi-dimensional spectra can be erroneous.
- The vector diagrams used to illustrate the BLOCH equations refer, strictly speaking, to first-order spectra only.

- Second order effects are mostly interpreted as *artifacts*, but they are real and even contain important information about the spin system, that can be useful when correctly interpreted [1.13], [1.14].

In conclusion, the understanding of coupled spin systems is the precondition to be able to deal with the interpretation of multi-dimensional data and a lack of understanding cannot be remedied by simply performing more multi-dimensional experiments.

The rapid development in computer hardware and of powerful operating systems with simple, graphical user interfaces have laid the foundations for the adaptation of the main frame iterative programs to run on personal computers but with a "user-friendly" interface. It is now possible to separate the acquisition of NMR data done on the spectrometer from the processing, analysis and interpretation of the spectra which may be performed on personal computers (PC's). This approach not only increases efficiency but also greatly simplifies the input into computer analysis programs.

Because the amount of NMR knowledge to be taught in universities and technical schools has increased enormously, the emphasis is usually placed on modern experimental techniques rather than on the detailed aspects of coupled spin systems. Students tend more to understand pulse sequences rather than to be able to analysis the spectrum obtained from a simple one-dimensional one pulse experiment and consequently the analysis of coupled spin systems is now a tool not frequently used in NMR laboratories. However, the NMR spectroscopist requires the help of all the tools at their disposal in solving structural and chemical problems and this also includes the analysis of coupled spin systems.

Modern Spectral Analysis is designed to fill the gap that has arisen in the understanding of spectrum analysis in an interactive way. The book contains the necessary theoretical background information to understand the links between NMR parameters and one-dimensional NMR spectra and also shows how to extract the various parameters from the experimental data. It has been assumed that you are familiar with the basics of NMR spectroscopy and with the processing of raw NMR data i.e. the FOURIER transformation of **F**ree **I**nduction **D**ecays (FID). However, it is recommended that before starting the spectrum analysis you "brush-up", if necessary, on your processing skills by practicing with some of the data supplied on CD ROM (see "Check its" in section 1.6.3). The spectra supplied on CD ROM will be used to generate the input data for the subsequent spectrum analysis that forms the major part of this book.

1.2 Organisation

Modern Spectral Analysis is composed of three main parts:
- The written book giving an introduction into the basic theory and presenting practical hints, examples and exercises – "Check its" – to enable you to apply directly what you have learnt in practice.
- Software tools including manuals (HELP routines) describing in detail the use of **1D WIN-NMR**, **WIN-DAISY**, **WIN-DR** and **WIN-DYNAMICS** supplied on CD-ROM.

- A large number of one-dimensional high-resolution NMR spectra and WIN-DAISY input files for a variety of compounds dealing with different spin systems plus T_1 and NOE data for glucose and an oligosaccharide taken from [1.15].

📖	**MODERN SPECTRAL ANALYSIS**	Basic Theory, Rules, Hints, Recommendations, Examples, Exercises
💿	**WIN-NMR SOFTWARE**	
	1D WIN-NMR	Manual Analysis of 1D NMR Data, Help Routine.
	WIN-DAISY	Simulation and iteration of coupled spin systems, Help Routine.
	WIN-DR	Simulation of double resonance NMR spectra.
	WIN-DYNAMICS	Simulation of exchange broadened NMR spectra.
💿	**DATA**	
	NMR data	1D high-resolution NMR spectra and FIDs (DEPT, T_1, NOE) are available on CD-ROM via the program setup.
	Input Files	WIN-DAISY input documents are also available from the setup, installed with the program WIN-DAISY.
	Solution*	The results for all example data are available on the CD ROM directory.

Fig. 1.1: Components of Modern Spectral Analysis (volume 3).

---

* The spectra can be loaded from the CD-ROM (read only). In you want to work on these files, copy them onto your local hard disk.

These three parts are combined in such a way that you may use all these educational tools interactively with your PC. In this way as you learn about the basics in theory, you directly apply what you have learnt in practice following the instructions given in the numerous "Check its". You experience step by step the scope of the powerful software modules, their advantages and limitations and you acquire the necessary skills to become proficient in NMR spectral analysis. It is essential at this stage to exploit and make full use of the powerful help routines if necessary. The help text explains the functionality of the pull-down menu entries, dialog boxes, buttons and entry field in the programs as a manual, in contrast to the introductory text which discusses the effects and the purpose of the various analysis options.

In working through this book and analyzing the various one-dimensional high-resolution NMR spectra, you will learn how spin systems are derived from the spectrum. You will also start with the suspected structure and learn how to apply the predicted spectrum to the experimental data. The analysis always starts with a manual, first order analysis in WIN-NMR where the spin system is defined and the values of the various spectrum parameters estimated. This parameter data set is then exported to WIN-DAISY and the spectrum simulated to establish whether the spectrum is truly a first order spin system. If the spectrum is not first order and cannot be easily analyzed, the WIN-NMR analysis provides the input data for the iteration document used for the parameter optimization in WIN-DAISY. Emphasis is laid on the fact that spectra are connected with structure and that the abstraction of the spin system helps considerably in evaluation of the extracted NMR parameters, i.e. which scalar coupling constants are defined by the spectrum and whether the signs can be determined from the spectrum. You will also be shown how to look for distinct signals in the spectrum which will confirm the analysis. In cases where the appearance of the spectrum is independent of particular NMR parameters, methods to determine these "hidden" parameters will be illustrated.

Theory has been kept to a minimum and is only briefly mentioned where it is necessary to understand the current discussion. A more detailed theoretical description is given in the appendix and in the references at the end of each chapter.

Some experimental aspects such as spectrum resolution, phasing, baseline correction, spectrum calibration and integration are also briefly discussed. To help in the analysis of the experimental data in various "Check its", the values of some NMR parameters in particular environments are given in the text. This information is very specific and is nether intended to be or is comprehensive enough for use in solving all NMR problems. Similarly, this book does not claim to contain examples of, or to discuss all of the relevant parameters of every type of problem that may occur in high-resolution NMR spectroscopy.

1.3 Personal Qualifications

It is assumed that you are familiar with the WINDOWS operating system either WINDOWS 3.1x, WINDOWS 95 or WINDOWS NT and are acquainted with the basic handling of the operating system and the features of the graphical user interface.

You should also have a basic knowledge of NMR spectroscopy and be competent in processing one-dimensional NMR spectra. A precondition for any successful analysis is that the experimental data has been correctly processed and it is strongly recommended that you study the first volume in this series, *NMR Spectroscopy: Processing Strategies* [1.15].

1.4 Contents

The book is divided into six chapters:

1. **Introduction**
2. **General Characteristics of Spin Systems**
3. **Structure and Spin System Parameters**
4. **NMR and the Periodic Table**
5. **Time Dependent Phenomena**
6. **Appendix**

Chapter 2 introduces the mystery called the 'spin system'. After a short introduction into the magnetic resonance effect the chapter is limited to the ^{1}H nucleus with spin quantum number I = ½. Starting with a single spin in the magnetic field the NMR parameters chemical shift and half-height line width forming the line shape are introduced. In this way you become familiar with the parameter handling in WIN-DAISY and the data exchange with 1D WIN-NMR (section 2.1). As more spins become involved, the concepts of scalar coupling constants and first order spin systems containing magnetically non-equivalent spins are introduced (section 2.2). You will learn how to use the first order analysis tool of 1D WIN-NMR to construct the spin system directly from the spectrum and how to transfer and process this data in WIN-DAISY. The spectral parameter describing the relationship between scalar coupling constant and chemical shift is introduced and second order spin systems and the concept of non-isochronous spins discussed (section 2.3). Spectral analysis using WIN-DAISY is compared with traditional 'by hand' methods. In the last section 2.4, magnetic equivalence is introduced starting with simple first order spin systems which are expanded to include second order effects. The principal signal pattern composition and decomposition is illustrated using a number of examples. Because the principal aim of this chapter is to focus on the spin system *characteristics*, the examples are mainly of aromatic type compounds where the scalar coupling constants are well defined and the sign of the scalar coupling constants do not have to be considered. Only in the section on magnetic equivalence will the examples have a different structural skeleton.

Chapter 3 starts by discussing the effects of symmetry in NMR spectra. The resulting spectra are strongly second order and cannot be analyzed on a first order basis. In the "Check its" the focus is on the relationship between structure and NMR parameters and it is necessary to apply the knowledge about spin systems learnt in chapter 2 in analyzing the spectra. Emphasis is laid on the definition of spin systems in order to determine which parameters can be identified in the spectrum and the recognition of spin systems which due to various parameter combinations may differ significantly in appearance. The sign of coupling constants and the methods used to determine sign are introduced. Double resonance techniques are discussed using the decoupling simulation program WIN-DR. Some advanced examples exhibiting special problems are included for you to analyze. The recognition of spin systems in overlapped spin systems and the occurrence of long-range coupling constants are also illustrated. Special emphasis is put upon the glucose and oligosaccharide spectra already discussed in volume 1 [1.15]. Finally the chapter closes with an analysis of the nicotine spin system.

Chapter 4 extends the concept of spin systems to other spin ½ isotopes. Starting with the pure, 100% natural abundance elements ^{19}F, ^{31}P and ^{103}Rh, both homonuclear and heteronuclear spin systems involving ^1H are discussed and the analysis of heteronuclear coupling patterns in WIN-NMR is outlined. The application of double resonance techniques in heteronuclear spectra is introduced. Organic and inorganic examples are used and the isotope effect on chemical shift outlined. Spin ½ isotopes with a natural abundance of less than 100% are then discussed and satellite spectra and the calculation of their intensities (probabilities) explained. Starting with ^{13}C spectra, (including the treatment of DEPT spectra starting from the FIDs) the determination of 'hidden' coupling constants is explained based on the analysis of satellite spectra. Only spin systems involving ^{15}N, ^{29}Si, ^{195}Pt and ^{77}Se are discussed in detail; some experimental spectra are provided for $^{115}Sn/^{117}Sn/^{119}Sn$. In the final section of this chapter quadrupolar nuclei and their basic features are introduced. The section starts with the most common quadrupolar nucleus in NMR spectroscopy, ^2H and discusses the isotope effect on coupling constants and the method of determination of 'hidden' proton-proton coupling constants by deuterium substitution. Other nuclei covered are ^{14}N, $^{11}B/^{10}B$, ^{17}O and finally one pure element ^9Be. Provided that you have worked consecutively through the text, these multi-nuclear spin systems will not present any problems.

Chapter 5 deals with the time relevant phenomena which have so far been ignored in the first chapters. Spin-lattice and (briefly) spin-spin relaxation are discussed using the glucose and oligosaccharide examples from reference [1.15]. NOE experiments are also discussed using the same examples. Because the data obtained from T_1 and NOE experiments is strongly influenced by the type of processing used, the data processing is regarded as an essential part of the analysis so that in these particular instances the FIDs are used as input data. Processing is performed in a semi- or fully automated way without any detailed discussion of the processing functionality's; detailed information regarding serial processing is given in volume 1 of this series [1.15]. The measured T_1 values and NOE enhancements are then related to the glucose and oligosaccharide structures. Finally a very brief introduction into dynamic NMR spectroscopy and the WIN-DYNAMICS program is given. Only the basic theory is discussed because to be treated properly dynamic processes would require its own volume. For a more detailed approach you are referred to the literature. The determination of the static parameters is outlined which directly picks up the thread of the analysis of spin systems discussed in chapters 2, 3 and 4. Fast exchanging spin systems analyzed in the previous chapters are reference to. The emphasis is on covering a few topics comprehensively rather than many topics superficially. Consequently only two site exchanging systems of mutual and non-mutual type are outlined, no intermolecular examples with complicated reaction scheme constructions are considered.

Chapter 6 is the Appendix. Section 6.1 contains additional quantum mechanical explanations and is subdivided into various sections. The first section contains the description of the AB spin system with extension to the general case and the second part explains the composite particle theory starting with the A_2 system and then extending the theory to the A_2B spin system. The section 6.2 contains the probability calculation of satellite spin systems using the example given in section 4.2.5. Section 6.3 gives some background information about the optimization algorithms implemented in WIN-DAISY and the glossary is found in section 6.4.

1.5 Technical Requirements

In order to install and run the programs 1D WIN-NMR, WIN-DAISY, WIN-DR and WIN-DYNAMICS to perform the exercises described in this book, the PC used should have the following minimum hardware requirements:

- IBM compatible 386, 486 or Pentium processor.
- at least 8 MB RAM.
- Arithmetic coprocessor.
- Monitor and graphic card for at least standard VGA resolution.
- Pointing device compatible with a Microsoft two button mouse.
- CD-ROM drive.
- Hard-disk with at least 35 MBytes spare capacity for standard installation - to perform all the "Check its" in the book about 100 MBytes are required.
- Printer (not indispensable).

Software requirements are as follows:

- Operating system/graphical user interface WINDOWS 3.1x, WINDOWS 95 or WINDOWS-NT are supported.
- Standard WINDOWS accessories NOTEPAD and CALCULATOR.
- The printer device should be correctly installed.

Remark:
Although all MS-WINDOWS versions starting from 3.1x are supported, it is strongly recommended to use either WINDOWS 95 or WINDOWS-NT, the setup will then install the 32-bit version of WIN-DAISY which on *identical hardware* is almost twice as fast as the 16-bit WINDOWS 3.1x version.

1.6 Software

1.6.1 General Remarks

The programs available on CD-ROM are special Teaching Versions of the programs 1D WIN-NMR, WIN-DAISY, WIN-DR and WIN-DYNAMICS. To avoid any problems with the full version of any of these programs already installed on your PC, select different paths for the teaching program version! Do not give path names where the complete release version of the programs are located. It is possible to run full and teaching tool versions on one PC at the same time without any interference. If you have already installed the CD-ROM delivered with volume 1 of this series [1.15] you may safely overwrite the teaching version of 1D WIN-NMR, all "Check its" from volume 1 work with the new version included with this book.

Special fonts and paragraph styles are used in this book to outline different actions and will help you to identify the relationship between the written text and the commands to execute on your PC. Although a detailed description will be given in section 1.6.4, to enable the installation exercises to be started three basic points are mentioned here:

- **Bold** printed expressions belong to pull-down menu commands, buttons, check boxes, radio buttons, their group description and descriptions of edit fields.
- CAPITAL letters are used for file names.
- The contents of the following type of <brackets> requires key board input.

1.6.2 Installation

If you wish to install the software tools and the examples make sure that there is at least 35 MByte free space available on your hard disk. If the desired hard disk is a network drive it should be mounted properly, you should have read and write access and it should be available any time you wish to run the "Check its" in this book. Only very few hard disk space is required from the system directory.

> **Check it in WINDOWS:**
>
> Insert the green CD ROM (stored inside the inner back cover of this book) into your CD-ROM drive. The operating systems WINDOWS 95 and WINDOWS NT will start the setup automatically. For WINDOWS 3.1x use the standard procedure to start the setup in a MS-WINDOWS environment (from the **Program Manager** use the **Run** command and type <[CD-ROM drive letter]:SETUP> as shown on the CD-ROM label. Consult the WINDOWS manual or Help File if necessary. After a few seconds the setup window will appear on the screen showing the Welcome dialog (Fig. 1.2). Click the **Next** button to proceed with the installation process which will open a number of dialog boxes one after the other. If possible, confirm the default options and entries for the whole installation process by simply pressing the **Next** button.

1.6 Software

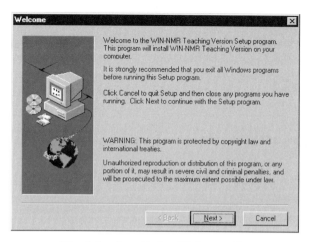

Fig. 1.2: Installation Welcome dialog.

The default options will install the Teaching Versions of the programs 1D WIN-NMR, WIN-DAISY, WIN-DR and WIN-DYNAMICS into the given destination directories (C:\TEACH\...). If selected, the example data for 1D WIN-NMR (spectra and FIDs in the default directory C:\TEACH\WIN1D\SPC\...) and WIN-DAISY (input parameter files, in the default directory C:\TEACH\DAISY\DATA\...) are also installed. This default installation will install on your PC the necessary data to complete all the "Check its" in this book (Fig. 1.3). The installation process is successfully performed if the dialog box shown in Fig. 1.4 appears on the screen.

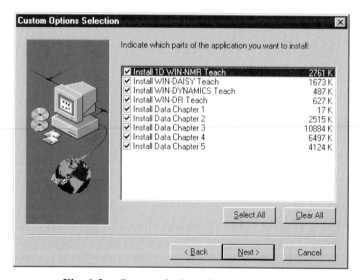

Fig. 1.3: Custom Options Selection dialog box.

10   *1 Introduction*

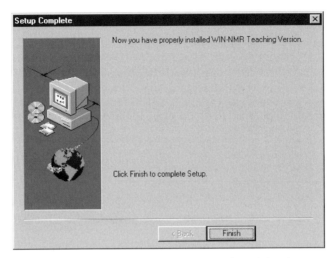

Fig. 1.4:   Setup Complete Installation dialog box.

Check it in WINDOWS:

If the installation process is initialized as described in the previous "Check it", all the required files will be copied to the destination directories, the initialization files in the WINDOWS system directories updated and icons to call the programs created. Accept the final message of a successful installation with **Ok**. The programs can be run from the WINNMR group from Program Manager (WINDOWS 3.1x, WINDOWS NT 3.51) by double clicking on the icon or from the **Start** button and select the **Programs I WIN-NMR Teaching Version ...** entries (WINDOWS 95 and WINDOWS NT 4.0). You may also create your own shortcut keys on the desktop.

1.6.3   Basic Program Knowledge

The basic program handling conforms with the standard WINDOWS conventions. Nevertheless the essential basics are mentioned here as the communication between the programs is tested. If one of these communication paths does not work as described below, the programs are not installed correctly and must be re-installed.

Check it in WINDOWS:

Run the program 1D WIN-NMR Teaching Version as described in the previous "Check it".

Check it in WIN-NMR:

As no data has been loaded in WIN-NMR most of the pull-down menu commands are disabled (grayed). Call the **Simulation I WIN-DAISY** command to start the program WIN-DAISY (Teaching Version) which is moved to the foreground and gets the input focus.

Check it in WIN-DAISY:

Again, most of the toolbar buttons and the corresponding pull-down menu entries are disabled because no data is loaded. Use either the pull-down menu command **File | Open** or the toolbar button **File Open** or the shortcut key <Ctrl + O> to display the File select box of WIN-DAISY input documents. The actual directory refers to the ...\DATA subdirectory of the WIN-DAISY program path where a number of standard input files are located. Enter the file name <A.MGS> or select the file using the mouse and then press the **Open** button. Immediately the corresponding WIN-DAISY document is displayed and a number of additional buttons / menu entries are enabled. Moving the mouse pointer over the toolbar displays tooltip and status line information about the various commands. Press the **SIM** button which corresponds to the **Simulation** run command to calculate the spectrum. The name of the simulated spectrum is listed in the status line **Calculated Spectrum:** of the WIN-DAISY document main window and the spectrum is usually stored in the temporary directory with the default name ...\999999.1R. After the simulation is finished the **Display** the **Calculated Spectrum** button (the button with a red peak) is enabled. Clicking this button will display the simulated spectrum in WIN-NMR.

Check it in WIN-NMR:

WIN-NMR is moved to the foreground with the simulated spectrum loaded. The corresponding file name is shown in the status line below the menu bar. Test the functions of the buttons in the button panel, e.g. the <> and **/2** buttons, **ALL** and **zoom** buttons. Now load an experimental spectrum using the **File | Open** command. The File selection box will be displayed for the installed spectrum directory ...\AX. With the mouse pointer select the first spectrum ...\AX\001001.1R and click on the **Ok** button. The spectrum is loaded into the display window and the file name shown in the status line. Remark: To test your own processing skills try your hand on the experimental FID ...\PROCESS\001001.FID. Process it in a way that the resulting spectrum has a sufficient digital resolution, is properly phased, base line corrected that it is comparable with the example spectrum ...\PROCESS\001999.1R.

Toggle back to **WIN-DAISY** using the **Simulation** pull-down menu command.

Check it in WIN-DAISY:

It is possible to export the contents of the current WIN-DAISY document to the programs WIN-DR and WIN-DYNAMICS. Press the **Export** to **WIN-DR** button in the panel bar (second from the right in the second toolbar row).

Check it in WIN-DR:

The program WIN-DR is started and the corresponding file A.WPR is loaded. In this Teaching Version it is only possible to export data from WIN-DAISY because loading input data files manually is not implemented. The toolbar buttons have the same meaning as in WIN-DAISY; however there is one additional button, the **Perturbation Parameters** for the double resonance experiments. Pressing the **SIM** button will run the **Simulation** and write the calculated spectrum with the default name …\999999.1R into the temporary directory. In this example the spectra calculated in WIN-DAISY and WIN-DR are identical because no parameter have been modified in WIN-DR. **Display** the **Calculated Spectrum** (red peak button) with WIN-NMR.

Check it in WIN-NMR:

WIN-NMR is moved to the WINDOWS foreground with the calculated spectrum displayed. Toggle back to **WIN-DAISY**.

Check it in WIN-DAISY:

Press the **Export** to **WIN-DYNAMICS** button.

Check it in WIN-DYNAMICS:

The program WIN-DYNAMICS is started and moved to the foreground and immediately the file A.DAT containing the parameter sent from WIN-DAISY loaded. Run a **Simulation** and **Display** the **Calculated Spectrum** (red peak button) with WIN-NMR.

Check it in WIN-NMR:

Use the shortcut <Alt + F4> to exit the program WIN-NMR (keep the Alt key depressed while pressing the function key F4).

Check it in WIN-DYNAMICS:

Because WIN-NMR has been closed WIN-DYNAMICS will now have the focus. Use <Alt + F4> to close WIN-DYNAMICS. (If there is no application program running and WINDOWS has the focus the <Alt + F4> command will shut down the system.)

Check it in WIN-DAISY:

Use the **File | Exit** command from the pull-down menu to close WIN-DAISY. This is equivalent to the shortcut used before to end the WIN-NMR session and is standard for all WINDOWS programs.

Check it in WIN-DR:

Now the last program WIN-DR gets the focus. Use the **File | Exit** command to close WIN-DR.

1.6.4 Book Formatting

A number of conventions have been adopted in the text:

- An indented paragraph with the string in the format shown below introduces the "Check it" exercise:

    Check it in PROGRAM:

- The program name(s) used in the "Check it" exercise refer to the following programs:

    1. WINDOWS   the operating system itself
    2. WIN-NMR   the program 1D WIN-NMR*
    3. WIN-DAISY   the program WIN-DAISY
    4. CALCULATOR   the WINDOWS Calculator
    5. WIN-DR   the program WIN-DR
    6. WIN-DYNAMICS   the program WIN-DYNAMICS

- **Bold** printed letter, words and expressions refer to commands and controls:

    in case of bold letters:

    1. In chemical structures, the bold letters refer to the nuclei forming the spin system which is currently being discussed.
    2. For certain abbreviations the complete phrase is given with the letters that appear in the abbreviation printed in bold.

    in WIN-NMR:

    1. Pull-down menu commands, e.g. **Simulation | WIN-DAISY**. The sub menu command is separated from the by a pipe. The example given above means: from the Simulation pull-down menu select the submenu entry WIN-DAISY. To improve the text flow in the *Check its* the various commands are often embedded into the sentence, e.g. for **Analysis | Multiplets**, the phrase 'Analyze the **Multiplets** in spectrum…' or 'Perform a **Multiplets Analysis**' or for **Analysis | Linewidth** simply 'measure a proper **Linewidth**' or enter the '**Linewidth** mode of'.
    2. Many menu commands change the mode of WIN-NMR and the buttons present in the button panel e.g.

---

* Because in this volume only 1D WIN-NMR will be used, no distinction has to be made between 1D and 2D WIN-NMR, thus the abbreviation WIN-NMR is used.

14   *1 Introduction*

> *2, **FT!** in initial display mode
> **Move Trace** in the **Display | Dual Display** mode
> **Calibrate** in the **Analysis | Calibration** mode
> Some of the buttons always appear in the button panel, e.g. *2, <>, **ALL**. The three buttons of the third line define the cursor mode and are referred to in the text as **zoom**, **maximum cursor** and **rectangular cursor** mode.

3. Buttons in dialog boxes, e.g.
   Execute in the **Options** dialog of the **Peak Picking** mode
   Export in the **Report...** dialog box of the **Multiplets** mode
4. Buttons in message boxes, e.g. **Yes**, **No**, **Ok**, **Cancel**.
5. Check boxes in dialog boxes, e.g. **Interpolation** in the **Peak Picking Options**.
6. Radio buttons and their group title, e.g. **ppm**, **Hz** and **Points** as the **Peak Labels** in the **Peak Picking Options**.
7. The text description for edit fields, e.g. to identify in the **Multiplet Mode** in the **Options** dialog box the **Min. Intensity** edit field.

in WIN-DAISY, WIN-DR and WIN-DYNAMICS:

1. These programs exhibit a toolbar button containing the most important pull-down menu commands. e.g. the pull-down menu command **Parameter | Main Parameters** in WIN-DAISY. When moving the mouse cursor over the toolbar buttons the corresponding pull-down menu commands are displayed as tooltips, and a more detailed description is given in the status line at the bottom of the main window. For the pull-down menu command **Parameter | Main Parameters** the tooltip text is abbreviated to the submenu entry **Main Parameters**, which is used throughout this book. The programs WIN-DR and WIN-DYNAMICS exhibit the same button layout and identical button symbols as WIN-DAISY and do not display tooltips. The menu-commands / toolbar buttons mostly open dialog boxes, execute calculations or call other programs.
2. Buttons in dialog boxes
   a) The standard buttons **Ok**, **Cancel**
   b) The buttons **Next** and **Previous** to display the corresponding **Parameter** dialog boxes.
   c) Buttons to display subdialog boxes, e.g. **Iter. Regions** in the **Spectrum Parameters** or to execute a function, e.g. **Auto Limits** in the same dialog of WIN-DAISY.
3. Buttons in message boxes, e.g. **Yes, No, Ok, Cancel**.
4. Check boxes in dialog boxes, e.g. **Disable** or **Iterate** in the **Frequencies** dialog box or in the **Lineshape Parameters** the **Nuclei specific linewidths** in WIN-DAISY.
5. Radio buttons and their group title, e.g. **ppm**, **Hz** in the **Frequencies** dialog box or in the **Symmetry Group** dialog box the **Select Symmetry** box containing the items **C1**, **C2**, **C3** etc. in WIN-DAISY.

6. The text description for edit fields, e.g. to identify in the **Spectrum Parameters** the **Number of spectral points** edit field.
7. Selections in list boxes or combo boxes and their description, e.g. **Single Simulation Mode** in the **Control Parameters**, the **Frequencies adapted to Spectrometer Frequency** of e.g. **500** MHz in the **Spectrum Parameters** or the list box on the main document window in WIN-DAISY.

- Key board strokes are given in <brackets>:

  1. Shortcut keys: keeping the control key depressed while another key is pressed. In WIN-NMR for some display options, e.g. <Ctrl + X> to switch the x-axis labels. In WIN-DAISY some shortcuts are shown in the pull-down menus, e.g. <Ctrl + W> for the run **WIN-NMR** command. <Ctrl + C> is used to copy the marked item into the WINDOWS Clipboard and <Ctrl + V> to paste contents of the Clipboard at the current cursor placement.
  2. Manual entries in edit fields in dialog boxes, e.g. in the WIN-NMR **Calibration** mode, clicking on the **Calibrate** button, enter a value <0> into the edit field to set the cursor position to zero ppm. In WIN-DAISY in the **Frequencies** dialog box enter another shift value, e.g. <2.5> ppm, or in the **Spectrum Parameters** set the **Number of spectral points** to <8K> Points for 8·1024 points. Most of the entries are of numerical type, an exception is the **Title** entry in the WIN-DAISY **Main Parameters** dialog box.

- File names are printed in CAPITAL LETTERS:
  1. WIN-DAISY input files have the extension 'MGS', e.g.: A.MGS and are found in subdirectory 'DATA' of the WIN-DAISY program path, by default: C:\TEACH\DAISY\DATA.
  2. WIN-DAISY simulation protocol files have the extension 'LST', e.g.: A.LST and are found in the subdirectory 'DATA' of the WIN-DAISY program path.
  3. WIN-NMR spectrum files with extension '1R' referring to the real spectrum part:
     a) For experimental WIN-NMR spectra e.g. ...\AX\001001.1R and the corresponding WIN-DAISY calculated spectra e.g. ...\AX\001999.1R the spectrum name consists of a six-digit number and a directory name which is located in the WIN-NMR spectrum path installed during the setup. For the default installation, '...' refers to the default path C:\TEACH\WIN1D\SPC.
     b) WIN-DAISY simulated spectra with no connected experimental data are written into the temporary directory of WIN-NMR e.g. ...\999999.1R. For the default installation, '...' refers to the default path C:\TEACH\ TEMP.
  4. WIN-DR input files have the extension 'WPR', e.g.: A.WPR and are found in the subdirectory 'DATA' of the WIN-DR program path., by default C:\TEACH\DR\DATA. This Teaching Version will only accept data exported from WIN-DAISY, it is not possible to store or open files.

16 1 Introduction

5. WIN-DYNAMICS input files have the extension 'DAT', e.g.: A.DAT and are found in the subdirectory 'DATA' of the WIN-DYNAMICS program path, by default C:\TEACH\DYNAMICS\DATA. Again, this Teaching Version will only accept data exported from WIN-DAISY, it is not possible to store or open files.

- Text written in *Italic* is used for equations, definitions, compound names below the corresponding structure in the "Check it" sections and to emphasize important statements, e.g.:

  1. Equation: $\lambda_{ik} = \left| \dfrac{J_{ik}}{v_i - v_k} \right| = \left| \dfrac{J_{ik}}{\Delta_{ik}} \right|$

  2. Definition: *Spin-active nuclei with identical resonance frequencies (chemical shifts) are called isochronous.*

  3. Compound names: *2-tert.-butyl-4,6-dinitro-phenol*

  4. Statements: 'A small *second order effect* is still present.'

  5. Keywords in the protocol files, e.g. *R-factor (%), correlation factor*.

- SMALL CAPITALS identify peoples names, e.g.
  LARMOR, FOURIER, EIGEN, HAMILTON, etc.

1.7 Recommended Reading and References

There are many excellent text books on NMR spectroscopy and references [1.16]-[1.22] are a representative selection which deal with the subject in varying degrees of detail. For more information about the practical aspects of NMR data processing the first volume of this series is recommended [1.15].

As a basis for the spectral analysis discussed in this book, the corresponding chapters in references [1.16]-[1.22] are adequate. Several monographs and reviews have been published containing detailed information about the quantum mechanics and analytical equations of individual spin systems [1.5]-[1.11], which are recommended only if you require further details about particular spin systems.

[1.1] Purcell, E.M., Torrey, H.C., Pound, R.V., *Phys. Rev.*, 1946, *69*, 37.
[1.2] Bloch, F., Hansen, W.W., Packard, M.E., *Phys. Rev.*, 1946, *69*, 127.
[1.3] Anderson, W.A., in *Encyclopedia of Nuclear Magntic Resonance*, Grant, D.M., Harris, R.K. (Ed.), Chichester: Wiley, 1996, *3*, 2126-2136.
[1.4] Ernst, R.R., in *Encyclopedia of Nuclear Magntic Resonance*, Grant, D.M., Harris, R.K. (Ed.), Chichester: Wiley, 1996, *5*, 3122-3132.
[1.5] Corio, P.L., *Structure of high-resolution NMR spectra*, London: Academic Press, 1966.

[1.6] Bishop, E.O., in *Annual Reviews of NMR Spectroscopy*: Mooney, E.F., Webb, J.A. (Eds.) London: Academic Press, 1968, *1*, 91.
[1.7] Hoffman, R.A., Forsén, S., Gestblom, B., in: *Basic Principles and Progress*: Diehl, P., Fluck, E., Kosfeld, R. (Eds.) Berlin: Springer, 1971, *5*, -.
[1.8] Abraham, R.J., *The Analysis of high resolution NMR Spectra*, Amsterdam: Elsevier, 1971.
[1.9] Swalen, J.D., *Progr. in NMR Spectr.*, 1966, *1*, 205.
[1.10] Haigh, C.W., in: *Annual Reports on NMR Spectroscopy*: Mooney, E.F., Webb, J.A. (Eds.), *4*, London: Academic Press, 1971, pp. 311–362.
[1.11] Diehl, P., Kellerhals, H., Lustig E., in: *NMR - Basic Principles and Progress*, Berlin: Springer, 1972, *6*, 1–90.
[1.12] Weber, U., Thiele, H., Spiske, R., Hägele, G., in: *Software Development in Chemistry*, Moll, R. (Ed.), Berlin: Springer, 1995, Vol. *9*, pp. 268–281.
[1.13] Radeglia, R., unpublished results.
[1.14] Traficante, D.D., *Abstracts*, XIX National NMR Symposium, Keuruu, Finland, June 1997. Jyväskylä, Finland, 1979 and
Traficante, D.D., Meadows, M.D., *Concepts in Magn. Reson.*, 1997, *9*, 359.
[1.15] Bigler, P., *NMR Spectroscopy: Processing Strategies*, Weinheim: Wiley-VCH, 1997.
[1.16] Kemp, W., *NMR in Chemistry - A Multinuclear Introduction*, London: Macmillan, 1986.
[1.17] Sanders, J.K.M, Hunter, B.K., *Modern NMR Spectroscopy - A guide for chemists*, Oxford: Oxford University Press, 2nd edition, 1994.
[1.18] Akitt, J.W., *NMR and Chemistry- An introduction to modern NMR spectroscopy*, 3rd ed., London: Chapman & Hall, 1992.
[1.19] Günther, H., *NMR Spectroscopy - Basic Principles, Concepts and Applications in Chemistry*, 2nd ed, Chichester: Wiley, 1995.
[1.20] Friebolin, H., *Basic One- and Two-Dimensional NMR Spectroscopy*, 2nd ed., Weinheim: VCH, 1993.
[1.21] Harris, R.K., *Nuclear Magnetic Resonance Spectroscopy*, Harlow: Longman, 1986.
[1.22] Canet, D., *Nuclear Magnetic Resonance - Concepts and Methods*, Chichester: Wiley, 1996.

2 General Characteristics of Spin Systems

Like any other spectroscopic method, Nuclear Magnetic Resonance (NMR) spectroscopy is based on the interaction between energy and matter.

The prerequisite for the detection of a nuclear magnetic resonance effect is the presence of a spin angular momentum (spin quantum number) I larger than zero arising from a magnetic moment µ. Only those isotopes with a nuclear spin quantum number I, greater than zero posses a nuclear magnetic moment µ and are NMR active, e.g. ^{1}H, ^{31}P. Isotopes with an even number of protons and an even number of neutrons, e.g. ^{12}C, ^{16}O have a zero nuclear magnetic moment and cannot be observed by NMR.

In absence of a magnetic field the spins are randomly oriented in matter so that the nuclear energy levels are degenerate. A static homogeneous magnetic field must be applied to abolish the degeneracy of the nuclear energy levels before it is possible to observe transitions according to the BOHR frequency condition $\Delta E = h\nu$. The stronger the field, the larger is the energy gap between the energy levels. The proportionality factor between the resonance frequency ν_0 and the magnetic field strength B_0 is the magnetogyric ratio. The magnetogyric ratio is a specific property for every isotope possessing a magnetic moment µ.

$$\nu_o = \frac{\gamma}{2\pi} \cdot B_0 \quad \textit{resonance condition / LARMOR frequency}$$

The differences between the energy levels correspond to the region of VHF radio waves (MHz) on the frequency scale. The number of levels depends on the magnitude of the nuclear spin quantum number. The presence or absence of a nuclear spin moment can be detected and bosons and fermions can easily be distinguished. Bosons are nuclei that have an odd number of protons and an odd number of neutrons and posses an integer nuclear spin quantum number I, e.g. ^{2}H, ^{10}B, ^{14}N while fermions have one number odd and the other even and have a half integer I value, e.g. ^{1}H, ^{11}B, ^{13}C, ^{15}N.

The phenomenon of nuclear resonance was predicted by PAULI on the basis of the hyperfine splitting of atomic spectra. The effect was discovered by PURCELL, TORREY and POUND [2.1] and BLOCH, HANSEN and PACKARD [2.2] in 1945. At that time the method seemed to be merely able to distinguish between different magnetic isotopes as described above; this would not explain the enormous field of applications available today. The chemical shift effect was found by PROCTOR and YU [2.3] and DICKINSON [2.4] in 1950; in the intervening years NMR has become a standard spectroscopic method in chemical laboratories.

2.1 The Chemical Shift

2.1.1 Spin Systems Without Coupling

The importance of NMR spectroscopy in chemistry is based on the fact, that the resonance frequency is dependent on the chemical environment in which the nucleus is situated. As discussed in the previous section, different isotopes show resonance frequencies differing by many MHz, while the differences between resonances of the *same* isotope caused by different chemical environments are of the order of Hz to kHz. These frequency differences are dependent on the strength of the applied magnetic field.

This dependence of the magnetic resonance frequency on the chemical environment of the nucleus is called the chemical shift. This phenomenon arises from the fact that in the applied magnetic field the nuclei are shielded (screened) by their electrons [2.5]. The electron density of the shell introduces a local magnetic field opposing the primary applied field B_0. Therefore the effective magnetic field operating on the nucleus is:

$$B_{local} = B_0 \cdot (1 - \sigma)$$

where σ is the shielding constant and is composed of two terms:

$$\sigma = \sigma_{dia} + \sigma_{para}$$

The diamagnetic screening σ_{dia} is reduced by a paramagnetic effect σ_{para} (of opposite sign as σ_{dia}) caused by bonding and the presence of positive centers. The constant σ is dimensionless and is normally given in **p**art **p**er **m**illion (ppm). The shielding constant cannot be detected directly, and chemical shifts δ are reported relative to a standard substance. For a constant magnetic field B_o the chemical shift is defined as follows:

$$\delta = \frac{v_{sample} - v_{reference}}{v_{reference}} \cdot 10^6 \ [ppm]$$

where $v_{reference}$ is the resonance frequency of reference standard.

The resulting chemical shift δ is only dependent on the shielding parameter σ_{sample} of the sample compound and is therefore a molecular property. This dimensionless parameter δ allows direct comparison of chemical shift measurements performed at different external magnetic field strengths. Table 2.1 gives an overview of the reference standards used for common NMR nuclei:

Table 2.1: Common reference standard compounds

nuclear isotope	NMR standard
$^1H, ^{13}C, ^{29}Si$	$(CH_3)_4Si$
^{31}P	External 85% H_3PO_4/H_2O
^{17}O	H_2O
^{19}F	CCl_3F
^2H	$(CH_3)_4Si\text{-}d_{12}$
^{11}B	$Et_2O\cdot BF_3$

Check it in WIN-NMR:

Open... the first 2 spectra from the examples directory ...\SHIFT\001001.1R and ...\SHIFT\002001.1R one after another into WIN-NMR. **Calibrate** the spectra using the corresponding **Analysis** mode. Mark the TMS peak with the maximum cursor and press the **Calibrate** button. Enter a new ppm value of <0.0> ppm and press the **Ok** button. **Save** the spectra after calibration.

Both spectra are obtained from the same compound but measured at different field strengths, one at 500 MHz the other at 80 MHz. As discussed the ppm scale allows the direct comparison independent on the field strength.

Check it in WIN-NMR:

Call the **Multiple Display** mode. Select the two calibrated spectra ...\SHIFT\001001.1R and ...\SHIFT\002001.1R with the mouse pointer. Click on the **Separate** button to see that the ppm scale allows the direct comparison of spectra recorded at different field strength. Switching the X-axis label into frequency units <Ctrl + X> illustrates the fact that in a 500 MHz spectrum 1 ppm represents 500 Hz while in a spectrum recorded at 80 MHz 1 ppm is 80 Hz.

The chemical shift depends on various factors which affect the electron density at the nuclei such as the electronegativity of substituents, solvent effects, concentration, pH and temperature. Nevertheless it is possible to define reliable intervals of chemical shifts for certain chemical structure fragments. Table 2.2 gives a rough overview of the regions where the resonance peaks for ^1H nuclei in various organic fragments can be expected.

Table 2.2: ^1H NMR shifts of some common organic fragments

structure fragment	chemical shift range [ppm]
C-H with sp^3-carbon	5.5 - -1.0
C-H with sp-carbon	3.0 - 2.5
C-H with sp^2-carbon	8.0 - 4.5
aromatic C-H (sp^2 carbon)	8.9 - 6.5
C-H of aldehydes (sp^2 carbon)	10.5 - 9.5
S-H of thioalcoholes	4.0 - 3.5
N-H of amines	4.8 - 3.7
O-H of sp^3 carbons	5.3 - 1.0
O-H of sp^2 carbons	10.0 - 4.0

More shielded protons appear at lower ppm, less shielded at higher ppm values. In discussion about such effects it is common to talk about high and low field shifts. This originates from early NMR years where field sweep experiments were the common technique. Nuclei appearing at high field are strongly shielded and show a lower frequency (low ppm value) in contrary to deshielded nuclei, resonating at higher frequency values (high ppm values). Although field has no concept in modern FT spectra this term is still used by many chemist, hence field will be used for clarity. However, remember that in reality a change to high field is a shift to low frequency.

Within the range of ^{1}H-NMR chemical shifts for the fragments in Table 2.2, the exact shift is determined by the effect of the substituents. Electronegative substituents cause a deshielding effect on the attached nuclei, so that proton resonance lines are shifted to higher frequency (higher chemical shift values) as the electronegativity of the substituent directly bonded to the carbon atom increases. For every NMR active isotope a correlation of chemical shifts with various structural parameters can be observed. Chemical shift tables are an important tool for the classification and structural assignments of compounds.

The theory of NMR chemical shift implies that all factors contributing to the electron density of the nucleus will influence its chemical shift. Chemical shifts can be based on the assumption empirical correlation's and additivity of substituent effects [2.6]–[2.8]. This empirical method takes into account only connectivities through bonds and other effects such as through-space contributions, while resonance effects are neglected. Prediction methods based on empirical substituent effects on chemical shifts are therefore not always satisfactory. Recently it has become possible to use quantum mechanical ab-inito programs to calculate the chemical shift such methods as the IGLO [2.9] and GIAO [2.10] approach.

2.1.2 The NMR Resonance Frequency

The basic resonance phenomenon of NMR can be explained using a single isolated spin active nucleus. By applying a homogenous external magnetic field B_o to the nucleus an energy difference within the nuclear energy levels is introduced. Each energy level is described by a distinct spin orientation in the magnetic field. The number of different energy levels is given by the *multiplicity g*, which is directly related to the *spin quantum number I*:

$$g = 2 \cdot I + 1$$

For some common NMR nuclei the multiplicities are listed in the following Table 2.3:

Table 2.3: Multiplicities of common NMR nuclei

Spin quantum number I	Multiplicity g	Spin active nuclei
½	2	^{1}H, ^{13}C, ^{29}Si, ^{19}F, ^{31}P
1	3	^{2}H (=^{2}D)
3/2	4	^{11}B
5/2	6	^{17}O
3	7	^{10}B

Each of the g spin orientations is characterized by its z component of the spin magnetization m_z as shown in Fig. 2.1 for $I = ½$ and in Fig. 2.2 for $I = 1$. The spin orientation is quantized by h (PLANCK's constant) and if we scale m_z in terms of h the levels are always separated by one unit. Without any perturbation by interaction with other spins the altogether g numbers of energy levels are equidistant in m_z scale.

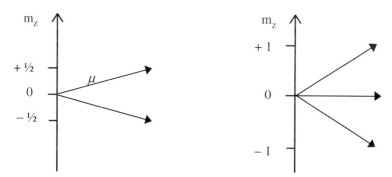

Fig. 2.1: Spin orientations for I = ½   **Fig. 2.2:** Spin orientations for I = 1

The range of values for the g numbers of spin orientations is defined as follows:

$$m_z \in [-I, +I] \text{ in steps of 1 unit: } \{-I, -I+1, \ldots, +I-1, +I\}$$

The selection rule defining the energy levels between which transitions can be detected using simple NMR experiments where only single quantum transitions give raise to detectable magnetization is:

$$\Delta m_z = \pm 1$$

The sign of the magnetogyric ratio determines whether the state with higher or lower m_z value corresponds to the higher energy. For positive magnetogyric ratio (e.g. ^{1}H, ^{2}H, ^{13}C, ^{19}F, ^{31}P) the antiparallel alignment of the spin (negative m_z) of the magnetic field

24   *2 General Characteristics of Spin Systems*

corresponds to the higher energy level. The exact selection rule for a positive magnetogyric ratio is as follows:

$$\Delta m_z = -1$$

Only transitions caused by absorption of one energy quantum (decreasing the m_z value by one unit) are possible.

In the following sections of chapter 2 and 3 we will focus on spin ½ nuclei, while higher spin values will be discussed in chapter 4. In section 2.4 the principle for understanding magnetic equivalence will be introduced using the concept of composite particles based on higher spin values.

2.1.2.1 The NMR Transition Frequency

Check it in WIN-DAISY:

Call the program **WIN-DAISY** directly from the program manager or from the WIN-NMR **Simulation** menu. **Open...** the WIN-DAISY document file A.MGS.* Examine the Spin System basic parameters in the **Main Parameters** dialog box using the pull-down menu **Parameter | Main Parameters** or the corresponding toolbar button or double-click the title **one spin system** in the list box of the document main window. The definition of the spin system contains only a single spin as to be seen from the **Number of spins or groups with magnetical equivalent nuclei**. Here some **Output Options** are already selected, which will be discussed later in this chapter. Quit the dialog box with **Ok** or **Cancel** (because no changes are made it does not matter). The nucleus **H** (proton) and its resonance frequency (v_0) is defined in the **Frequencies** dialog box. You may switch between **Hz** and **ppm** units. In the **Spectrum Parameters** the data for the digitized lineshape to be calculated are defined. Execute a synthesis of the NMR spectrum based on the parameters defined by pressing the **SIM** button or using the **Run | Simulation** menu command. The **calculated spectrum** can be **display**ed in WIN-NMR by pressing the toolbar button marked with a red peak. (This button is enabled only after the calculation is finished or if the spectrum with the corresponding name given in the WIN-DAISY document main window **Calculated Spectrum** status line already exists).

Check it in WIN-NMR:

Change the scaling to Hertz <Ctrl + X> and you will see the resonance peak at the given resonance frequency.

This rather simple demonstration of a one spin system is intended to introduce the simple relation between resonance frequency of the nucleus and the transition frequency appearing in the spectrum. The quantum mechanical formalism behind will not be discussed here in detail. For complete theory the reader is referred to textbooks [2.11]–

---

*   The data are found in the subdirectory DATA of the WIN-DAISY installation path, by default: C:\TEACH\DAISY\DATA\.

[2.15]. Here simply the basic concept which forms the basis for understanding the signal pattern of complicated spin systems will be mentioned.

If the m_z scheme in Fig. 2.1 is redrawn as an energy diagram (Fig. 2.3) it has to be considered, that the antiparallel alignment of the spin relative to the magnetic field ($m_z=-½$) corresponds to the higher energy level:

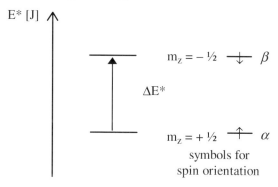

Fig. 2.3: Energy level diagram for a single spin

In all the following chapters energies are always written in units of HERTZ and not in JOULES. The conversion factor between both units is PLANCK's constant h. In this way the NMR transition frequency, which is represented by the difference between both energy levels, is directly obtained in frequency units.

$$E * [J] = \frac{E*}{h}[Hz] = E[Hz]$$

For a single spin ½ nucleus in an external magnetic field only the two energy levels shown in Fig. 2.3 are present. The magnetic field causes the nuclei to align their spins, which would otherwise be randomly oriented in the absence of the magnetic field. The observation of a transition is possible due to the population differences as given by BOLTZMANN statistics. Radio frequency that corresponds to the exact value of the energy gap between the two energy levels induces the transition.

The parallel and antiparallel orientations of the nuclear spin with respect to the applied magnetic field can be expressed using the m_z values (+½, −½), the spin up (↑) and spin down (↓) notation or the corresponding *spin basis functions* (α, β).

Check it in WIN-DAISY:

Load the **Simulation Protocol** A.LST into an editor using the corresponding toolbar button for displaying: **Show Simulation Protocol**. The **Output Option Linear combinations** selected in the **Main Parameters** lists the basis functions used to represent the spin system. So that any editor may be used to display the functions, Latin letters *(a)* and *(b)* are used instead of Greek 'α' and 'β'.

In quantum mechanical terms the m_z value is the EIGENvalue of the spin operator $\hat{I}_z$ when the latter is applied to the one spin basis function |φ⟩ ('ket' in DIRAC's notation):

26 2 General Characteristics of Spin Systems

$$\hat{I}_z|\phi\rangle = m_z|\phi\rangle$$

Two equations corresponding to two different one-spin wave functions are obtained for a single spin with $I = \frac{1}{2}$:

$$\hat{I}_z|\alpha\rangle = +\frac{1}{2}|\alpha\rangle$$
$$\hat{I}_z|\beta\rangle = -\frac{1}{2}|\beta\rangle$$

As we are only considering chemical shifts in this section the HAMILTONian simply consist of the negative I_z operator, so that the corresponding EIGENvalues are given by the so-called ZEEMAN term. The nuclear energy levels correspond to the EIGENvalues given by the following equation for zero order ZEEMAN energies:

$$E_j = -m_z(j) \cdot \nu_0$$

for every basis function $j \in \{\alpha, \beta\}$

As a result one EIGENvalue is obtained for every basis function, so that the basis function refers to an EIGENstate of the spin system. For a one spin system with spin quantum number $I=\frac{1}{2}$ the basis functions (α, β) are EIGENfunctions describing EIGENstates of the spin system and the values $(-m_z \cdot \nu_0)$ are the corresponding EIGENvalues. In this case the m_z values are called *good quantum numbers*. We will expand this concept of good quantum numbers to higher spin systems in following sections.

Check it in WIN-DAISY:

The **Output Option EIGENvalues** in the **Main Parameters** dialog box writes the determined *EIGENvalues* (energy levels) into the **Simulation Protocol** file after the basis functions are written. Each energy level is assigned a number to identify the origin and destination levels for the transitions.

According to the selection rule the NMR transition frequency is the difference between the destination and origin energy levels. The higher energy level for nuclei with positive magnetogyric ratio γ belongs to the wave function β (nuclear spin anti parallel to the applied field) as follows:

$$E_{destination} = E_\beta = +\frac{1}{2}\nu_0$$
$$E_{origin} = E_\alpha = -\frac{1}{2}\nu_0$$

The transition frequency in the resulting NMR spectrum is then given by:

$$\Delta E = E_{destination} - E_{origin} = \nu_0$$

So in the present case of one spin interacting situated in a homogenous magnetic field will resonate at its LARMOR frequency ν_0.

Check it in WIN-DAISY:

The **Output Option Transitions and energy levels** of the **Main Parameters** lists the transition frequency, the intensity and the corresponding energy level indices. The single peak in the spectrum is caused by the single possible transition.

2.1.2.2 The NMR Transition Intensity

The NMR signal is not only described by its frequency but also by its intensity which is determined by the value of the square of the transition probability M^2. For a single spin system the operator consists of the lowering operator $\hat{I}_-$.

The lowering operator changes the spin function into the function with a m_z value *one* unit lower than the original function according to the selection rule: function $|\alpha>$ ($m_z=+\frac{1}{2}$) is changed into the function $|\beta>$ ($m_z=-\frac{1}{2}$). If this operator is applied to the function $|\beta>$ whose m_z value cannot be lowered any further, the function vanishes.

We will not go into detail for calculation of the transition intensities, for further reading the user is referred to the corresponding textbooks [2.11]–[2.15] and the appendix, section 6.1.

Check it in WIN-DAISY:

The calculated intensity for the transition is reported in the file A.LST as *1.0*. Here the quantum-mechanical calculated transition probability is listed.

2.1.2.3 The NMR Lineshape

The signals detected in NMR spectroscopy are not "infinitely sharp" due to the HEISENBERG uncertainly principle. The transition is broadened due to relaxation processes. For more detailed information about relaxation phenomena refer to section 5.1. NMR lines normally display a LORENTZian lineshape $f_L(\nu)$ [2.11]:

$$f_L(\nu) = \frac{I_i \cdot T_2^*}{1 + 4\pi^2 (\nu_i - \nu)^2 T_2^{*2}}$$

where ν_i: transition frequency
$\quad\quad\quad I_i$: transition amplitude
$\quad\quad\quad T_2^*$: effective transversal relaxation time

Therefore the intensity of an NMR signal refers to the *integral* of the corresponding peak, not to the *amplitude* (peak height). The integral is defined as the area under the LORENTZian lineshape and refers to the transition probability M^2, discussed in the previous section. The relation between the NMR linewidth at half-height $\nu_{1/2}$ and the effective transversal relaxation time T_2^* is as follows:

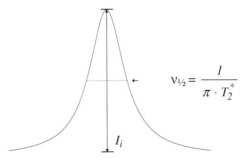

Fig. 2.4: Relation between linewidth at half-height and LORENTZian lineshape.

The clear consequence of this is as follows: the longer the relaxation time T_2^* the sharper the NMR signal will be.

Check it in WIN-NMR and WIN-DAISY:

From the **Analysis** pull down menu chose the **Linewidth** option and determine the linewidth at half-height of the calculated LORENTZian signal: Mark the peak top with a mouse click in the **maximum cursor** mode and press the right hand mouse button. The location where the linewidth is measured is displayed and the value is shown in the header status line.

Return to WIN-DAISY by selecting the **WIN-DAISY** option in the **Simulation** pull-down menu of WIN-NMR. Then WIN-NMR will automatically return to the normal display mode. Compare the determined linewidth with the given linewidth in the WIN-DAISY **Lineshape Parameters** dialog box. Change the global linewidth parameter, e.g. to <1.5> Hz, quit the dialog box with **Ok** to accept the new value, execute a **Simulation**, **display** the **calculated** signal in WIN-NMR and determine the linewidth in the **Analysis Linewidth** mode again.

If the linewidth parameter value is decreased to a very low value, such as *0.01 Hz*, WIN-NMR cannot determine the linewidth correctly because of insufficient digital resolution. Toggle back to **WIN-DAISY** and open the **Lineshape Parameters** again to change the global linewidth to <0.01> Hz and **Simulate** and **Display** the **Spectrum**. You may visualize the digital resolution of an NMR peak by changing the WIN-NMR display from the default line mode to the line-and-cross mode <Ctrl + O> and expand the x-scale with the <> button of the panel bar. Switch into the **Analysis Linewidth** mode and determine the linewidth parameter and compare it with the value you entered in WIN-DAISY. Toggle back to **WIN-DAISY**.

To avoid problems due to insufficient digital resolution the number of points for the total spectrum can be increased or the limits of the lineshape calculation for the spectrum can be changed in the WIN-DAISY **Spectrum Parameters**. The easiest way to obtain a reasonable digital resolution is to use the **Auto Limits** option. Press the button, quit the dialog with **Ok**, execute a **Simulation**, **Display** the **Calculated Spectrum** and determine the linewidth of the peak in the WIN-NMR **Analysis Linewidth** mode.

Close the WIN-DAISY document using the **File | Close** pull-down menu command.

2.1.3 Two Non-Coupled Spins

In this section NMR spin systems consisting of two spin active nuclei without any interaction between them will be introduced.

For a single spin systems the single-spin wave functions α, β are used to describe the basis functions. An easy way to define a functional basis for multiple spin systems is the creation of *basis product functions (bpf)*. For this purpose all possible combinations of the single spin functions are set up.

For a system, consisting of two nuclei with spin $I=½$ four combinations are possible (to be read vertically):

nucleus 1: |α>, |α>, |β>, |β>;   or symbolized by ↑, ↑, ↓, ↓.
nucleus 2: |α>, |β>, |α>, |β>;   or symbolized by ↑, ↓, ↑, ↓.

If N is the number of spins in the spin system, the maximum number of combinations and therefore the number of basis product functions is 2^N.

The functional basis of a two spin system is:

$$|αα>, |αβ>, |βα> \text{ and } |ββ>.$$

To characterize these new functions the total spin value m_T can be determined based on the total spin operator F_z:

$$\hat{F}_z = \sum_{i=1}^{N} \hat{I}_{z,i}$$

Every operator $\hat{I}_{z,i}$ effects only the corresponding basis function of spin i and leaves all other spins of the basis product function unaffected, so that the EIGENvalue which corresponds to the energy is built as the sum of all m_z values of the functions involved. This sum of the m_z values is called the total spin value m_T.

The total spin value is discussed here, because the field contribution to the HAMILTONian is composed of the F_z operator and the resonance frequencies of the member spins. On condition that the spins 1 and 2 do *not* interact (i.e. there is no coupling between them) all four basis product functions are EIGENstates of the two-spin system and we can obtain directly the energy levels by the following formula:

$$E_j = -\sum_{i=1}^{N} m_{z,i} \cdot v_i \quad \text{for every EIGENfunction } j$$

Consequently the sum of the m_z values, the total spin values m_T, are here as well good quantum numbers.

Applied to the two-spin system we obtain:

$$E_{\alpha\alpha} = -\tfrac{1}{2}\nu_1 - \tfrac{1}{2}\nu_2$$
$$E_{\alpha\beta} = -\tfrac{1}{2}\nu_1 + \tfrac{1}{2}\nu_2$$
$$E_{\beta\alpha} = +\tfrac{1}{2}\nu_1 - \tfrac{1}{2}\nu_2$$
$$E_{\beta\beta} = +\tfrac{1}{2}\nu_1 + \tfrac{1}{2}\nu_2$$

There are two possibilities:
- either both nuclei have the same resonance frequency $\nu_1 = \nu_2$,
- or both spins exhibit different resonance frequencies $\nu_1 \neq \nu_2$.

2.1.3.1 Two Isochronous Spins

Spin active nuclei with identical resonance frequencies (chemical shifts) are called isochronous.

This definition does not distinguish between spins that are in the same chemical environment and spins who's chemical shifts are identical by chance.

In the case where $\nu = \nu_1 = \nu_2$ the expressions for the energy levels can be written as follows:

$$E_{\alpha\alpha} = -\nu$$
$$E_{\alpha\beta} = E_{\beta\alpha} = 0$$
$$E_{\beta\beta} = +\nu$$

Energy levels with the same energy value are referred to as degenerate.

The energy level diagram given in Fig. 2.5 illustrates the degeneracy of the both energy levels with total spin value $m_T = 0$.

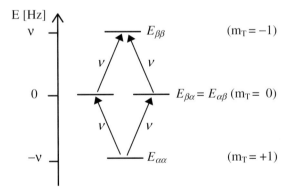

Fig. 2.5: Energy level diagram for two uncoupled isochronous spins.

The four transitions arising from the selection rule $\Delta m_T = -1$ all have the same frequency ν. In analogy to the EIGENstates they are called *degenerate transitions*.

2.1 The Chemical Shift

As in the case of the one-spin system, the two-spin system discussed here refers to a non-perturbed system, so that none of the basis functions mixes with the others. All the basis functions describe directly EIGENstates of the system (corresponding m_T values are good quantum numbers). The only difference is that the EIGENfunctions are represented by product functions. To calculate the transition probability the one-spin lowering operator is replaced by the total lowering operator in the same way as for the z component of the spin momentum. No details are discussed here, just the result shall be mentioned.

Four non-zero transition probabilities result due to the selection rule $\Delta m_T = -1$. Neither zero quantum transitions $\Delta m_T = 0$ nor double quantum transitions $\Delta m_T = -2$ are allowed.

All non-zero transition probabilities are listed below:

$$M^2_{\beta\alpha-\alpha\alpha} = 1, \quad M^2_{\beta\beta-\alpha\beta} = 1$$
$$M^2_{\alpha\beta-\alpha\alpha} = 1, \quad M^2_{\beta\beta-\beta\alpha} = 1$$

The overall intensity in relative quantum mechanical units for the two-spin system is '4'. We should remember that the intensity connected with a single spin is '1'. For a spin system consisting of N spins the total intensity for the whole system is:

$$N \cdot 2^{N-1}$$

The one-spin contribution to the total intensity of the spectrum is given by:

$$2^{N-1}$$

It has to be mentioned that these rule is valid for all nuclei possessing spin quantum number $I = \frac{1}{2}$.

Check it in WIN-DAISY:

Re**open**... the input document A.MGS of the single-spin system and execute a **Simulation** with the original parameters again. Call the **Main Parameters** and increase the **Number of spins or groups with magnetic equivalent nuclei** to <2> and select the **Use all spins** option for the absolute intensity norming. Open the **Frequencies** parameter and set the resonance frequency of the second nucleus to the same value as the first nucleus <7.2> **ppm**. Click on the **Change** button on the WIN-DAISY document window or use the **Edit | Change Spectrum Name** command and alter the simulated spectrum name e.g. the number from ...\999999.1R to ...\999998.1R. Simulate the spectrum using the new parameter set by pressing the **SIM** button. Click on the **red peak button** to load the two-spin simulation into WIN-NMR.

Check it in WIN-NMR:

From the **Display** pull-down menu choose the **Dual Display** option and load the one-spin simulation as the **Second Filename** ...\999999.1R. The only difference visible is the absolute intensity of the peak. Change the intensity scaling in the Dual Display from **absolute** to **relative** using the **Options** button or simply press <Ctrl + Y>. No difference in the lineshape of the peak can be seen. Click on the **Separate** button to display the two simulated spectra in separate windows. Toggle back to **WIN-DAISY**.

Check it in WIN-DAISY:

Inspection of the **Simulation Protocol** list A.LST shows the difference. The basis functions are listed, four energy levels and four transitions, all with the intensity '1.0'. As there are no other peaks for comparison we cannot determine from the spectrum how many spins are responsible for the signal, because no absolute intensity measurement is possible in the experimental NMR spectrum of isolated systems.

2.1.3.2 Two Non-Isochronous Spins Without Coupling

For two spins with different resonance frequencies ($v_1 > v_2$) the resulting energy level diagram is constructed as shown in Fig. 2.6:

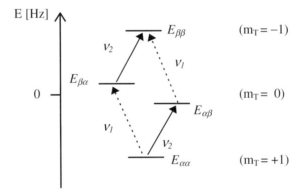

Fig. 2.6: Energy level diagram of a two spin system with different resonance frequencies but without coupling. Legend in the text.

The two energy levels for $m_T = 0$ are no longer degenerate. Four transitions are observed, as in the isochronous case. The transitions involving spin flip of nucleus 1 are denoted by dashed lines, those for a flip of nucleus 2 by full lines. The values of the transition frequencies are listed below. The symbols T_{1i} and T_{2i} are used to distinguish between the degenerate transitions of v_1 and v_2 respectively.

$$T_{11} = E_{\beta\beta} - E_{\alpha\beta} = v_1$$
$$T_{12} = E_{\beta\alpha} - E_{\alpha\alpha} = v_1$$
$$T_{21} = E_{\beta\beta} - E_{\beta\alpha} = v_2$$
$$T_{22} = E_{\alpha\beta} - E_{\alpha\alpha} = v_2$$

In the NMR spectrum two signals are observed, one at frequency v_1 and the other at v_2. Both signals consist of two degenerate transitions.

2.1 The Chemical Shift

The same equation for the transition probabilities as used for the isochronous spin system is applicable for the transition intensity, because only the EIGENstates and not the resonance frequencies of the nuclei, influence the transition moment. We may assign the four transition moments to the two transition frequencies v_1 and v_2:

$$\text{transition } v_1: \quad M^2_{\beta\alpha-\alpha\alpha} = 1, \quad M^2_{\beta\beta-\alpha\beta} = 1$$

$$\text{transition } v_2: \quad M^2_{\alpha\beta-\alpha\alpha} = 1, \quad M^2_{\beta\beta-\beta\alpha} = 1$$

The result shows that both signals in the spectrum have the same relative intensity of '2' at the position corresponding to the resonance frequencies v_1 and v_2. The one-spin contribution to the total intensity of the spectrum for a spin system of size of N is given by $2^{N-1}=2$ as discussed in the previous section.

Check it in WIN-DAISY:

Open... the input document AX.MGS or modify the file A.MGS again. You can do both, because WIN-DAISY can load more than one document at the same time: For the latter way call the **Main Parameters** of A.MGS and increase the **Number of spins or groups with magnetic equivalent nuclei** to <2> – if this is not still present from the last "Check it". In the **Frequencies** dialog box enter a value for the second resonance frequency different from the first nucleus, e.g. <7.3> **ppm**. Increase the limits of the calculated spectrum in the **Spectrum Parameters** dialog box to cover both resonance frequencies or simply press the **Auto Limits** button to set appropriate values automatically. Execute a **Simulation** and **Display** the **Calculated Spectrum** with WIN-NMR. You may compare the EIGENvalues listed in the **Simulation Protocol** file with the formulae derived above. **Close** the WIN-DAISY documents.

2.1.4 Basic Notation of Spin Systems

The convention of labeling nuclei according to their chemical shift values was introduced by POPLE, SCHNEIDER and BERNSTEIN (PSB-Notation) [2.16]. This notation uses letters of the Latin alphabet to label nuclei with different chemical shifts. Nuclei showing identical chemical shifts (isochronous nuclei) are assigned the same letter and if necessary they may be labeled with primes to distinguish between them (see below).

Letters which are further apart in the alphabet are used when there are larger differences in chemical shifts. The assignment of spins with letters (from the beginning of the alphabet) starts with the most deshielded nuclei (i.e. with highest chemical shift) by convention.

Applied to the two spin examples of the last section, the case with two isochronous spins would be labeled AA' or A_2. (The latter notation with subscript numbers is only applicable because no other spin active nuclei are present in the spin system; this will be explained in section 2.4.) The two-spin system with different chemical shifts is labeled AX. The detailed notation and labeling scheme also depends on the interaction between the nuclei and will be discussed in the sections 2.2 and 2.3. Further notational details will be discussed in these sections.

2.1.5 Summary and Examples

In this chapter the following basics of simple one-dimensional NMR spectra have been discussed:
- For non-interacting nuclei the resonances occur unperturbed in the spectrum.
- The NMR spectrum shows LORENTZian lineshapes.
- The integral of the NMR peaks represent the relative number of spins responsible for the signal.
- The chemical shift is dependent on the shielding of the nucleus by its electrons.

Based on these features the assignment of structures using NMR spectra is possible in conjunction with chemical shift correlation charts like Table 2.2 and where applicable the integral information from the spectra.

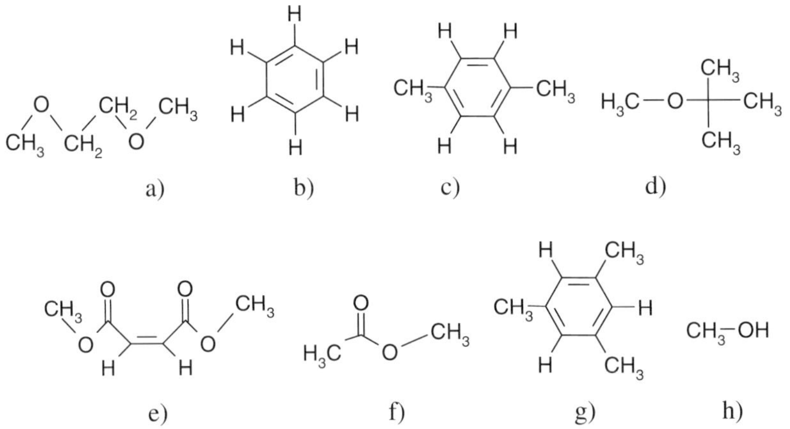

Fig. 2.7: a) *1,2-dimethoxyethane*, b) *benzene*, c) *p-xylene*, d) *t.-butyl methyl ether*, e) *maleic acid dimethyl ester*, f) *acetic acid methyl ester*, g) *mesitylene*, h) *methanol*.

Check it in WIN-NMR:

Open... all the ^{1}H spectra of the compounds shown in Fig. 2.7 one after another with WIN-NMR: The corresponding spectra are found in the directory ...\SHIFT\002001.1R up to ...\SHIFT\009001.1R. Use the **Integration** option from the **Analysis** pull-down menu and use in the **rectangular cursor** mode the left mouse to place the cursor on the spectrum. Fix the border of the integral region with the right-hand mouse button click. Define the second border of the integral in the same way. Integrate all the signals of the compound in each spectrum to determine the relative numbers of protons responsible for the signals. For that purpose you may call the **Options...** to normalize the integral values: For **Manual Calibration** you can enter either a value for the **Selected Integral** (default) or for the sum over **All Integrals**.

Take into account that TMS is added as reference peak. After having integrated all eight spectra decide which of the given structures corresponds to which spectrum.

Note: Some of the spins of these compounds do interact with each other (coupling). The spectra are recorded at 80 MHz, so that the spin-spin interactions can be neglected at this stage of analysis. These features will be discussed in the next sections. The aim of this exercise is to be able to assign spectra to structures only using information about chemical shift and intensity. Take into account, that the signals can show different linewidths, especially OH-protons often show broader lines due to exchange process with water residue (refer to section 3.3.1.1 and 5.3).

2.2 First Order Spin Systems

In addition to chemical shift effects, many high-resolution NMR spectra determined in liquids show an additional feature, which contributes to structural information: this is the phenomenon of spin-spin interaction, or spin-spin coupling. In the isotropic liquid phase the direct spin-spin interactions through space are averaged to zero due to BROWNian motion. Direct spin-spin interactions must be taken into account if the liquid phase exhibits anisotropic properties what is not the point of discussion in this book.

NMR spectra obtained from isotropic solutions show indirect interactions via bonds, the so-called *scalar couplings* (*J-couplings*). These interactions result in the splitting of the signals. To demonstrate their effect on the quantum mechanical description of a spin system we will modify the AX spin system introduced earlier to take into account the spin-spin coupling interaction.

In this section only spin systems consisting of single spins $I = \frac{1}{2}$ are discussed. Every coupled spin splits the signal of the coupled partner nucleus into two lines (multiplicity rule g=2). The distance between the lines corresponds to the absolute value of the coupling constant. For the most common nuclei in NMR spectroscopy (^{1}H, ^{13}C, ^{19}F, ^{31}P) the spin quantum number is $I=\frac{1}{2}$. Weakly coupled spin systems will be discussed in this section and strongly coupled systems in section 2.3. Up to this point only aromatic proton spin systems are discussed. In Section 2.4 the theory of *magnetic equivalence* will be introduced shortly. Chapter 3 deals with important factors relating to spin systems such as *diastereotopy*, *symmetry* and linewidths and also explains how these factors may be used for the effective generation of the input parameters required to simulate the spectra of large spin systems. The transfer of the spin system properties to other parameter values and the influence of coupling constant signs is discussed.

The transfer of the spin systems to other nuclei is part of chapter 4 as well as spin systems involving nuclei with spin quantum number $I > \frac{1}{2}$.

2.2.1 Introduction to Spin-Spin Coupling

In cases of weak coupling interactions in spin systems the so-called first order rules apply. The decision as to whether these rules are applicable or not can be made on the basis of the differences between the resonance frequencies and the coupling constants J_{ik} involved.

> *If the absolute value of the scalar coupling constant J_{ik} can be regarded as **small** compared with the difference between the resonance frequencies ν_i and ν_k the first order rules may be used to describe the spin system.*

$$|\nu_i - \nu_k| >> |J_{ik}|$$

The perturbation parameter λ_{ik} is defined as a measure of the spectral relation between the nuclei i and k:

2.2 First Order Spin Systems

$$\lambda_{ik} = \left| \frac{J_{ik}}{\nu_i - \nu_k} \right| = \left| \frac{J_{ik}}{\Delta_{ik}} \right|$$

The spin system is of first order type, when $\lambda_{ik} \ll 1$. The borderline to so-called second or higher order spin systems is not well defined. The effect of the coupling in a first order spin system is the splitting of the signals, what can be expressed by the *rule of repeated spacing* (first order splitting rule):

> When a single nucleus i is coupled by the coupling constant J_{ik} to a single spin k, each signal (i.e. of i and of k) is split into two lines, which are separated by the absolute value of the coupling constant. These pairs of lines are symmetrically centered at the resonance frequencies ν_i and ν_k respectively and are of equal intensity. This pattern is called a doublet.

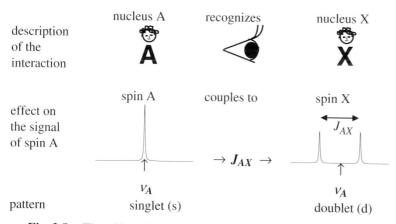

Fig. 2.8: The effect of spin-spin coupling interaction in a two-spin system AX.

Spin systems in which the nuclei belong to different NMR active isotopes (heteronuclear spin systems) always show coupling patterns in which the first order rules can be applied. For different heteronulcei the resonance frequencies usually differ by several MHz or more so that the condition $\lambda_{ik} \ll 1$ is always fulfilled.

Check it in WIN-DAISY:

Open... the input file AX_HETER.MGS and execute a **Simulation** or create your own AX system by modifying the input document A.MGS. To do the latter, **Open...** the document A.MGS. In the **Main Parameters** dialog box increase the **Number of spins or groups with magnetic equivalent nuclei** to <2> and in the **Frequencies** dialog box press the **PSE** (Periodic System of Elements) button of the second row (spin 2, initially shows a **H**) and select any spin ½ isotope, e.g. ^{19}F (Press the **F** button). Switch with **Next** into the **Scalar Coupling Constants** dialog box and enter a coupling constant, e.g. <50> Hz. Press the **Auto Limits** button in the **Spectrum Parameters** dialog box to make sure that they cover all transitions for the selected **isotope ¹H**. Only those transitions that fall inside the spectrum limits are used for lineshape generation, if the limits are incorrectly set only part of the spectrum will be calculated. The **Auto Limits** option will only cover the signals for the isotope displayed on the **PSE** button! **Simulate** and **Display** the **Spectrum. Close** the WIN-DAISY document.

The first order rule is treated quantum mechanically as the so-called X *approximation*. If the nuclei on the PSE buttons are different the X approximation is made. This method may be used within homonuclear spin systems as well, if the isotope number (**ISO value**) of the nuclei is changed. Changing this number for individual spins of a homonuclear spin system causes WIN-DAISY to use the X approximation. Either the atom (**PSE**) or the isotope number (**ISO value**) must be different before the X approximation is applied by the program.

Check it in WIN-DAISY:

Open... the input file AX_HOMO.MGS or create your own input file by modifying the input document A.MGS. To do the latter, proceed in a similar manner to the previous "Check it", but ensure that in the **Frequencies** dialog box the **PSE** (Periodic System of Elements) for the second spin is not altered. Enter a *any* coupling constant value in the **Scalar Coupling Constants** dialog box. Edit the resonance frequencies in the **Frequencies** dialog box of the two spins so that λ_{ik} is much smaller than 1, for example <0.0017>. To calculate a proper resonance frequency you may consult the **Calculator** to be called from the WIN-DAISY toolbar and switch the units in the **Frequencies** to **Hz**. Make sure that **Auto Limits** option is selected in the **Spectrum Parameters** dialog box. Run a **Simulation** and inspect the **Calculated Spectrum** in WIN-NMR. **Close** the WIN-DAISY document.

The spin-spin coupling affects the nuclear energy and thus the scalar spin-spin coupling constant contributes to the nuclear HAMILTON operator as a perturbation. The parameter λ_{ik} indicates the strength of the perturbation the scalar coupling constant produces on the ZEEMAN energies $E_0(j)$. The first order perturbation contribution $E_1(j)$ is added to the zero order ZEEMAN part resulting in $E_j = E_0(j) + E_1(j)$:

$$E_j = -\sum_{i=1}^{N} m_{z,i}(j) \cdot v_i + \sum_{i=1}^{N} \sum_{i<k}^{N} m_{z,i}(j) \cdot m_{z,k}(j) \cdot J_{ik}$$

for every basis product function j.

2.2.2 The AX Spin System

The characteristic first order spin system is based on weak interactions ($\lambda_{ik} \ll 1$), so that truncation of the HAMILTONian after the second term $E_1(j)$ is allowed and the mixing term can be neglected (*X approximation*). The effect of the whole HAMILTON operator on a spin system without this approximation (i.e. when strong coupling interactions are present) will be discussed in section 2.3. The energy of the EIGENstates of the two-spin system can be summed up by the ZEEMAN energies and the first order perturbation:

2.2 First Order Spin Systems

$$E_{\alpha\alpha} = -\tfrac{1}{2}\nu_1 - \tfrac{1}{2}\nu_2 + \tfrac{1}{4}J_{12}$$
$$E_{\alpha\beta} = -\tfrac{1}{2}\nu_1 + \tfrac{1}{2}\nu_2 - \tfrac{1}{4}J_{12}$$
$$E_{\beta\alpha} = +\tfrac{1}{2}\nu_1 - \tfrac{1}{2}\nu_2 - \tfrac{1}{4}J_{12}$$
$$E_{\beta\beta} = +\tfrac{1}{2}\nu_1 + \tfrac{1}{2}\nu_2 + \tfrac{1}{4}J_{12}$$

It becomes clear that for spin-½ nuclei the absolute value of the first order coupling contribution is always ¼ J_{AX}. The sign depends on the sign of the m_z values of the spin orientations in the basis product function j. If we set $\nu_1 = \nu_A$, $\nu_2 = \nu_X$ and $J_{12} = J_{AX}$ for the corresponding transition frequencies we obtain:

$$T_{A1} = E_{\beta\beta} - E_{\alpha\beta} = \nu_A + \tfrac{1}{2}J_{AX}$$
$$T_{A2} = E_{\beta\alpha} - E_{\alpha\alpha} = \nu_A - \tfrac{1}{2}J_{AX}$$
$$T_{X1} = E_{\beta\beta} - E_{\beta\alpha} = \nu_X + \tfrac{1}{2}J_{AX}$$
$$T_{X2} = E_{\alpha\beta} - E_{\alpha\alpha} = \nu_X - \tfrac{1}{2}J_{AX}$$

In the following the transition frequencies will be identified by symbols T and resonance frequencies by ν. If we compare the energy levels with those obtained for a two-spin system without coupling, we can describe the coupling interaction as a perturbation of the basis energies derived from the ZEEMAN term. The resulting transition frequencies T are called the *effective LARMOR frequencies*. Dependent on the sign of the coupling constant the corresponding energy levels are stabilized (energy lowered, negative sign of coupling contribution) or destabilized (energy raised, positive sign of coupling contribution).

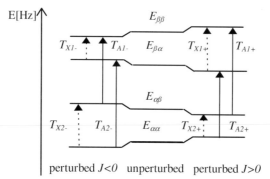

Fig. 2.9: Energy level diagram for an AX spin system $\nu_A > \nu_X$.

The transitions are marked with the sign of the coupling constant. It is clear that the following relations connect the transitions:

$$T_{A2-} = T_{A1+}, \qquad T_{A1-} = T_{A2+}$$
$$T_{X2-} = T_{X1+}, \qquad T_{X1-} = T_{X2+}$$

2 General Characteristics of Spin Systems

The appearance of the spectrum is independent of the sign of the coupling constant J_{AX}, because the transition frequencies are identical, although the energy values are modified. Changing the sign of J_{AX} interchanges the transition labels.

Check it in WIN-DAISY:

Open... the input file AX_HETER.MGS and run a **Simulation**. **Change** the name of the calculated spectrum, e.g. to ...\999998.1R and check if the **Output Options Transitions and energy levels** in the **Main Parameters** are selected. Change the sign of the coupling constant in the **Scalar Coupling Constants**, run a **Simulation** and **Display** the actual **Calculated Spectrum**.

Check it in WIN-NMR:

Use the **Dual Display** option of the **Display** menu and select the original simulation ...\999999.1R as the second **Filename**. Compare both spectra. You may use the **Separate** option to display the spectra in different windows. Toggle back to **WIN-DAISY** using the corresponding **Simulation** command. In this way WIN-NMR returns automatically to the normal display mode and it is open to receive data from other programs.

Check it in WIN-DAISY:

Load the **Simulation Protocol** into the editor (command **Display I Show Simulation Protocol**). Compare the transitions with the formulae given above with the **Calculator** and the data given in the protocol file.

Interchange the resonance **Frequencies** A (index 1) and X (index 2) and/or alter the sign of the **Scalar Coupling Constant**. Run a **Simulation** and inspect the **Simulation Protocol**. Only the labeling of the energy levels and transitions can change, while the appearance of the spectrum is identical in all cases. **Close** the WIN-DAISY document.

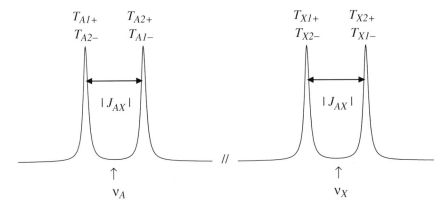

Fig. 2.10: Calculated AX Spectrum (zoomed) showing the transition frequencies associated with a positive and negative coupling constant.

2.2.2.1 Heteronuclear AX Spin Systems

Let us take a look at the perturbation parameter λ_{AX} in a heteronuclear spin system. If for example we take an AX spin system consisting of one fluorine and one proton, the qualitative estimation of the resonance frequencies is that based on the magnetogyric ratios, so that on a 300 MHz spectrometer (which defines the proton basic frequency) the ^{19}F nuclei have a resonance frequency of about 282 MHz. Assuming a proton-fluorine coupling constant of approximately 50 Hz, the λ_{AX} value is given as follows:

$$\lambda_{AX} \approx \left| \frac{50\,Hz}{300\,MHz - 282\,MHz} \right| = 2.8 \cdot 10^{-6}$$

Thus even when the resonance frequencies of different isotopes differ only by 1 MHz and the coupling constant is rather large, the first order precondition for the X approximation is always satisfied. Therefore the selection of a different isotope from the periodic table in WIN-DAISY is always sufficient to define the X approximation.

Check it in WIN-DAISY:

Open... the input file AX_HETER.MGS and execute a **Simulation**. Looking at the **Display** of the **Calculated Spectrum** in WIN-NMR shows the doublet arising from the A nucleus (proton).

Toggle to **WIN-DAISY**, change the **Isotope** in the **Spectrum Parameters** from ^{1}H to ^{19}F and press the **Auto Limits** button to determine appropriate spectrum limits for the fluorine spectrum. Execute a **Simulation** and look at the **Calculated** fluorine **Spectrum Display**. It would make no sense to simulate both signals arising from ^{1}H and ^{19}F within the same spectral width, because in practice they must be measured by switching the transmitter frequency.

Remark: You must only make sure that the **Frequencies** in units **Hz** are different. If the transition frequencies overlap or the spectrum limits by accident cover both frequency regions, the unwanted spin(s) have to be **Disabled** in the **Frequencies** dialog box (refer to section 2.3.2.2) or the **Upper** and **Lower frequency limits** in the **Spectrum Parameters** have to be modified manually.

2.2.2.2 Homonuclear AX Spin Systems

Check it in WIN-DAISY:

Open... the input document AX_HOMO.MGS. If you check the λ_{AX} value for it with the **Calculator** and the values given in the **Frequencies** and **Scalar Coupling Constants** dialog boxes, you will obtain the value $\lambda_{AX} = 0.0017$. On checking the intensities of the transitions in the **Simulation Protocol** AX_HOMO.LST you will find that they are not exactly '1.0' for every signal, but *0.9983* and *1.0017* respectively. The EIGENvalues of the two bpf's with $m_T = 0$ differ slightly from the first order estimated values.

2 General Characteristics of Spin Systems

From the transition intensity values reported in the simulation protocol the meaning of the perturbation parameter is clear: The first order intensity values are disturbed by the perturbation parameter and in this special two-spin system the perturbation parameter can be seen from the intensity disturbance directly. One transition intensity is raised the other decreased by the perturbation parameter value, so that the sum remains identical. Thus in spite of a quite small perturbation parameter value (λ_{AX} = 0.0017) the spin system is not totally first order. A small - with the naked eye not visible - *second order effect* is still present. This phenomenon is named the *roof effect* of spin systems, which are close to but not fully identical with first order systems. The $m_T = 0$ value is no longer a good quantum number, because the corresponding basis product functions correspond not exactly to the EIGENfunctions.

Check it in WIN-DAISY:

You may alter the perturbation parameter λ_{AX} either

- by changing the **Spectrometer frequency the frequencies are adapted to** in the **Spectrum Parameters** to a *lower* value. On the basis of this value and the magnetogyric ratio of the selected **Isotope**, the **Reference Transmitter Frequency** for the conversion of ppm/Hz is determined. Select the value of **60** MHz value from the combo box **Frequencies adapted to Spectrometer Frequency** and quit the dialog box with **Ok**. Answer the question 'Do you wish to adjust the data to the new spectrometer type?' with **Yes** to change the perturbation parameter. If the box is accepted with **No**, the perturbation parameter is kept constant, thus no spin system change is made. Change the name of the calculated spectrum by using the **Change** button, execute the **Simulation** and compare both spectra in the WIN-NMR Dual Display by **Display** of the **Calculated Spectrum** and loading of the original simulation as the second file. **Calculate** the perturbation parameter for this parameter set (0.0083). To decrease the value further in another way, go on as follows:

- by changing the resonance frequencies to reduce the corresponding difference in the **Frequencies** dialog box, e.g. the first frequency to <4.0> ppm (<240> Hz). Check afterwards if all transitions still remain with the **Auto Limits** function within the **Spectrum Parameters**. **Simulate** and **Display** the **Spectrum** in WIN-NMR (Fig 2.11). **Calculate** the perturbation parameter. Afterwards change the first chemical shift in the **Frequencies** dialog box back to <7.2> ppm. The roof effect is numerically symbolized in the signal intensities listed in the **Simulation Protocol** (intensities). To obtain the *same* perturbation parameter ($\lambda_{AX} \approx 0.019$) go an as follows:

- by changing the coupling constant to a higher absolute value in the **Scalar Coupling Constants** dialog box. **Calculate** the coupling constant required for the same perturbation, enter the value into the edit field, run a **Simulation** and **Display** the **Calculated Spectrum**. The latter two spectra differ only in the linewidth of the signals, the signal intensities in the **Simulation Protocol** indicating the roof effect are identical with the previous calculation (for J_{AX} = 6.5 Hz). **Close** the document.

2.2 First Order Spin Systems 43

Fig 2.11: Spectrum calculated by changing the first resonance frequency to 240 Hz (4.0 ppm) at 60 MHz. The dotted lines illustrate the roof effect.

The ridge of the roof is lying in between the A and X signals, independently of the sign of the coupling constant. The slope of the roof always points to the position of the signal of the coupled partner spin.

Check it in WIN-NMR:

Open... the spectrum ...\AX\001001.1R. Determine the linewidth of one peak using the **Analysis | Linewidth** mode. The spectrum contains only the signals of the two protons due to the AX spin system. Analyze the first order multiplet structure using the **Analysis | Multiplets** mode. Use the **maximum cursor** mode to mark one peak top with the spectrum cursor.

2-tert.-butyl-4,6-dinitro-phenol

Switch into the **Free Grid** mode either via the panel bar button or by clicking the right mouse button within the spectrum window which opens up a context menu. Holding down the left mouse button, drag the grid over the spectrum. Release the left mouse button when the second line of the doublet is found. A tick mark is placed on top of the peak. Using the **Define Mult.** option (panel bar or context menu) will define the doublet with its shift and coupling constant. The corresponding values are displayed in the header status line. Repeat this procedure with the other doublet (first mark the peak top with the arrow...). Open the **Report...** box where all values are listed. To connect the coupling partners automatically press the **Auto Connect** button and in the new dialog box change the **Maximum of Difference between Couplings** value, e.g. to <0.3>.

There are other ways to connect coupling partners. Double click on an entry in the **Report...** list box and in the new dialog box select the connection displayed to establish the connection. Another way to connect the coupling partners is manually in the spectrum window. Using the **rectangular cursor** mode mark one of the multiplets, the color changes from green to red. If necessary, use the <page down> key to select the appropriate coupling constant within the multiplet. Click the right mouse button and select the option **Designate Multiplet**. Mark the other multiplet with the arrow cursor and again, if necessary, use the <page down> key to select the appropriate coupling constant. The designated multiplet color changes into black. Pressing the right mouse button, selecting **Connect Multiplets**, the connected multiplet color changes into pink color. The result will appear as shown in Fig. 2.12. For complete description of the Multiplet mode refer to the WIN-NMR **Help** File.

Check it in WIN-NMR:

Using the **maximum cursor** mode mark one peak of the spin system signals. **Export...** the data from the WIN-NMR **Report...** dialog box to **WIN-DAISY**. Answer the questions WIN-NMR asks with **Yes** in order to send also predefined spectral regions defined for the actual multiplet signals. Then the WIN-DAISY program is moved to the WINDOWS foreground.

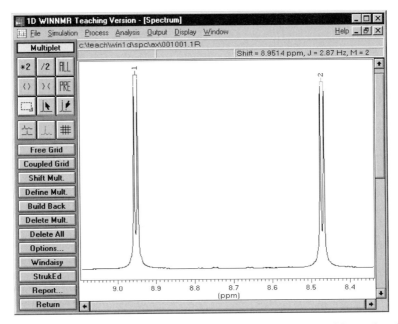

Fig. 2.12: WIN-NMR multiplet analysis of the aromatic protons of *2-tert-butyl-4,6-dinitrophenol*.

Check it in WIN-DAISY:

A new WIN-DAISY document is created. Check the values in the WIN-DAISY **Parameter** dialog boxes. The coupling constant J_{AX} displayed in the **Scalar Coupling Constants** dialog box is the average of both values estimated from the two doublets in the experimental spectrum. Execute a **Simulation** with WIN-DAISY. **Calculate** the perturbation parameter for the spins A and X.

The result can be visualized in two different ways from WIN-DAISY:

- using the **Export** to **WIN-NMR exp.** feature will draw the calculated transition frequencies as a stick spectrum below the experimental spectrum and shows the first order multiplet pattern based on the data taken from the WIN-DAISY dialog boxes at the top of the spectrum window.

- with the **Dual Display** command both spectra, the experimental and calculated will be shown (Fig. 2.13). The experimental spectrum is displayed as the first trace (by default blue color) and the second trace refers to the calculated lineshape (by default green color – you may change it in the WIN-NMR **Display | Colors | Trace B** combo box).

2.2 First Order Spin Systems 45

Only a small roof effect is visible for the spectral parameter value $\lambda_{AX} \approx 0.019$, as approximately determined in a recently done "Check it". A careful comparison of the experimental and calculated spectrum will reveal a slight shift of the calculated lines with respect to the experimental spectrum (Fig. 2.13).

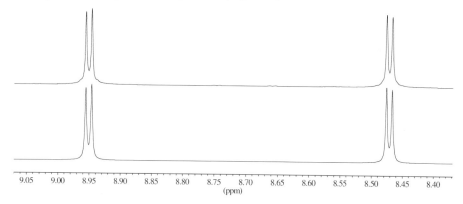

Fig. 2.13: WIN-NMR Display of the experimental (upper) and calculated (lower) lineshape.

Check it in WIN-DAISY:

Modify the spin system to be exactly first order (using the X approximation) by changing the **ISO** value of either spin A or spin X, enter e.g. <2>. Execute a **Simulation** and compare the result with the experimental spectrum in WIN-NMR **Dual Display**. Use the **Separate** option to visualize the intensity difference.

The correct line positions obtained by the WIN-NMR first order multiplets analysis are reproduced in the simulation, but the line intensities do not show the roof effect observed in the experimental spectrum. Thus the first order analysis leads to a satisfactory estimation of the NMR parameters even in cases where a roof effect is present. Such spin systems displaying a roof effect are not strictly first order, so that the notation using the letters A and X is not quite correct, because the notation implies the application of the X approximation. The second order effect is not negligible in such spectra, so in calculations for homonuclear spectra it is important *not* to assign different ISO values manually, because the X approximation will lead to an error in the calculated result. A spectrum iteration can be executed in order to obtain the optimized parameters represented by the measured data. This procedure modifies the estimated NMR parameters in an iterative process until the calculated lineshape gives the best fit to the experimental data. The parameter optimization has no need of line assignments, because it uses a total-lineshape fitting procedure, described in the appendix, section 6.3. In the context only necessary information is given.

46 2 General Characteristics of Spin Systems

Check it in WIN-DAISY:

Using the data file exported from WIN-NMR with the unchanged **ISO values**, (open the **Frequencies** dialog box and reenter **ISO Value** <1>), execute a spectrum **Iteration**. After the calculation is finished look at the result in the **Dual Display** of WIN-NMR and the data in the WIN-DAISY dialog boxes **Frequencies** and **Scalar Coupling Constants**. The **Iteration Protocol** gives at the end statistical information about the optimization. The initial parameter values were already very close to the best parameter values.

2.2.2.3 Summary

- For each spin two signals appear in the spectrum. If the spin system is homonuclear, all signals are situated in the same spectrum.
- The two signals are symmetrically positioned with respect to the resonance frequencies v_A and v_X.
- The splitting between the two signals corresponds to the absolute value of the coupling constant $|J_{AX}|$.
- The coupling constant is found twice due to the rule of repeated spacings, at v_A and v_X.
- The sign of the coupling constant cannot be determined from the spectrum, because the appearance of the spectrum is independent of sign.
- If a roof effect is visible in the signal intensities this indicates a second order influence. In this case the first order analysis is not correct and an optimization of the parameters describing the spectrum should be completed. Stronger second order effects and their treatment will be discussed explicitly in section 2.3.

The contribution of the coupling term to the energy is always the scalar coupling constant J_{ik} multiplied with the product of the $m_{z,i}$ and $m_{z,k}$:

$$J_{ik} \prod_{j=\{i,k\}} m_{z,j} \quad \text{for each bpf}$$

2.2.3 The AMX Spin System

All the derivations worked out in the previous section can be directly adapted to a first order three-spin system $I = \frac{1}{2}$ (and also for larger spin systems).

Every single spin splits the signal of the coupled partner into a doublet.

Applied to a three spin system (when all spins interact with each other) every multiplet is a doublet of doublets (dd): The coupling to one partner results in a doublet (two lines separated by the absolute value of the coupling constant between the two nuclei) while the coupling to the other partner results in each line of the original doublet being split further into a doublet (Fig. 2.14, two lines separated by the absolute value of the coupling constant between the original nucleus and its new partner).

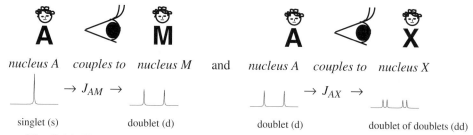

Fig. 2.14: Signal of nucleus A due to interaction with both spin M and spin X.

If the value of a spin-spin coupling is zero, no splitting is visible in the spectrum. In this introduction to AMX spin systems only rigid compounds (of aromatic type) will be discussed, non-rigid species and other rigid types will be subject of chapter 3.

Check it in WIN-NMR:

Open... the spectrum ...\AMX\001001.1R of *1,2,4-trichlorobenzene* with WIN-NMR. Determine the signals of the three protons via **Integration**. For that purpose you may **Load...** the predefined **Regions...** later on used for iteration in WIN-DAISY. Use the **Split. Int.** option to divide the first region into two: Place the spectrum cursor in the **rectangular cursor** mode in the middle between the two signals and press the right hand mouse button.

1,2,4-trichlorobenzene

Start the analysis of the multiplet structure using the **Analysis | Multiplets** mode as described in a previous "Check it".

One of the signals shows an additional splitting, because it is coupled to both the other protons. To analyze this multiplet start with the first doublet splitting as usual. To duplicate the doublet change the cursor into the **Coupled Grid** mode. By holding the left mouse button the doublet pattern can be dragged as a grid over the whole spectrum. Release the mouse button when the pattern matches. These two distances relating to the two coupling constants can be found again in one of the other multiplets according to the rule of repeated spacing. **Auto Connect** the coupling constants in the **Report...** box to assign the coupled nuclei.

Number	Chem. Shift [ppm]	J [Hz]	Multiplicity	Connections
1	7.4347	2.4326	2	J(1,3)
2	7.3471	8.5140 ?	2	J(2,3)
3	7.1643	8.6245 ?	2	J(3,2)
		2.4326	2	J(3,1)

Fig. 2.15: 1D WIN-NMR Multiplets Analysis Report box of the AMX spin system of *1,2,4-trichlorobenzene* with $J_{AM} = 0$.

The result obtained will be similar to the WIN-NMR spectrum displayed in Fig. 2.16 and the Report box in Fig. 2.15.

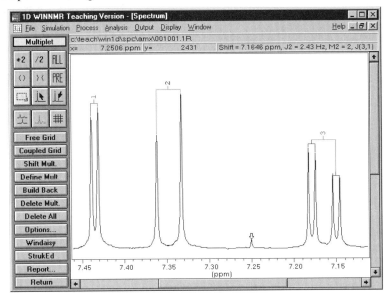

Fig. 2.16: WIN-NMR Multiplets Analysis of the AMX spin system of *1,2,4-trichlorobenzene* with J_{AM}=0. The arrow indicates the solvent signal.

Usually the coupling constant values measured in the different parts of the spectrum are not exactly identical. This difference can result from the line overlap of the signals. If two LORENTZian lines do overlap, the position of the maximum is shifted towards the overlapped neighbor signal. The multiplets mode takes the data point at the peak top position frequency for calculation of the distance. The *question marks* beside the coupling constant J_{23} indicate a second order effect between spins 2 and 3 (Fig. 2.15).

Check it in WIN-DAISY:

After having measured a linewidth in the **Analysis | Linewidth** mode **Export...** the multiplet structure from WIN-NMR to **WIN-DAISY**. Check the values of the coupling constants in the **Scalar Coupling Constants** dialog box. Run a **Simulation** and compare the calculated and experimental spectrum in WIN-NMR **Dual Display** (Fig 2.17).

Open the **Frequencies** dialog box, therein the chemical shifts are ordered according to their coupling paths. Because the low-field proton is coupled to the high-field proton the latter one appears as the second nucleus in the dialog. Open the **Scalar Coupling Constants** dialog box. One coupling constant (J_{13}) was not determinable by first order splitting in the spectrum, thus its value remains zero. Place the cursor in the corresponding edit field and check the **iterate** option in the bottom line. Execute an **Iteration** with WIN-DAISY and look at the result with WIN-NMR **Dual Display** and in the dialog boxes for **Scalar Coupling Constants** and **Frequencies**. The

2.2 First Order Spin Systems

optimized values are very close to the starting parameters determined with WIN-NMR as to be seen from the **Iteration Protocol**.

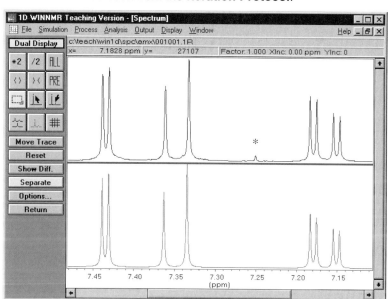

Fig 2.17: WIN-NMR Dual Display of the experimental (upper) and calculated (lower) lineshape, *: CHCl$_3$.

The protons in aromatic systems are geometrically fixed. The magnitude of the coupling constants depends mainly on the number of bonds separating the coupled hydrogen atoms. The aromatic J-couplings can be classified by the following rule:

$$^3J_{ortho} > {^4J_{meta}} > {^5J_{para}}$$

Dependent on the chemical environment not only the chemical shift, but also the values of coupling constants will vary. The signs of aromatic proton-proton coupling constants are known as to be positive.

Table 2.4: Range of coupling constants for aromatic compounds.

Coupling Constant	Coupling range [Hz]
$^3J_{ortho}$	≈ 6 to 10
$^4J_{meta}$	≈ 1 to 4
$^5J_{para}$	≈ 0 to 1

Check it in WIN-DAISY:

Look at the **Scalar Coupling Constants** and assign the labels A, M and X to the structure using the values of the coupling constants (J_{AM} = 0 Hz, J_{AX} = 2.4 Hz and J_{MX} = 8.6 Hz) and Table 2.4. **Close** the WIN-DAISY document.

50   2 General Characteristics of Spin Systems

This assignment based on coupling constant values is also supported by estimating the relative chemical shifts based on the environment of the individual protons (refer to Fig. 2.18).

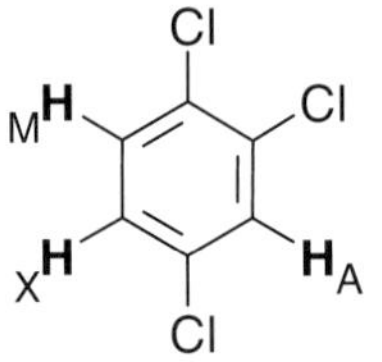

Fig. 2.18: Spin notation of *1,2,4-trichlorobenzene*.
$^5J_{AM} = 0$ Hz *(para)*, $^4J_{AX} = 2.4$ Hz *(meta)*, $^3J_{MX} = 8.6$ Hz *(ortho)*

Check it in WIN-NMR and WIN-DAISY:

In a similar manner to the previous "Check it" **Analyze** the aromatic three-spin systems of 2,4-dichloro-phenol ...\AMX\002001.1R and 2-(2,4-dichloro-phenoxy) propionic acid methylester ...\AMX\003001.1R. This includes the measurement of the **Linewidth**, which will be not always explicitly mentioned from this point on. In case you want to optimize the parameters, always determine a linewidth initially. Then enter the **Multiplets** mode.

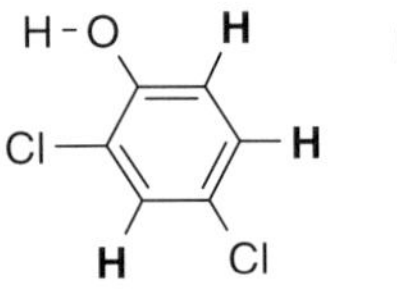

2,4-dichlorophenol 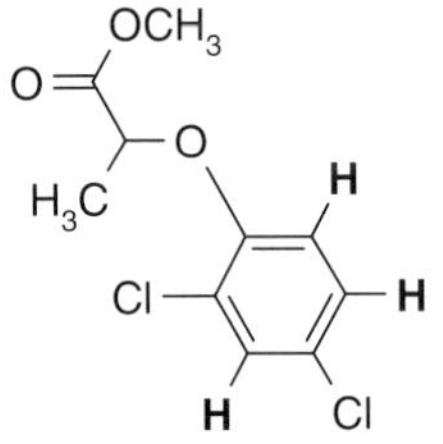   *2-(2,4-dichloro-phenoxyl) propionic acid methyl ester*

Label the aromatic protons in the structure by evaluating the magnitudes of the coupling constants. You may assign the labels to the Multiplets via the context menu option **Define Identifier** by clicking the right mouse button when the desired coupling tree is highlighted and substitute the question mark by the label, e.g. <A>. **Export...** the data to **WIN-DAISY** and perform spectrum **Iteration**s. Do not forget to take into account as well the para coupling constant.

The assignments show how strong the substituent influence at position 1 in the benzene ring is. Note the absence of the question marks in the **Report...** dialog for the latter spectrum analysis. Note also the different quality of the result, reported in the *R-factor (%)* in the **Iteration Protocol**. This value refers to the deviation between experimental and calculated spectrum in per-cent. For detailed information about the iteration routines refer to the appendix, section 6.3.

Check it in WIN-NMR:

Use the **Multiple Display** mode of WIN-NMR to compare the first three AMX spectra (either experimental or calculated) and to illustrate the differences between them. Use the **Separate** button and select the spectrum with the largest spectral width with a left hand mouse click, use the **Equal X** button and all three spectra will be displayed using the same spectral range. Fig. 2.19 shows a multiple display of calculated spectra. Keep one of the WIN-DAISY document open - do *not* close it.

2.2 First Order Spin Systems 51

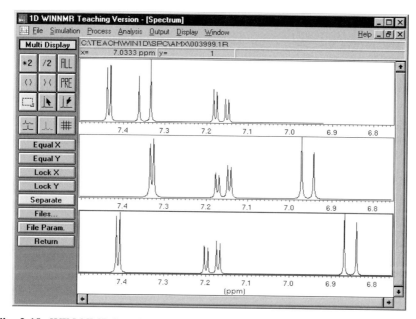

Fig. 2.19: WIN-NMR Multiple Display of the three calculated AMX spectra.

The lower trace in Fig. 2.19 is a good example for illustrating of the roof effect. The nucleus M shows coupling to both the A and X spins and its signal is found in between the others. In Fig. 2.20 the multiplets are displayed showing quite clearly the roof effect.

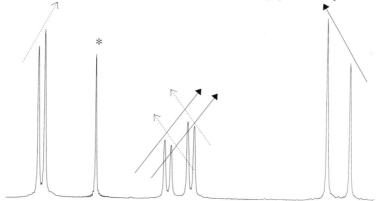

Fig. 2.20: Experimental spectrum of *2-(2,4-dichloro- phenoxyl) propionic acid methyl ester* with roof effects; * $CHCl_3$ impurity of the solvent $CDCl_3$.

The number of bpf's for a spin system consisting of N spins is given by 2^N. The systematic generation of all the possible bpf's for spin systems consisting of only single spins with $I = \frac{1}{2}$ is very simple: They are contained in the binary code of the 2^N integer

2 General Characteristics of Spin Systems

numbers from zero up to 2^N-1. Binary numbers consist solely of the elements '0' and '1' and can be used to represent the spin orientation.

Check it in the Calculator:

While an input document is still loaded start the **Calculator** from the WIN-DAISY toolbar and chose the **scientific view**. For the three-spin system AMX *(N=3)* we will generate all of the bpf's starting with the βββ spin state. Enter the sequence $2^N-1= 7$ in the following way: Click on the buttons **2 x^y 3 - 1** and **=**. Change the display from **Dec.** *(decimal)* to **Bin** *(binary)*. The binary number '*111*' is shown, every digit representing a one-spin function. A '*0*' can symbolize the α spin state and a '*1*' the β spin state. Toggle back to the **dec**imal display and go on generating the binary codes of the integer number 0 up to 7. Note that preceding zeros are neglected in the binary display. The result of the assignments is shown in Table 2.5.

This method works for any integer number not exceeding the number of bits for the integer type used. The program WIN-DAISY uses 8-byte integer numbers (1 byte = 4 bit) resulting in 32 bit numbers.

Check it in the Calculator:

Enter in the **dec**imal mode the integer number "2^{32}" (click the buttons **2, x^y, 32** and **=**) and switch over to the **bin**ary display. As you can see, this number cannot be represented. Press the button **C** and try the **dec**imal number "$2^{32}-1$" (click on **2, x^y, 32 – 1** and **=**) and turn into the **bin**ary display.

Table 2.5: Systematic formation of basis product functions from integer numbers.

integer number	binary number	bpf	Σβ	m_T value
0	000	ααα	0	1½
1	001	ααβ	1	½
2	010	αβα	1	½
3	011	αββ	2	– ½
4	100	βαα	1	½
5	101	βαβ	2	– ½
6	110	ββα	2	– ½
7	111	βββ	3	–1½

In this manner the bpf's for a maximum of 32 spins in a spin system may be represented. In practice other factors will limit the number of spins that can be successfully handled in a calculation. However, the theoretically maximum for the number of spins per spin system using this type of function formalism based upon integer numbers is dependent on the number of bits used to represent integer numbers and ultimately depend upon the computer system itself.

2.2 First Order Spin Systems

Check it in WIN-DAISY:

Open the **Main Parameters** and check the **Output Options Linear combinations**, **EIGENvalues** and **Transitions and energy levels**. Run a **Simulation** and open the **Simulation Protocol**. Therein the bpf's are given ordered according to the m_T values and the transitions between the energy levels are listed. Construct a m_T diagram for an AMX spin system. Label the energy levels with the bpf's symbols. Using different types of arrows, draw the transitions of A, M and X spins respectively.

For the present three spin system the resulting diagram can be described as a cube in which the eight corners represent the energy levels and all parallel edges (three types for the same number of spins) correspond to the transitions of the *same spin* of the system as visualized in Fig. 2.21. The basis functions within the bpf's are always written in the spin order: first A, M and then X.

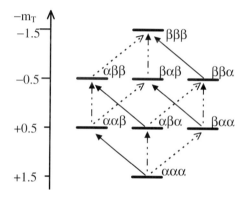

Fig. 2.21: m_T scheme for the AMX spin system
↗ A transitions, ↑ M transitions, ↖ X transitions.

In contrast to the energy level diagram, the m_T diagram is clearly arranged. The same rules for the calculation of the first order energy levels (involving ZEEMAN and Coupling term as discussed in the previous section) are applicable.

Check it in WIN-DAISY:

Force the last AMX example into the X approximation by entering three different **ISO values** for the three chemical shifts. Select the **Output Options EIGENvalues**, **Transitions and energy levels** and **Frequencies after degeneracy** and **Simulate** the spectrum. Manually **Calculate** the first order energy levels and transition frequencies (refer to Table 2.6 and

Table **2.7** and Fig. 2.21) and compare the values with the **Simulation Protocol** information.

Table 2.6: Energy levels according to X approximation for an AMX spin system.

bpf	energy level
ααα	$-\frac{1}{2}\nu_A - \frac{1}{2}\nu_M - \frac{1}{2}\nu_X + \frac{1}{4}J_{AM} + \frac{1}{4}J_{AX} + \frac{1}{4}J_{MX}$
ααβ	$-\frac{1}{2}\nu_A - \frac{1}{2}\nu_M + \frac{1}{2}\nu_X + \frac{1}{4}J_{AM} - \frac{1}{4}J_{AX} - \frac{1}{4}J_{MX}$
αβα	$-\frac{1}{2}\nu_A + \frac{1}{2}\nu_M - \frac{1}{2}\nu_X - \frac{1}{4}J_{AM} + \frac{1}{4}J_{AX} - \frac{1}{4}J_{MX}$
βαα	$+\frac{1}{2}\nu_A - \frac{1}{2}\nu_M - \frac{1}{2}\nu_X - \frac{1}{4}J_{AM} - \frac{1}{4}J_{AX} + \frac{1}{4}J_{MX}$
αββ	$-\frac{1}{2}\nu_A + \frac{1}{2}\nu_M + \frac{1}{2}\nu_X - \frac{1}{4}J_{AM} - \frac{1}{4}J_{AX} + \frac{1}{4}J_{MX}$
βαβ	$+\frac{1}{2}\nu_A - \frac{1}{2}\nu_M + \frac{1}{2}\nu_X - \frac{1}{4}J_{AM} + \frac{1}{4}J_{AX} - \frac{1}{4}J_{MX}$
ββα	$+\frac{1}{2}\nu_A + \frac{1}{2}\nu_M - \frac{1}{2}\nu_X + \frac{1}{4}J_{AM} - \frac{1}{4}J_{AX} - \frac{1}{4}J_{MX}$
βββ	$+\frac{1}{2}\nu_A + \frac{1}{2}\nu_M + \frac{1}{2}\nu_X + \frac{1}{4}J_{AM} + \frac{1}{4}J_{AX} + \frac{1}{4}J_{MX}$

Table 2.7: First order transition frequencies for an AMX spin system.

transition label	destination EIGENstate	origin EIGENstate	transition frequency
T_{A1}	βββ	αββ	$\nu_A + \frac{1}{2}J_{AM} + \frac{1}{2}J_{AX}$
T_{A2}	ββα	αβα	$\nu_A + \frac{1}{2}J_{AM} - \frac{1}{2}J_{AX}$
T_{A3}	βαβ	ααβ	$\nu_A - \frac{1}{2}J_{AM} + \frac{1}{2}J_{AX}$
T_{A4}	βαα	ααα	$\nu_A - \frac{1}{2}J_{AM} - \frac{1}{2}J_{AX}$
T_{M1}	βββ	βαβ	$\nu_M + \frac{1}{2}J_{AM} + \frac{1}{2}J_{MX}$
T_{M2}	αββ	ααβ	$\nu_M - \frac{1}{2}J_{AM} + \frac{1}{2}J_{MX}$
T_{M3}	ββα	βαα	$\nu_M + \frac{1}{2}J_{AM} - \frac{1}{2}J_{MX}$
T_{M4}	αβα	ααα	$\nu_M - \frac{1}{2}J_{AM} - \frac{1}{2}J_{MX}$
T_{X1}	βββ	ββα	$\nu_X + \frac{1}{2}J_{AX} + \frac{1}{2}J_{MX}$
T_{X2}	αββ	αβα	$\nu_X - \frac{1}{2}J_{AX} + \frac{1}{2}J_{MX}$
T_{X3}	βαβ	βαα	$\nu_X + \frac{1}{2}J_{AX} - \frac{1}{2}J_{MX}$
T_{X4}	ααβ	ααα	$\nu_X - \frac{1}{2}J_{AX} - \frac{1}{2}J_{MX}$

The transition labeling depends on the size and sign of the coupling constants in the actual spin system. The latter two experimental AMX examples exhibits an almost non-detectable para coupling constant (J_{AX}) that results in a nearly degeneracy in pairs of transition frequencies for A and X. Whether the transitions are resolved or not depends on the relation between the size of the small coupling constant and the linewidth.

$$T_{A1} \approx T_{A2} \quad \text{and} \quad T_{A3} \approx T_{A4}$$
$$T_{X1} \approx T_{X2} \quad \text{and} \quad T_{X3} \approx T_{X4}$$

In three-spin systems more than the twelve transitions are possible according to the selection rule $\Delta m_T = -1$ (refer to Fig. 2.21). If we only look at the net change in m_T, three additional transitions may be possible due to the selection rule:

ααβ → ββα	$\Delta m_z(A)=-1, \Delta m_z(M)=-1, \Delta m_z(X)=+1;$	$\Delta m_T=-1$
αβα → βαβ	$\Delta m_z(A)=-1, \Delta m_z(M)=+1, \Delta m_z(X)=-1;$	$\Delta m_T=-1$
βαα → αββ	$\Delta m_z(A)=+1, \Delta m_z(M)=-1, \Delta m_z(X)=-1;$	$\Delta m_T=-1$

These three transitions are the so-called *combination lines*. Here all three nuclei change their spin orientation (and m_z value) at the same time, so that the net change in m_T is also minus one. In first order spin systems these kind of transitions are forbidden.

Check it in WIN-DAISY:

Calculate the three transition frequencies for the combination lines. Enter the **Minimum intensity** of zero: <0> in the WIN-DAISY **Main Parameters** and reenter *ISO* values of <1> for each chemical shift in the **Frequencies** dialog box. **Simulate** the spectrum. Compare your manually calculated transition frequencies with the data recorded in the WIN-DAISY **Simulation Protocol**. Note also the deviation in the intensities from "1.0". Examine the transition frequencies carefully and you will see beside the fact that each line of the A doublet and each line of the X doublet is nearly degenerate the three forbidden transitions frequencies with zero transition intensity.

The general pattern of 1,2,4-trisubstituted benzene compounds has been discussed previously. Occasionally the $^5J_{para}$ coupling constant is resolved, so that a small additional splitting can be seen. The spectra for these types of compounds always show one ortho, one meta and one para coupling. 1,2,3-trisubstituted benzene derivatives (substituents 1 and 3 different) can be easily distinguished from the 1,2,4-trisubstituted compound, because two ortho and one meta couplings are involved. If the values of the ortho couplings are quite similar two lines of the doublet of doublets (dd) overlap and result in the appearance of a pseudo-triplet signal (Fig. 2.22).

Check it in WIN-NMR and WIN-DAISY:

Analyze the aromatic region of 4-chloro-2-oxobenzothiazoline-3-yl acetic acid ...\AMX\004001.1R. Take into account that the highfield signal is supposed to be a doublet of doublets! Mark one signal with the spectrum cursor and **Export...** the data to **WIN-DAISY**.

Run a WIN-DAISY **Iteration** and **Export** the result back to **WIN-NMR exp.** spectrum (refer to Fig. 2.22).

4-chloro-2-oxobenzothiazoline-3-yl acetic acid

2 General Characteristics of Spin Systems

Fig. 2.22: Experimental lineshape of the X-nucleus with calculated stick spectrum.

The assignment of spins A and M to the protons in the structure is not as easy as it is for systems with very different coupling constants and additional information is needed about chemical shifts of such compounds. Compilations about proton NMR chemical shifts and coupling constants are given in the references [2.22].

Check it in WIN-NMR and WIN-DAISY:

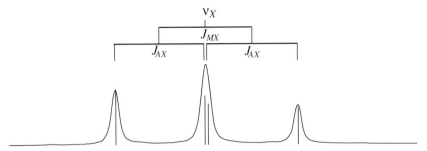

7-bromo-5-chloro-8-hydroxy-quinoline

Analyze the AMX spin system of the substituted quinoline ...\AMX\005001.1R. The spectrum shows a small amount of a different isomer that shall be ignored in analysis. **Export...** the data to **WIN-DAISY**, run an **Iteration** and try to assign the signals to the protons in the structure.

The best result in the **Iteration Protocol** will refer to a R-factor of about 1.65 % due to the impurity lines which are not taken into account but contribute to the lineshape.

The shifts and coupling constants of the aromatic protons of the pyridine skeleton differ considerably from those found in the benzene skeleton discussed above. The dipole moment of the pyridine caused by the electronegativity of the nitrogen deshields ortho (2) and para (4) position of the ring and shields the meta (3) position. The two possible ortho coupling constants have very different values because the nitrogen reduces significantly the coupling involving the ortho protons.

In almost all aromatic spin systems the roof effect is visible indicating a significant second order influence. However, a preliminary analysis may be performed on the basis of the first order rules to estimate the shifts and coupling constants. A complete analysis using WIN-DAISY requires the second order effects to be taken into consideration. Three-spin systems with a stronger relation between the spins will be explained in Section 2.3. Not all aromatic spin systems are easily analyzable using the first order rules.

2.2.4 Extended First Order Spin Systems

The method of repeated spacings can be extended to bigger first order single spin systems without any limitation. Every additional spin (which is coupled to the others) gives rise to an extra doublet splitting.

In homonuclear proton spin systems the spin-spin couplings over more than five bonds are usually very small and are not usually resolved. Thus extended spin systems do not generally exhibit all possible couplings. In many cases there exist substructures separated from substructures either by a non-spin active nucleus or by many chemical bonds. In these situations the subsystems can be regarded as isolated spin systems within the total spin system. Large fully coupled spin systems are often found in non-rigid aliphatic structures. These spin systems are normally not first order and there are many examples where at least two spins show a strong second order effect. Most frequently fully coupled spin systems are connected either with long-chained aliphatic or condensed ring systems.

Heteronuclear spin systems containing natural abundance spin active nuclei e.g. including ^{19}F, ^{31}P, generally have coupling constants J_{XH} which are larger than homonuclear J_{HH} and thus long-range couplings over many bonds are detectable. The most effective way treating such types of spectra will be subject of Section 4.1 (^{19}F) and 4.2 (^{31}P). The effect of a stronger second order effect will be introduced on the basis of pure proton NMR spectra in section 2.3 and in section 3.2 many practical examples will be given.

2.2.4.1 AMRX Spin Systems

The compounds to be discussed here are actually symmetrical eight-spin systems arranged in two identical 4-spin subsystems of the AMRX type. Due to the fact that there is no coupling interaction between the two subsystems (both aromatic ring structures), only one AMRX spin system needs to be calculated. If there were at least one non-zero coupling constant between the two rings, the spectrum would be much more complicated (these type of spectra will be explained in section 3.1 treating spin system symmetry).

Even though there is no long-range coupling, the presence of a duplicate subsystem in the total spin system must be taken into account in any full analysis. Duplicate subsystems will be discussed fully in chapter 3.

Check it in WIN-NMR and WIN-DAISY:

Analyze the **multiplet** signals of the proton spectrum of 2,2'-bipyridine ...\AMRX\001001.1R on the basis that there is no coupling between the two pyridine rings. **Export...** the data to **WIN-DAISY**. Note that no question marks appear in the **Report...** box.

2,2'-bipyridine

Adjust the NMR parameters by **Iteration** and display experimental and calculated spectrum in WIN-NMR **Dual Display**. Obviously the solvent signal is positioned next to the high-field multiplet which is excluded from the iteration region in the **Interval** regions defined.

Label the protons in the structure taking into account the structural effects on shifts and coupling constants in pyridinic structures as discussed in Section 2.2.3.

Check it in WIN-NMR and WIN-DAISY:

Analyze the **Multiplet** signals of the proton spectrum ...\AMRX\002001.1R on the basis that the two phenyl rings are identical and independent and **Export...** the data to **WIN-DAISY**.

3,6-bis(2-chlorophenyl)-1,2,4,5-tetrazine

Include the **Scalar Coupling Constant** J_{13} into the **iterat**ion as well, which is not determinable as a splitting. If you **Simulate** a spectrum with the exported data and compare the calculation with the experimental spectrum in the **Dual Display** it is obvious that the experimental data exhibit some lineshape problems; a hump is visible under the measured transitions. In spite of this, reliable results may be obtained by fixing the global linewidth value in the **Lineshape Parameters** to about <0.8> Hz for example and disable the **iterate** option for the linewidth. Run a spectrum **Iteration** and display the result in WIN-NMR **Dual Display**.

2.2.4.2 Fragmentation of Spin Systems

Analyzing the spin systems in section 2.2.4.1 it was already taken into account that for the eight-spin system only half (AMRX) need to be defined, because there is no interaction between both equivalent parts present. There are many examples of connected aromatic systems which represent independent spin systems, the so-called *spin islands*.

They total spin system may be subdivided into independent quantum mechanical systems not interfering each other. In such cases WIN-DAISY can treat the subsystems within one document as different *fragments*. All Parameter values that have already been used to define the spin systems such as:

- Main Parameters
- Frequencies
- Scalar Coupling Constants
- Lineshape Parameters

are available for each individual fragment.

The simulated NMR spectrum to be displayed with WIN-NMR will contain the transitions of all the fragments. Thus the **Spectrum Parameters** (and also the Calculation **Control Parameters**) belong to the whole WIN-DAISY document - they are defined for the total lineshape.

2.2 First Order Spin Systems

The Parameters not mentioned here (e.g. **Symmetry Group, Symmetry Description**) will be explained in section 3.1, the **Dipolar Couplings** are not subject of discussion in this book at all.

When exporting data from the WIN-NMR Multiplets analysis and exporting data, WIN-DAISY will check the coupling connections in order to determine whether the definition fragments is possible.

Check it in WIN-NMR:

Analyze the proton spectrum **Multiplets** of the substituted benzamide ...\AMX\006001.1R assuming that there is no inter-ring coupling between the two phenyl fragments. Mark one signal with the **maximum cursor**. **Auto Connect** the couplings in the **Report**... box in order to build two three spin systems. **Export**... the data and answer WIN-DAISY question "Do you wish to subdivide your spin system into independent fragments?" with **Yes**.

5-chloro-N-(2-chloro-4-nitro-phenyl)-2-hydroxy-benzamide

Check it in WIN-DAISY:

The WIN-DAISY document main window now shows two lines in the list box corresponding to the two different three-spin systems. Double clicking on either line in the **Title** line of the list box will display the corresponding **Main Parameters** of the fragment. In the header line of each fragment-specific dialog box the **Next** and **Previous** buttons are enabled to offer access to the parameters of the other fragment(s). There is no internal program limit for the number of fragments, so that WIN-DAISY can be used to calculate the spectrum of any molecule which can be subdivided into independent fragments. Applications of this type of approach will be given in section 3.2 and 3.3. Select with one mouse click or the arrow key <↓> the second fragment of the actual WIN-DAISY document, the second title line is highlighted. Then open the **Frequencies** dialog box. The header line shows that the displayed shift values correspond to the second fragment. Press the **Next** or **Previous** button of the header line in order to update the displayed data for the first fragment.

Practice switching between fragments using the buttons in the header dialog box line and switching between the different Parameter windows using the buttons in the bottom line of the dialog box.

Change the **Title** in the **Main Parameters** of one of the fragments and close the dialog box with **OK**. The list box in the WIN-DAISY document main window is directly updated and the header information in the **Parameters** dialog boxes show the new title.

Execute a spectrum **Iteration** and inspect the result in the WIN-NMR **Dual Display** mode. Return to **WIN-DAISY**; from the **Export** pull-down menu choose the **WIN-NMR exp.** option or use the appropriate blue button in the WIN-DAISY toolbar to export the iteration result as a multiplet tree into WIN-NMR. These data will be used later on again.

2 General Characteristics of Spin Systems

Check it in WIN-NMR:

The WIN-NMR **Multiplets** mode is called with **Windaisy** button depressed. Open the **Report...** box and note that the coupling trees representing the WIN-DAISY iteration result are ordered in a different way from those obtained by the manual WIN-NMR multiplet analysis. WIN-DAISY orders the multiplets according to their subsystem membership while WIN-NMR simply numbers the multiplets from left to right. It is possible to export the WIN-DAISY results stored with the experimental spectrum back to WIN-DAISY what will be exercised later on.

In contrast to all the previous WIN-DAISY documents the **Edit** pull-down menu now offers another option **Delete fragment**. The edit commands are used to manage the fragments that make up the complete WIN-DAISY document, unlike the **Parameter** commands which operate on the individual fragments. Only the **Spectrum** and **Control Parameters** belong to the complete WIN-DAISY document and not individual fragments.

The WIN-DAISY input hierarchy is shown in Fig. 2.23.

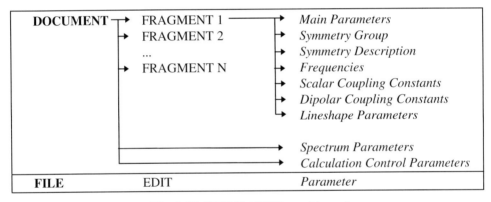

Fig. 2.23: WIN-DAISY input hierarchy.

The **Edit** commands that may be used in the management of the fragments within the same document are:

- New
- Copy
- Merge
- Delete

Because a WIN-DAISY document must consist of at least one fragment, the **Delete fragment** option is only available when more than one fragment is defined. Further details including the subdivision of spin systems into fragments will be given in chapter 3. There as well the fragment copy *between* different documents **Copy fragment to...** is discussed.

Check it in WIN-DAISY:

Display the previously **Calculated Spectrum** with WIN-NMR (button with red peak). Toggle back to **WIN-DAISY**. Then **Change** the name of the calculated spectrum (e.g. ...\006998.1R). Using the arrow keys <↑> and <↓> the different fragments can be selected in the WIN-DAISY fragment list box in the main window. Select the second line. From the **Edit** pull-down menu choose the **Delete fragment** option or use the appropriate button in the WIN-DAISY toolbar (recycle bin) to delete the fragment selected on the main document window. After the delete command is confirmed with **Ok**, only the other (first) fragment title line remains. Execute a **Simulation** and compare both calculated spectra in the WIN-NMR **Dual Display** (refer to Fig. 2.24). For that purpose toggle with **Run WIN-NMR** to the display of the previously loaded first simulation and load the second calculation (...\006998.1R) manually into the WIN-NMR **Dual Display**. Click on the **Separate** and **Show Difference** button (Fig. 2.24).

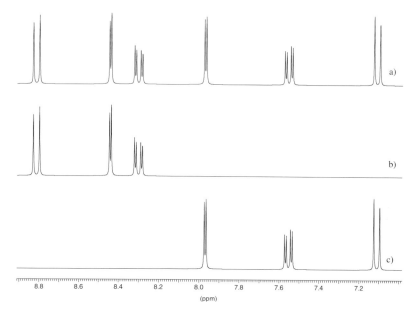

Fig. 2.24: a) Total simulation of both AMX spin systems.
            b) Simulation of the 'left' AMX spin system.
            c) Difference spectrum a) – b) showing the 'right hand side' AMX system.

Check it in WIN-DAISY and WIN-NMR:

Close the WIN-DAISY document, return to **WIN-NMR** and enter the **Multiplets** mode with depressed **Windaisy** button. Open the **Report...** box and **Export...** the WIN-DAISY iteration result to **WIN-DAISY** again with fragmentation (**Yes**). Toggle back to **WIN-NMR** and **Export...** the data again but this time answer the fragmentation question with **No**. Inspect the WIN-

DAISY dialog boxes for the differences between exporting the multiplet analysis with and without fragmentation. Execute a **Simulation** with both WIN-DAISY documents and note the calculation times. (The second calculation will overwrite the first spectrum file, because the destination names of the spectra are identical. But for the moment we are only interested in the protocol files). The time needed for the six-spin system is slightly longer than for the two three-spin systems (it is also dependent on other WINDOWS processes running).

There are disadvantages in treating both AMX subspectra as a single fragment instead of two independent fragments with their own parameter sets. The spectrum simulation of such a single fragment is much more time consuming because the matrices to be diagonalized are much bigger. In addition, because many of the coupling constants are zero, every transition appearing in the spectrum has an eight fold degeneracy so that the 192 calculated transitions are reduced to 24 after checking for degeneracy.

Check it in WIN-DAISY:

Open the **Simulation Protocol** for both simulations and look at the calculated transition list to verify the statements made. The high number of degenerate lines in the six-spin system simulation is obvious.

When treated as two independent, three spin fragments only 24 transitions (12 from each fragment) are calculated. Therefore in cases where no long range coupling is observed (or expected) it is an advantage to subdivide a molecule into independent fragments. In case later on long range couplings shall be included the **Edit | Merge fragments** command may be used to combine two independent fragments.

2.3 Second Order Spin Systems

The pattern described in the previous section allows the chemical shifts and coupling constants to be determined directly from the spectrum. In these cases spin simulation is simply used as a tool to either confirm or disprove the manual analysis. The real need for programs that can simulate and interactively adjust NMR parameters arise with spectra that exhibit strong second order effects.

In section 2.2.1 the effect of the scalar coupling constant has been introduced with respect to the perturbation parameter λ_{ik} to determine whether the relation between two nuclei is first order or second order. For first order spin systems the rule of repeated spacing was introduced. The mathematical treatment of this phenomenon as the X approximation can be used to simplify the calculation by neglecting of the HAMILTONian mixing (flip-flop) term. Nevertheless it was pointed out that the borderline between first and second order systems is not well defined. Homonuclear spin system should be always calculated *without* using the X approximation to avoid biased results. However, the X approximation may always be used for spectra that display coupling between different NMR active isotopes, refer to chapter 4.

The theoretical background of second order spin systems is included in every textbook [2.12]-[2.15] in various depth and in [2.18] the theory with stress on the AB spin system is given. Here the necessary information to proceed and understand the analysis is given and the appendix, section 6.1 contains more detailed theoretical discussions.

2.3.1 Notational Aspects

Based on the PSB notation [2.16] described in section 2.1.4 the labeling of spins using letters of the alphabet can be extended to lay stress upon the second order relation between nuclei. If a strong homonuclear coupling exists between two spins, they are represented by consecutive letters of the alphabet.

For example a two spin system with first order character is named AX and with second order character AB. A three spin system, in which all spins are weakly coupled is labeled AMX; in the case where the two low field spins (A and M) are strongly coupled, but their relation to spin X is of first order type, the system is labeled ABX. When all three spins show distinct second order influence the correct notation is ABC.

2.3.2 The AB Spin System

As an introduction to second order spin systems you should start with the analysis of an AX spin system as discussed in section 2.2.2.

Check it in WIN-NMR and WIN-DAISY:

Analyze the two spin system **Multiplets** of the proton spectrum ...\AX\002001.1R. The **Interval** regions file contains only the doublet signals. The central part of amino and phenyl signals will be discussed in chapter 3. Optimize it by **Iteration**. We will neglect for the moment, that the linewidths of the two signals are different.

2-chloro-6-nitro-3-phenoxy-aniline

Use the **Dual Display** command to compare the spectra and toggle back to WIN-DAISY. Open the **Control Parameters** and change the **Mode** into **Single Simulation**. Click on the **Previous** button to open the **Spectrum Parameters** dialog box. In the **Frequencies adapted...** parameter box select the lowest field strength available (**60 MHz**). Quit the dialog box by clicking the **Ok** button and answer **Yes** in the new box that pops up. (The 60MHz simulation data obtained in this way can be compared with the input document AB_60.MGS.) **Change** the name of the simulated spectrum to e.g. ...\AX\002998.1R and click on the **Simulation** button. Use the WIN-DAISY command **Dual Display** to compare the low magnetic field calculation with the high-field experimental data by pressing the button. In WIN-NMR use the **Multiple Display** to compare experimental, iterative and calculated data (...\002001.1R, ...\002999.1R, ...\002998.1R). Expand the x-region with <>, click the **Separate** button and switch the X-axis into Hertz-units <Ctrl + X> (refer to Fig. 2.25) to accentuate the difference in the chemical shift scale.

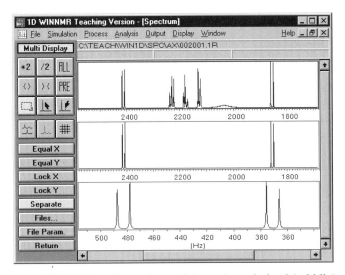

Fig. 2.25: Multiple Display of experimental (upper), optimized (middle) and calculated spectra at different field strength (lower).

Clearly the second order effect visible by the roof effect has been intensified by switching to a lower magnetic field strength.

Check it in WIN-DAISY:

Call the **Calculator** from the WIN-DAISY toolbar and evaluate the perturbation parameter λ_{AB} for both calculated spectra. Take the values from the **Frequencies** and **Scalar Coupling Constants** dialog boxes for the 60 MHz spin system. To restore the 300 MHz data, you simply open the **Control Parameters** and an **Iteration mode** (either **Standard** or **Advanced**) and quit the dialog with **Ok**, accepting the message box with **Yes**. When selecting an iteration mode the spectrometer type of simulation and experimental data must be identical! Then change the data back to **60** MHz.

The values obtained for the present spectra are:

$$300 \text{ MHz}: \lambda_{AB} \approx 0.018 \qquad 60 \text{ MHz}: \lambda_{AB} \approx 0.09$$

Based on the fact, that the perturbation parameter is 0.018 at 300 MHz, it is obvious, that by dividing the basic transmitter frequency by 5, the perturbation parameter at 60 MHz is five times larger than its value at 300 MHz.

This is the easiest way to modify the order of a spin system involving a single change in the parameter set. If you want to increase the second order effect further, the spectrometer type selection doesn't offer any lower magnetic field strengths, but the effect may be obtained by either modifying the chemical shift difference or the magnitude of the coupling constant as exercised already basically in section 2.2.2.2.

Check it in WIN-DAISY:

Use the 60 MHz AB spectrum as the basic data set (either the input document still opened, or **Open...** the file AB_60.MGS). Open the **Control Parameters** dialog box and change the **Calculation Mode** to **Sequence Simulation**. Typically, set the **Default Parameter Increment [Hz]** to <10> Hz and **Default Number of Steps** to <19>. Click on the **Scalar Coupl.** button and check the available box for the coupling constant. Return with **Ok** to the **Control Parameters** dialog box. A total number of **20** spectra will be simulated as mentioned in the dialog; the first calculated spectrum refers to the original reference simulation while in the other 19 spectra the scalar coupling constant is incremented by +10 Hertz. Close the dialog with **Ok**. Conversely, you may also use the input file AB_JSEQ.MGS. Run a **Simulation Sequence** to perform the calculations. The spectra are stored consecutively with decreasing process number. After the sequence is finished the original data is restored in the dialog boxes. Display the reference calculated spectrum by pressing the red peak button. In this way WIN-NMR changes already into the correct directory.

Check it in WIN-NMR:

Switch to the **Multiple Display** mode. Starting with the highest process number (move the slider to the bottom), select all the twenty calculated spectra (starting with the number in the WIN-NMR header status line, e.g. ...\002998.1R and use then the arrow key <↑>) and compare them in the **Stacked Plot** mode (refer to Fig. 2.26). You may need to change the Y-gain to relative units <Ctrl + Y> when experimental data are connected.

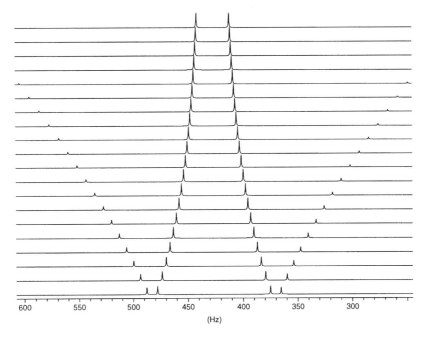

Fig. 2.26: Stack plot of 60 MHz simulations varying the scalar coupling constant from 10 Hz (bottom spectrum) to 200 Hz (top spectrum).

The visual impression of Fig. 2.26 is, that starting from the doublet of doublet in the bottom spectrum the roof effect is increased until the outer lines of the doublets almost disappear.

Carrying on the sequence of simulations would finally lead to the inner signals collapsing to a single line containing the total spectrum intensity. As the intensity of the outer lines decreases, the magnitude of the coupling constant can no longer be accurately determined and consequently it is not possible to determine the true chemical shifts of spin A and spin B. Thus the top trace in Fig. 2.26 might be misinterpreted as belonging to two non-coupled spins. (The transitions are shifted about 40 Hz from the true resonance frequency of the nuclei.) In fact the two protons are so strongly coupled ($\lambda_{AB} \approx 1.82$), that the second order effect prevents the reliable determination of the shifts and coupling constant. This particular example of a deceptively simple AB system arises when either the outer lines disappear or the inner lines collapse:

2.3 Second Order Spin Systems

- If half of the reciprocal of the square of the perturbation parameter is smaller than the minimal detectable intensity, the outer signals (*weak outer lines*) are not detectable and the spectrum seems to consist only of two single lines:

$$\tfrac{1}{2} \cdot \lambda_{AB}^{-2} < i_{min}$$

- When limes (λ_{AB}) = ∞ a singlet is obtained. However before this limit is reached, another condition causes the inner lines to collapse, when the value given below is smaller than the half-height linewidth $v_{1/2}$:

$$\tfrac{1}{2} \cdot \Delta_{AB} \cdot \lambda_{AB}^{-1} < v_{1/2}$$

Check it in WIN-DAISY:

The other way to decrease the ratio λ_{AB} is to shift one resonance frequency towards the other. This may be illustrated by simulation a set of spectra keeping the size of the coupling constant fixed and changing, in approximately 10 Hz steps one of the resonance frequencies. Either **Open...** the input document AB_FSEQ.MGS or modify your actual input: Proceed in a similar manner to the previous 'Check it'. In the **Control Parameters** dialog box click the **Scalar Coupl.** button to ensure that the coupling option is switched off. Click the **Frequencies** button and select **Spin 2** as the chemical shift to be varied and the **No. of steps** to <13>. To keep the other calculations, you may **Change** the spectrum name (e.g. ...\002899.1R). After **Sequence Simulation** the result of 14 calculations is shown in the WIN-NMR Multiple Display (Fig. 2.27).

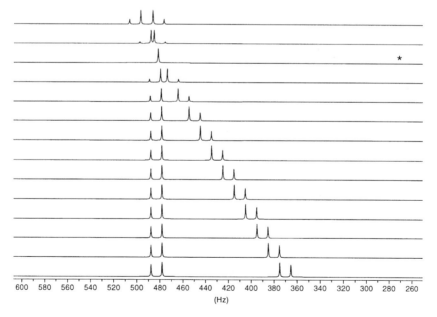

Fig. 2.27: Stack plot of 60 MHz simulations shifting the resonance frequency of nucleus B in 10-Hertz-steps towards and beyond shift A (lowest to uppest spectrum), * deceptively simple spectrum.

Because λ_{AB} is inversely proportional to the chemical shift difference, it converges much faster towards infinity compared with varying J_{AB}. In the marked spectrum (*) in Fig. 2.27 the chemical shifts are identical ($\Delta_{AB} = 0$), the perturbation parameter becomes infinite ($lim(\lambda_{AB}) = \infty$) and the signals collapse.

In this case (*) the resonance frequencies are approximately identical and the information about the shifts and coupling constant is lost because nuclei A and B are isochronous. In the trace above the shift of nucleus B lies across shift A, the absolute value of the resonance frequency difference and therewith λ_{AB} is not identical to that of the spectrum beyond trace (*). The AB system occupies a special position within the category of second order spin systems, because the number of transitions is identical with those of the first order AX spin system. No additional transitions appear.

Check it in WIN-NMR and WIN-DAISY:

Measure a **Linewidth** and analyze the aromatic AB spin system in the spectrum 2,3,4-trichloroanisole ...\AB\001001.1R, using the **Multiplets** mode of WIN-NMR. Mark one of the peaks with the spectrum cursor belonging to the spin system when **export**ing the data to **WIN-DAISY**.

2,3,4-trichloroanisole

Perform an **Iteration** and compare the lineshapes in WIN-NMR **Dual Display**. Due to a long range coupling with the methoxy group (refer to section 2.4) the A and the B signals have different linewidths. Toggle back to **WIN-DAISY** and open the **Lineshape Parameters**. Select the option **Nuclei specific linewidths** and press the **Iterate All** button. Run an **Iteration** and look at the result in WIN-NMR **Dual Display**. The best result obtainable is an R-factor of ca. 0.7% in the **Iteration Protocol**. Use the linewidth information to assign the signals to the structure. **Export** the result as a coupling tree back to the **exp**erimental spectrum in **WIN-NMR**. The second order effect in the present example is not strong. Keep the document open.

Check it in WIN-NMR and WIN-DAISY:

Analyze the aromatic AB spin system in the proton spectrum of 2,3,6-trichloroanisole ...\AB\002001.1R, using the **Multiplets** mode of WIN-NMR (note the question mark in the **Report...** box). Mark one of the signals with the spectrum cursor and **Export...** the data to **WIN-DAISY**, run first a **Simulation** and call the **Dual Display**.

2,3,6-trichloroanisole

The comparison of the simulation based on the first order analysis is erroneous in resonance frequencies, not in the determination of the coupling constant: Use the **Move Trace** option and move the calculated trace with depressed <Shift> key over the experimental. Toggle back to **WIN-DAISY** and perform an **Iteration**. **Export** the WIN-DAISY result as a coupling tree back to **WIN-NMR exp.** and compare the coupling tree with the peak positions (refer to Fig. 2.28). **Calculate** the λ_{AB} values for both the current and the previous AB example and relate this to the chemical environment of the protons.

2.3 Second Order Spin Systems 69

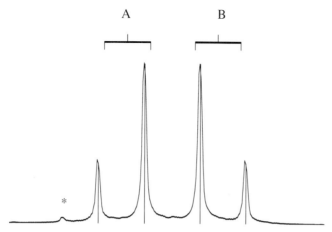

Fig. 2.28: Experimental. AB spectrum with calculated transitions and coupling-tree, *: CHCl$_3$ impurity signal.

In Fig. 2.26, Fig. 2.27 and Fig. 2.28 it evident that for an AB spin system the absolute value of the coupling constant appears twice in the signal pattern as the distance between the first pair and the second pair of lines. But the first order coupling tree displayed above the spectrum in Fig. 2.28 clearly illustrates, that the resonance frequencies v_A and v_B are *not* the arithmetic mean of the transition frequencies. However the WIN-DAISY iteration is able to extract the correct parameters on the basis of the Multiplet analysis as input data. Another way to extract the NMR parameters v_A, v_B and J_{AB} is discussed in the following section. A detailed description of second order spin system type with priority to the AB spin system is given in [2.18], [2.11] and [2.14].

2.3.2.1 The Graphical Spectrum Analysis

Another special feature of AB spin systems is the possibility to obtain the resonance frequencies using graphical methods.

Check it with WIN-NMR:

Using the aromatic AB spin system in the proton spectrum of *2,3,6-trichloroanisole* ...\AB\002001.1R, perform a **Mouse PP.** with **Interpolation**, units **Hz**, in the **Analysis Peak Picking** mode of WIN-NMR. **Undo** the automatic Peak Picking before the manual. Use <Ctrl + X> to switch the X-axis into Hz units. **Preview** the AB spectrum using a well defined Hertz-per-cm ratio, e.g. <2> Hz/cm in the **Output Page Layout** and **print...** it.

With the distance between T_{A1} and T_{B1} as the radius (refer to Fig. 2.29), use a pair of dividers to draw a circle using T_{A1} as the origin. The height the circle cuts the peak T_{A2} is used to define a right-angle triangle with its hypotenuse equal to the radius of the circle. Measure the cathetus height Δ_{AB} where the circle cuts the peak T_{A2} and convert this distance into frequency units. To get the resonance frequencies measure on the X-axis ½Δ_{AB} at either side from the arithmetic center of the four line signal pattern.

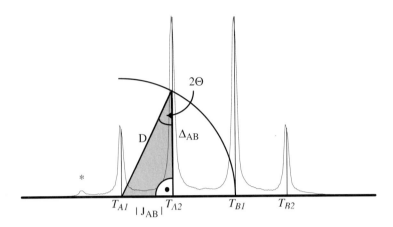

Fig. 2.29: Graphical solution of an AB signal pattern, *: CHCl$_3$

The equations used in the following analysis are based on PYTHAGORAS theorem:

$$D^2 = \Delta_{AB}^2 + J_{AB}^2$$

$$|J_{AB}| = T_{A1} - T_{A2} = T_{B1} - T_{B2}$$

using the auxiliary parameters: $D = T_{A1} - T_{B1} = T_{A2} - T_{B2}$

and

$$v_z = \tfrac{1}{2}\cdot(T_{A1} + T_{B2}) = \tfrac{1}{2}\cdot(T_{A2} + T_{B1})$$

to result in

$$\Delta_{AB} = \sqrt{D^2 - J_{AB}^2}$$

$$v_A = v_z + \tfrac{1}{2}\cdot \Delta_{AB}$$

$$v_B = v_z - \tfrac{1}{2}\cdot \Delta_{AB}$$

Two other ways are possible to yield the resonance frequency difference Δ_{AB}:
Based on the angle 2Θ, the frequency difference can be calculated using trigonometric functions. The angle Θ will occur again in other sections. 2Θ is obtainable from the values D and $|J_{AB}|$:

$$2\Theta = \arcsin\left(\frac{|J_{AB}|}{D}\right) \quad \text{(can be extracted by protractor)}$$

$$\Delta_{AB} = D \cdot \cos 2\Theta$$

Classifying the four transition lines into 'inner' (more intense: T_{A1} and T_{B2}) and 'outer' (less intense: T_{A2} and T_{B1}) lines offers a third way to calculate the chemical shift difference based on the separation of outer and inner lines:

$$S_{inner} = T_{A1} - T_{B2}$$
$$S_{outer} = T_{A2} - T_{B1}$$
$$\Delta_{AB} = \sqrt{S_{outer} \cdot S_{inner}}$$

Check it in WIN-DAISY:

Run the **Calculator**. Use a protractor to measure the angle 2Θ; also calculate 2Θ using trigonometric functions. Use the S_{inner} and S_{outer} method to determine v_A and v_B. Mark the position of the calculated resonance frequencies on your plotted spectrum. As you can see they are shifted towards the inner lines of the AB pattern.

The auxiliary parameter D will appear in other second order cases and will occur also within the quantum mechanical formalism used by WIN-DAISY to calculate the NMR transitions based on the NMR parameters v_A, v_B and J_{AB} in the theory sub section 2.3.2.3. Here only the results will be discussed. The following Table 2.8 gives a stepwise introduction of the different perturbation levels included in the calculation of the energy levels of $m_T = 0$ in an AB spin system: The ZEEMAN energy only takes into account chemical shifts (without coupling perturbation) and the effect of the coupling perturbation can be regarded as either first order or second order coupling effect. In former section the EIGENvalues (energy levels) E of the HAMILTONian were discussed, here the operator $\hat{H}$ will be used, because for second order spin systems the corresponding energy levels cannot be determined in the easy way described (for explanation refer to the appendix, section 6.1). The corresponding total spin value $m_T = 0$ is called a *bad quantum number* in contrast to the *good quantum number* $m_T = +1, -1$.

Table 2.8: Comparison of the EIGENvalues obtained for $m_T = 0$ in an AB system. $\hat{H}_0$ refers to energy levels E_j^0, $\hat{H}_1$ to E_j^1 and $\hat{H}_2$ to E_j^2 (mixing term).

HAMILTONian	$\hat{H}_0$	$\hat{H}_0 + \hat{H}_1$	$\hat{H}_0 + \hat{H}_1 + \hat{H}_2$
perturbation level	no perturbation (only chemical shifts)	first order (X approximation)	second order (full Hamiltonian)
description	ZEEMAN energy	weakly coupled	strongly coupled
energy levels	$\pm \Delta_{AB}$	$-\frac{1}{4} J_{AB} \pm \Delta_{AB}$	$-\frac{1}{4} J_{AB} \pm \frac{1}{2} D$

Check it in WIN-DAISY:

Using the formulae given in Table 2.8, **calculate** the two energy levels E_2 and E_3 corresponding to the HAMILTONian submatrix with $m_T = 0$. For the input parameters use the values of v_A, v_B and J_{AB} obtained from the analysis of the aromatic AB spin system in the proton spectrum of *2,3,6-trichloroanisole*. The parameters you use may either be extracted directly from the spectrum or obtained using WIN-DAISY.

Select the output option **EIGENvalues** in the **Main Parameters** dialog box and repeat a spectrum **Simulation**. Compare your manually calculated energy levels with the values determined by the program shown in the **Simulation Protocol**.

Check the second order perturbation effect on these energy levels by changing the **ISO** value in the **Frequencies** parameters dialog box. **Change** the calculated spectrum name and **Simulate** the spectrum. Use the Dual Display mode of WIN-NMR to compare both simulated spectra. **Calculate** manually the corresponding EIGENvalues using the first order energy level equations (with X approximation) given in Table 2.8 (middle column).

The mixing of the two states $\alpha\beta$ and $\beta\alpha$ ($m_T = 0$) results in the modification of the separation between the two energies for this total spin value (compare last row in Table 2.8). (The value of D given below results from the p-q formula to extract the EIGENvalues of 2x2 matrices, see appendix, section 6.1).

$$\Delta_{AB} \quad \text{turns into} \quad D = \sqrt{\Delta_{AB}^2 + J_{AB}^2}$$

Table 2.9 illustrates the transition energies for the different levels of perturbation discussed:

Table 2.9: Comparison of the transition frequencies obtained for an AB system.

transition frequency	$\hat{H}_0$	$\hat{H}_0 + \hat{H}_1$	$\hat{H}_0 + \hat{H}_1 + \hat{H}_2$
T_{A1}	v_A	$v_A + \frac{1}{2} J_{AB}$	$v_z + \frac{1}{2}D + \frac{1}{2} J_{AB}$
T_{A2}	v_A	$v_A - \frac{1}{2} J_{AB}$	$v_z + \frac{1}{2}D - \frac{1}{2} J_{AB}$
T_{B1}	v_B	$v_B + \frac{1}{2} J_{AB}$	$v_z - \frac{1}{2}D + \frac{1}{2} J_{AB}$
T_{B2}	v_B	$v_B - \frac{1}{2} J_{AB}$	$v_z - \frac{1}{2}D - \frac{1}{2} J_{AB}$

Comparing the transition frequencies for the second order case AB the graphical solution of the spin system described in Section 2.3.2.1 becomes evident, e.g.:

$$T_{A1} - T_{A2} = v_z + \tfrac{1}{2}D + \tfrac{1}{2}J_{AB} - (v_z - \tfrac{1}{2}D + \tfrac{1}{2}J_{AB}) = D$$
$$T_{A1} - T_{B1} = v_z + \tfrac{1}{2}D + \tfrac{1}{2}J_{AB} - (v_z + \tfrac{1}{2}D - \tfrac{1}{2}J_{AB}) = J_{AB}$$

The determination of the transition intensities will be discussed in the appendix, section 6.1 but essentially it is the calculation of the transition probability, in which the angle 2Θ appears again:

Table 2.10: Transition frequencies and intensities obtained for an AB system.

transition	energy levels	frequency	intensity
T_{A1}	$1 \to 3$	$v_z + \tfrac{1}{2}D + \tfrac{1}{2} J_{AB}$	$1 - \sin 2\Theta$
T_{A2}	$2 \to 4$	$v_z + \tfrac{1}{2}D - \tfrac{1}{2} J_{AB}$	$1 + \sin 2\Theta$
T_{B1}	$1 \to 2$	$v_z - \tfrac{1}{2}D + \tfrac{1}{2} J_{AB}$	$1 + \sin 2\Theta$
T_{B2}	$3 \to 4$	$v_z - \tfrac{1}{2}D - \tfrac{1}{2} J_{AB}$	$1 - \sin 2\Theta$

2.3 Second Order Spin Systems

Check it in WIN-DAISY:

Using the printout of the aromatic AB spin system in the proton spectrum of *2,3,6-trichloroanisole*, analyze the spectrum and extract the angle Θ. **Calculate** the transition intensities. In the **Main Parameters** dialog box of WIN-DAISY, select the output option **Transition and energy levels** and **Simulate** the spectrum. Compare the manually calculated intensities with those given in the WIN-DAISY **Simulation Protocol** file.

The ratio of the distance between outer lines (S_{outer}) and the inner lines (S_{inner}) of the four transitions is inversely proportional to their intensity ratio (I_{outer}/I_{inner}):

$$\frac{S_{outer}}{S_{inner}} = \frac{I_{inner}}{I_{outer}}$$

Check it in WIN-NMR:

Open the proton spectrum of *2,3,6-trichloroanisole* ...\AB\002001.1R with WIN-NMR, and perform a **Peak Picking** over the aromatic AB spin system. Click on the **Report...** button and print out the Report Peaks table. Using the intensities in this table, calculate the intensity ratio and compare it with the value obtained using the distance ratio.

The intensity ratio determined using the calculated intensities coincides well with the distance ratios. In contrast the experimental peak picking intensity ratio is smaller than expected.

One important point is, that the intensity of NMR transitions is not represented by the peak height but by the peak area - the integral. In cases where the transition peaks overlap, even at the peak fringes, the values extracted are biased. The integration of the individual peaks will not help to obtain better results, because only frequency regions can be defined to detect the integral information. To obtain more reliable intensity data, the deconvolution of the signals is recommended. In this technique, LORENTZian line profiles with variable position, intensity and linewidth are decomposed from the spectrum and the integral of each individual line is listed.

Check it in WIN-NMR:

Open... the proton spectrum of *2,3,6-trichloroanisole* ...\AB\002001.1R with WIN-NMR. From the **Analysis** pull-down menu choose the **Deconvolution 1** option and perform a deconvolution over the aromatic region. Click the **Region...** button; use the left mouse button to position the frequency markers and the right mouse button to define the frequency region. Click on the **Options...** button and define **LORENTZian lineshape** and **Number of Peaks** <4>. Click the **Execute** button. After the calculation is finished the individual lines are shown in different colors on the screen. The numerical values are accessible from the **Report...** button.

Although the values determined by the spectrum decomposition give a more accurate intensity ratio than peak-picking, the fact remains that the intensity data is more susceptible to perturbation than frequency data. As discussed in section 2.2.3, the effect of signal overlap on frequency data is to simply shift the peak maxima. In the present case the effect of the signal overlap influences the intensity data but does not effect the frequency data perceptively.

Another point that has to be considered is that the noise in experimental data influence the accuracy of values extracted from such data.

Check it in WIN-NMR:

Open... the calculated AB spectrum after iteration with WIN-NMR ...\AB\002999.1R and repeat the **Peak Picking** and **Deconvolution** using this noise-free lineshape for the analysis.

The problems arising from noise and line overlap will become more critical in spin systems of higher complexity.

2.3.2.2 Suppression of Particular Transitions

In the case of first order spin systems the nuclei A, M, ... X responsible for the individual transitions is easy to determine, because the bpf's are identical with the stationary states. On this basis it is evident, which spin causes which transition as shown in Table 2.11:

Table 2.11: Spin responsibility for particular transitions in an AX spin system.

transition	energy levels	origin bpf's	destination bpf's	spin flip of
T_{A1}	$1 \to 3$	$|\alpha\alpha>$	$|\beta\alpha>$	A
T_{A2}	$2 \to 4$	$|\alpha\beta>$	$|\beta\beta>$	A
T_{X1}	$1 \to 2$	$|\alpha\alpha>$	$|\alpha\beta>$	X
T_{X2}	$3 \to 4$	$|\beta\alpha>$	$|\beta\beta>$	X

In the case of second order spin systems the situation is slightly more complicated as illustrated in Table 2.12:

Table 2.12: Spin responsibility for particular transitions in an AB spin system.

transition	energy levels	origin bpf's	destination bpf's	spin flip of
T_{A1}	$1 \to 3$	$|\alpha\alpha>$	$|\alpha\beta>$ and $|\beta\alpha>$	?
T_{A2}	$2 \to 4$	$|\alpha\beta>$ and $|\beta\alpha>$	$|\beta\beta>$	?
T_{B1}	$1 \to 2$	$|\alpha\alpha>$	$|\alpha\beta>$ and $|\beta\alpha>$	?
T_{B2}	$3 \to 4$	$|\alpha\beta>$ and $|\beta\alpha>$	$|\beta\beta>$	?

The reason why is the mixing of the bpf's in order to form the EIGENstates for the quantum mechanical spin system. Based on the calculation of spectra at the beginning of this section we know, that the low-field spectral part arises from spin A and the high-field part is caused by nucleus B. A closer look at the coefficients for the linear combination requires the evaluation of the definition range for Θ.

As illustrated by the graphical solution of the AB spin system and the fact it is a right-angled triangle, it can be concluded, that:

2.3 Second Order Spin Systems

$$\Theta \in [\,0°\,;\,45°\,]$$

The two limiting cases $\Theta = 0°$ describes the AX situation and $\Theta = 45°$ is connected with the AA' or A_2 conditions of identical resonance frequencies (see section 2.1.2.1 and section 2.4). The corresponding coefficients of the bpf's in the EIGENfunctions used to define the linear combinations

$$c_{1k}\,|\alpha\beta\rangle + c_{1j}\,|\beta\alpha\rangle = E_1\,|\alpha\alpha\rangle \qquad \text{(an example for i=1)}$$

are shown in Table 2.13 (for detailed information refer to the appendix, section 6.1):

Table 2.13: Coefficient values for the linear combinations.

c_{ik}	$\Theta = 0°$ (AX)	$\Theta = 45°$ (AA')
$\sin \Theta$	0	$\dfrac{1}{\sqrt{2}}$
$\cos \Theta$	1	$\dfrac{1}{\sqrt{2}}$

For the AX spin system calculated with second order effect ($\Theta = 0°$) the contribution of one term of the linear combination vanished, so that the remaining bpf represents the EIGENstate.

The other limiting case ($\Theta = 45°$) appears for two isochronous spins. Inspecting the coefficients of the bpf's to form the EIGENfunctions it can be concluded that both spins are equally involved in every transition ($2^{-\frac{1}{2}}$). Consequently it is not possible to distinguish between spin A and A' and therefore WIN-DAISY cannot decide which spin is responsible for the transition. A complete discussion of this type of problem can be found in the section about magnetic equivalence (section 2.4).

The conclusion of these explanations is, that for all cases with $\Theta \in [0°;45°[$ (the latter bracket means 45° excluded!) the nucleus giving the main contribution to the transition can be determined on the basis of the coefficients of bpf's forming the linear combinations of the corresponding EIGENfunctions.

Check it in WIN-DAISY:

Use the input document of the analyzed AB spin system in the proton spectrum of *2,3,6-trichloroanisole*. **Simulate** and **Display** the **Calculated Spectrum** with **WIN-NMR**. Toggle back to **WIN-DAISY** and open the **Frequencies** dialog box and select the **Disable** option for Spin 2. **Change** the calculated spectrum name (e.g. to ...\002998.1R to keep the former calculation) and execute a **Simulation**. Toggle to **WIN-NMR**.

Check it in WIN-NMR:

Load the new calculated spectrum as the second file in the **Dual Display**. Press the **Separate** and **Show Difference** button to display in the third screen part the difference spectrum representing the suppressed transitions (refer to Fig. 2.30).

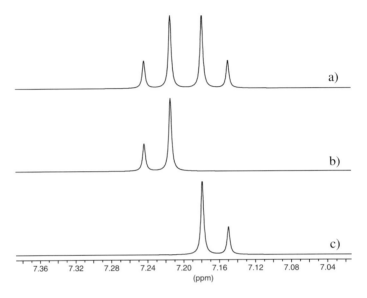

Fig. 2.30: a) Calculation of the whole AB spin system.
　　　　　b) Calculation of the A part of the AB system under suppression (disable of spin B).
　　　　　c) Difference spectrum a) – b) showing the suppressed B spectrum part.

2.3.3 The ABX Spin System

Second order spin systems consisting of three nuclei can be divided into different types:
- Either all three nuclei are strongly coupled (notation: ABC).
- Or only two spins show a distinct second order effect and are weakly coupled to the third member of the spin system (notation: ABX or AXY respectively).

The ABX case occurs quite frequently and displays the second order effects not only in intensity contributions and slight frequency shifts but also in the presence of additional transitions. The introduction into this type of spin system is achieved by modifying one resonance frequency of an AMX spin system to an ABX type system (actually the example refers to an AXY spin system, the spins X and Y representing the second order effect).

Check it in WIN-DAISY and WIN-NMR:

Open... the input document ABX_SEQ.MGS. Calculate the spectrum using the initial parameters by simply pressing the **Simulation** button and **Display** the **Calculated Spectrum** in WIN-NMR. **Analyze** the spectrum in the **Multiplets** mode using the first order rules as an AMX spin system, each signal as a doublet of doublets (dd). The high-field part of the spectrum (AB-type) can be analyzed as a doublet of doublets with slightly distorted intensities. But the main roof effect already indicates a strong second order effect (Fig. 2.31). Toggle back to **WIN-DAISY**.

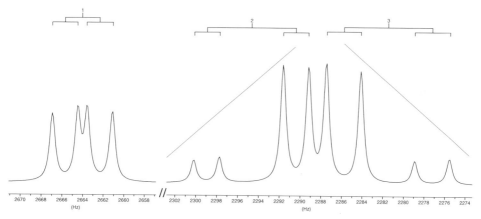

Fig. 2.31: Multiplet analysis of the single simulation of ABX_SEQ.MGS.

Check it in WIN-DAISY and WIN-NMR:

Click in the **Control Parameters** dialog box the **Frequencies** button. You can see that the second frequency, which has already a strong second order relation to spin 3, will be shifted in 1 Hertz-steps towards shift 3. Calculate the **Simulation Sequence** and display the 10 spectra using the Multiple Display mode of WIN-NMR. Fig. 2.32 shows the 'AB' type part (right) and the 'X' type part (left) of the spin system simulation.

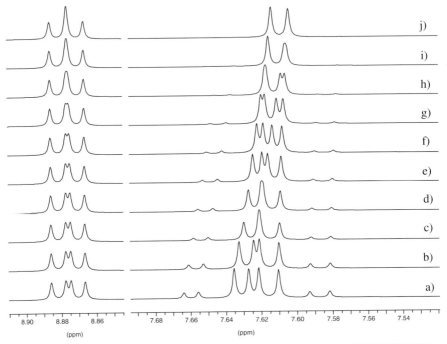

Fig. 2.32: Zoomed X (left) and AB (right) part of the sequence ABX_SEQ.MGS.
a) Reference spectrum. b) - j) Shift 2 decreased by 1 Hz respectively.
X-part intensities increased.

The identification of the doublets of doublets in the 'AB' part of the spectrum becomes more difficult the closer the resonance frequencies of 'A' and 'B' move together. We will take for example trace g) of Fig. 2.32 to look closer at the second order pattern.

Check it in WIN-NMR and WIN-DAISY:

Open... trace g) of Fig. 2.32 - the seventh simulation ...\999993.1R - in WIN-NMR. From the **Output** pull-down menu choose the **Title...** option to display the frequency of spin 2 (2287 Hz) used in the actual simulation.

Toggle back with **Simulation | WIN-DAISY** and either modify the actual input document by changing the frequency of Spin 2 to <2287> Hz or simply **Open...** the input document ABX_2.MGS. **Change** the calculated spectrum name (to keep the sequence simulations) e.g. to ...\999001.1R, execute a **Simulation** and **Display** the **Calculated Spectrum** in WIN-NMR. This spectrum should be identical to trace g) in Fig. 2.32.

Go back to **WIN-DAISY** and in the **Frequencies** dialog box **disable** spin 3. **Change** the calculated spectrum name again, e.g. to ...\999002.1R and execute a **Simulation**. Toggle to **WIN-NMR** and load the new calculated spectrum as the second trace in the **Dual Display**. Click the **Separate** and **Show Difference** buttons to display the difference spectrum representing the suppressed transitions (Fig. 2.33).

2.3 Second Order Spin Systems   79

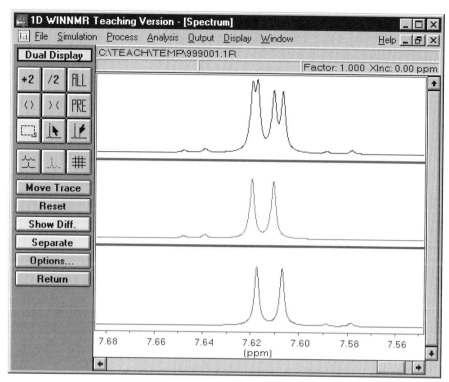

Fig. 2.33: 'AB' part of ABX system. Upper trace shows the total simulation. Second trace with suppression of spin 3, lower trace difference spectrum.

Fig. 2.33 shows in the A and B multiplets that the 'doublets of doublets' overlap and that the overall pattern looks very different. Despit4 displaying a strong roof effect, the intensities do not follow the rule of the direction of the coupling partner any more. This is also true for all the spectra in the simulated sequence where this distorted doublet-of-doublet pattern in the overlapping AB part is clearly visible until the outer lines disappear.

Thinking about the AB spin system discussed in the previous section, the suppression of the transitions of one spin resulted in the display of one half of the AB signal pattern. In the present ABX type system we have an additional nucleus X coupled weakly to spins A and B which results in a duplication of each half of the AB signal pattern (Fig. 2.33). To complete this argumentation based on the AB system, the combination of one part of the A and one part of the B simulation results in one *ab subspectrum*, and the combination of the remaining parts leads to the second *ab subspectrum*, as shown in Fig. 2.34. Subspectra will be always labeled with small letters instead of capitals. In the present example both ab subspectra display different AB characteristics and are easily distinguished.

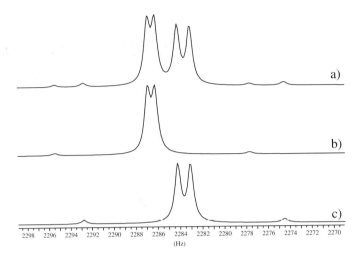

Fig. 2.34: a) Complete AB part of ABX spectrum.
b) Low-field AB: subspectrum (ab)$_1$.
c) High-field: AB subspectrum (ab)$_2$.

Because spin X is only weakly coupled to nuclei A and B, the subspectra result from the two different possible spin orientations of the X nucleus (↑ and ↓). Thus one (ab) subspectrum corresponds to the subsystem ABX with $m_z(X) = + \frac{1}{2}$ and the other (ab) subspectrum to the subsystem with $m_z(X) = -\frac{1}{2}$. Here again the concept of *effective LARMOR frequencies* is applicable [2.19], [2.20]. In every (ab) subsystem the spins are therefore labeled with small letters *a* and *b* to identify them as effective LARMOR frequencies in contrast to capital letters *A* and *B* for resonance frequencies.

In every (ab) subsystem the coupling J_{AB} is found twice like in every normal AB spin system. The difficulty in many ABX spectra is rather the correct assignment of the a- and b-parts to form the two (ab) subspectra. This problem occurs in cases where the AB characteristic of both subspectra is quite similar. Before this problem will be discussed any further, the X-part of the ABX system will be considered. Fig. 2.32 show the X transition lines of the calculation sequence. The main pattern is a doublet of doublet that moves closer together. As spins A and B become isochronous the inner lines of the X part collapse into a single line. But this is not the whole truth, by zooming in the base line of the X spectra in an overlay plot (Fig. 2.35) it becomes obvious that *additional lines* appear in the X spectrum part. Consequently an ABX spectrum consists of 14 lines instead of 12 lines as in an AMX spectrum; 8 transitions in the AB part (two (ab) subspectra with 4 lines each) and 6 lines in the X part. The two additional lines are so-called *combination lines* and due to their weak intensity they are also called *weak outer lines*.

2.3 Second Order Spin Systems 81

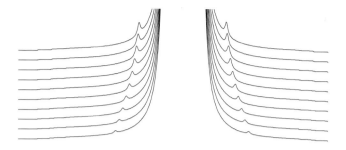

Fig. 2.35: X spectrum part of ABX sequence with zoomed base line.

2.3.3.1 Analytical Solution for the ABX Spin System

To understand this phenomenon reconsider the energy levels and the transitions in the AMX case. In Section 1.2.3. three additional transitions were considered but for first order spin systems these combination lines are forbidden. In an ABX spin system however, two of these transitions are allowed. In Fig. 2.36 two different representations are given. X-transitions are shown as normal arrows, A and B transitions in dotted lines. The diagram on the left-hand side shows the normal scheme for a three spin system using the cube standing on one edge and includes the combination lines (X transitions). To accentuate the appearance of two (ab) subsystems the right-hand side illustration may be chosen. This diagram emphasizes that the method of effective LARMOR frequencies can be applied to the two (ab) sub systems of an ABX spin system and not just to individual lines. Thus the coupling constants between the weakly coupled spins J_{AX} and J_{BX} may be incorporated into effective LARMOR frequencies which do not represent the true resonance frequencies.

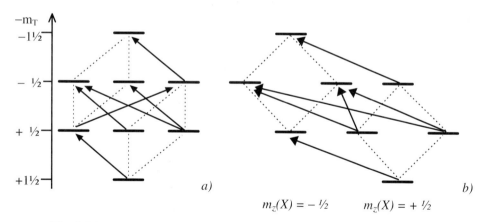

Fig. 2.36: m_T scheme for the ABX spin system including transitions
⋯⋯ A and B transitions,   ← X transitions.
a) Cubic representation.   b) Illustration of (ab) subsystems.

Analyzing the two (ab) subsystems using the rules for the AB spin system given in section 2.3.1. leads to two sets of effective LARMOR frequencies v_{a1}, v_{b1} and v_{a2}, v_{b2} respectively:

$$v_{a1} = v_A + \frac{J_{AX}}{2} \qquad v_{a2} = v_A - \frac{J_{AX}}{2}$$

$$v_{b1} = v_B + \frac{J_{BX}}{2} \qquad v_{b2} = v_B - \frac{J_{BX}}{2}$$

Check it in WIN-NMR or WIN-DAISY:

Either execute a Peak Picking in WIN-NMR (**Analysis | Peak Picking** mode with options Peak Labels in **Hz** and **Interpolation**) or take the transition frequencies from the WIN-DAISY **Simulation Protocol**. Analyze the two (ab) subspectra using the formulae given in Table 2.14.

Table 2.14: Analytical equations for analysis of (ab) subspectra.

Parameter	$(ab)_1$	$(ab)_2$
$|J_{AB}|$	$T_{a11}-T_{a21} = T_{b11}-T_{b21}$	$T_{a12}-T_{a22} = T_{b12}-T_{b22}$
v_z	$v_{z1} = \frac{1}{2}(T_{a11}-T_{b21})$	$v_{z2} = \frac{1}{2}(T_{a12}-T_{b22})$
S_{inner}	$S_{i1} = T_{a21}-T_{b11}$	$S_{i2} = T_{a22}-T_{b12}$
S_{outer}	$S_{o1} = T_{a11}-T_{b21}$	$S_{o2} = T_{a12}-T_{b22}$
Δ	$\Delta_1 = \sqrt{S_{i1} \cdot S_{o1}}$	$\Delta_2 = \sqrt{S_{i2} \cdot S_{o2}}$
D	$D_1 = T_{a11}-T_{b11}$	$D_2 = T_{a12}-T_{b12}$
v_a	$v_{a1} = v_{z1} + \frac{1}{2}\Delta_1$	$v_{a2} = v_{z2} + \frac{1}{2}\Delta_2$
v_b	$v_{b1} = v_{z1} - \frac{1}{2}\Delta_1$	$v_{b2} = v_{z2} - \frac{1}{2}\Delta_2$
v_a^*	$v_{a1} = v_{z1} + \frac{1}{2}\Delta_1$	$v_{a2}^* = v_{z2} - \frac{1}{2}\Delta_2$
v_b^*	$v_{b1} = v_{z1} - \frac{1}{2}\Delta_1$	$v_{b2}^* = v_{z2} + \frac{1}{2}\Delta_2$

In Table 2.14 two solutions for the effective LARMOR frequencies are given. The solution marked with an asterisk is the so-called *conjugated solution*. Evaluating solely from the AB part of an ABX spectrum, the result is ambiguous regarding the relative assignment of effective LARMOR frequencies. Both solutions lead to the same signal pattern in the AB spectrum part (Fig. 2.37 but without taking into account the X part, the resonance frequencies v_A and v_B cannot be determined. In order to finish the analysis both normal and conjugated parameter sets must be taken into account.

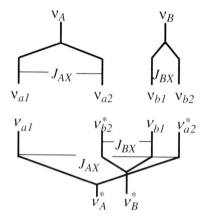

Fig. 2.37: Relative combination of effective LARMOR frequencies in ABX spectra.

The X spectrum part of an ABX spin system consists of six lines. Two distinct transitions appear in this multiplet: the so-called *N-lines*. Together these two lines contain 50 % of the total intensity in the X spectrum part (each N line is 25 %) and they are separated by the modulus of the sum of the weak coupling constants:

$$N = |J_{AX} + J_{BX}|$$

The remaining four lines are labeled as *inner* and *outer* lines.

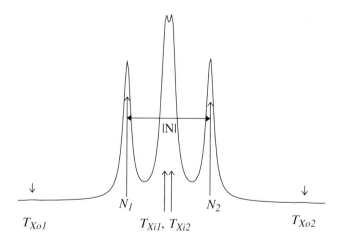

Fig. 2.38: Line identification in the X spectrum part.

Table 2.15: Analytical equations for analysis of X part of ABX spectra.

Parameter	normal solution	conjugated solution
N	$N_1 - N_2$	
S_{iX}	$|D_1 - D_2| = T_{Xi1} - T_{Xi2}$	
S_{oX}	$|D_1 + D_2| = T_{Xo1} - T_{Xo2}$	
v_X	$\frac{1}{2}(N_1 + N_2) = \frac{1}{2}(T_{Xi1} + T_{Xi2}) = \frac{1}{2}(T_{Xo1} + T_{Xo2})$	
J_{AX}	$J_{AX} = v_{a1} - v_{a2}$	$J^*_{AX} = v_{a1} - v^*_{a2}$
J_{BX}	$J_{BX} = v_{b1} - v_{b2}$	$J^*_{BX} = v_{b1} - v^*_{b2}$
v_A	$v_A = \frac{1}{2}(v_{a1} + v_{a2})$	$v^*_A = \frac{1}{2}(v_{a1} + v^*_{a2})$
v_B	$v_B = \frac{1}{2}(v_{b1} + v_{b2})$	$v^*_B = \frac{1}{2}(v_{b1} + v^*_{b2})$

Check it in WIN-DAISY:

Analyze the X part of the simulated ABX spectrum and determine the NMR parameters v_A, v_B, v^*_A, v^*_B, v_X, J_{AX}, J_{BX}, J^*_{AX} and J^*_{BX}. $|J_{AB}|$ has already been determined from the AB part of the spectrum. Perform a **Simulation** using the 'normal' and 'conjugated' (*) solution as the input data for WIN-DAISY. **Change** the calculated spectrum name to keep both simulations. Compare both spectra using the Dual Display mode of WIN-NMR (Fig. 2.39). Which is the correct solution?

In many cases the values obtained for the coupling constants may be used to distinguish between the right and wrong solution.

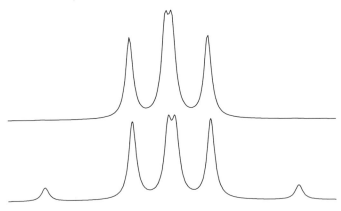

Fig. 2.39: a) X part of the normal solution of the ABX analysis.
b) X part of the conjugated ABX solution.

2.3 Second Order Spin Systems

The difference between both solutions (normal and conjugated) affects only the *transition intensities* in the X spectrum part. The AB part and the line frequencies of the X nucleus remain unchanged.

Check it in WIN-DAISY:

The conjugated solution includes one *negative* coupling constant. In the **Scalar Coupling Constants** dialog box change the sign of this coupling constant. (You may also compare your results with the input document ABX_1.MGS which contains the parameters for the conjugated solution.) **Change** the calculated spectrum name and **Simulate** the spectrum. Compare all three simulated spectra using the Multiple Display mode of WIN-NMR.

The spectrum obtained by changing the sign of either J_{AX} or J_{BX} results in a totally different signal pattern for the A, B and the X transitions. Altering the sign of the strong coupling constant J_{AB} leaves the spectrum unchanged. The ABX spin system is the first example in which the second order effect allows the determination of the relative signs of certain coupling constants.

The purpose of the following 'Check it' is to illustrate that it is possible to obtain the correct NMR parameters by iteration even when the starting values are poor: The most extreme situation conceivable is to start the optimization with the 'wrong' solution.

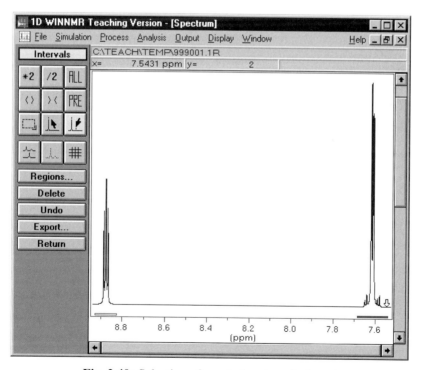

Fig. 2.40: Selection of spectral regions for iteration.

Check it in WIN-DAISY and WIN-NMR:

Open... the WIN-DAISY input document ABX_2.MGS containing the parameters for the normal ABX solution. **Simulate** the spectrum and **Display** the **Calculated Spectrum** (red peak button). Return to **WIN-DAISY** and **Close** the input document ABX_2.MGS.

Open the input document ABX_1.MGS containing the parameters for the conjugated ABX solution. Toggle to **WIN-NMR**. The simulated spectrum of the normal solution is still displayed. From the **Analysis** pull-down menu choose the **Interval** option and define two regions containing the AB and the X part of the spectrum. Do not make the intervals too large, but ensure that the outer lines are included in the defined regions (refer to Fig. 2.40). Click on the **Regions...** button and **Save...** the interval file using the default name for further use. Click on the **Export...** button and send this 'pseudo experimental data' to WIN-DAISY ensuring that **No** Multiplet analysis data is transferred, in case any exist.

WIN-DAISY is moved into the foreground. In the message box click the **Ok** button. The **Connect** button in the document window of the WIN-DAISY input document ABX_1.MGS is enabled. Click this button and make sure that the correct spectrum name appears in the status line **Experimental spectrum**. In case the experimental name features the default simulated spectrum name, WIN-DAISY will choose an appropriate name for the calculated spectrum name. If this is not possible a message box with the request to change the calculated spectrum name is displayed. Click on the **Options** button to change the default limits for **Scalar Coupling Constants** to $\pm$<10> Hz with **allow sign change** option checked. Quit the dialog box with **Ok**.

Open the dialog box for **Control Parameters** and change the **Mode** to **Advanced Iteration**. The default smoothing parameter **Broadening medium** and **No. of Cycles medium** will be automatically selected. Click the check box **Starting Frequencies poor**. Click on the **Frequencies** button and press the **Select All** button to select all three frequencies for iteration. Switch to the **Scalar Coupl.** dialog box and click on the **Select All** button to select all three J-couplings for iteration. You may check the range of the limits: They should be $\pm$<10> Hz. The bottom line will be updated when the check boxes are clicked once.

Execute an **Iteration** and inspect the result using the **Dual Display** mode of WIN-NMR. Examine the **Iteration Protocol** list - the statistical information implies that the exact solution has been achieved (*R-factor* 0%), because the pseudo-experimental spectrum does not contain any noise.

Not every possible iteration method is able to find the correct result defining the 'pseudo-experimental' ABX spectrum. The Smoothing parameters are necessary, because of the following reason: Every time a calculated and experimental LORENTZian line overlap a minimum results in the error hyperspace. In fact the initial parameters to start the iteration already define a local minimum, because the AB part of the 'pseudo-experimental' and calculated spectrum are identical. The only difference is found in the intensities of the inner and outer lines in the X-spectrum part. In spite of this the parameters v_A, v_B, J_{AX} and J_{BX} have to be changed drastically. Without any smoothing of the spectra the iteration algorithm is not able to escape from this local minimum to find the global minimum.

2.3 Second Order Spin Systems   87

Up to now smoothing has not been necessary because in first order spectra the correct lines in the experimental and calculated spectrum already display line overlap. As a result of this signal overlap, the local minimum obtained using the original starting parameters is the same as the global minimum describing the correct solution.

In the present example the case is different. Many local minima appear in the error hyperspace due to line overlap on the way to the final solution. These minima are sharp and narrow but not very deep. The deepest minimum corresponds with the global minimum (the correct solution), but even this is not much deeper than the initial simulation. To reach the correct solution the program must be able to escape from the sharp, narrow local minima and overcome the higher error values induced by shifting lines over the spectral region on the way to the minimal error. This is done by the Selection of the Smoothing parameters **Broadening** and **No. of Cycles**.

Simple iterative algorithms such as steepest descent or Gauss-Newton methods have no chance to escape from these kind of local minima, because they only consider frequency data and ignore intensity information. Consequently these types of program cannot distinguish between the normal and conjugated solution. For NMR spectra minima resulting from overlapping LORENTZian lines (including intensity information) are narrow and very sharp. The WIN-DAISY lineshape fitting procedures represent a combination of iterative methods (steepest descent and Gauss-Newton) combined with a smoothing of the error hyperspace. The initial level of the starting smoothing is defined by the **Broadening** parameter. It smoothes the surface of the error hyperspace in the sense that it looses the fine-structure. This will guide the iteration into the direction of the global minimum. In every iteration cycle (smoothing step) the parameter vector representing the lowest error value is determined until the relative change in the root mean square value undergoes the **Convergence Criterion** parameter value. When no further error reduction is possible the smoothing is taken back (controlled by the parameter **No. of Cycles**). The Advanced iteration method offers an additional control parameters: **Starting Frequencies poor**. Selecting this option enables the program to be more sensitive to frequency distances.

The present optimization problem - starting the ABX iteration - with the incorrect solution represents one of the most difficult tasks: The transition frequencies are already at the correct positions, but the intensities of four of the lines are incorrect. The program must be able to drastically modify the parameters describing the quantum mechanical problem until the global minimum is reached as well as being capable of overcoming the rather high error values that will be induced as it iterates to find the correct solution. The best way to solve such problems is the **Advanced Iteration mode** with **poor Starting Frequencies**. For some theoretical background of the WIN-DAISY iteration method the user is referred to [2.24]–[2.23]. and the appendix, section 6.3.

Check it in WIN-DAISY:

Close the input document ABX_1.MGS and **Open...** it again. Open the **Options** and check **allow sign change** and modify the default **Scalar Coupling Constants Parameter limits for iteration** again to ±<10> Hz (because in the Teaching Version they cannot be stored). Open the **Control Parameters** dialog box and try out different iteration parameters; **Iteration**

Mode, different values for **Broadening** and **No. of Cycles**. Which combination of iteration parameters gives the correct result? Afterwards **Edit | Disconnect exp. data**.

In cases where WIN-DAISY is not able to reach the correct result from a distinct set of NMR parameters and Iteration Control parameters two ways are open to overcome the problem: One way is to try different Iteration Control parameters and the other is to modify the NMR parameters defined in the starting set. However with this approach, you must keep in mind the expected range for each NMR parameter and adjust the iteration limits accordingly.

2.3.3.2 First Order Based ABX Solution

The analysis of an ABX spectrum as described in the previous section based on analytical equations seems to be rather cumbersome and tedious. In addition it can also give the wrong result if the (ab) subspectra are assigned incorrectly or if the wrong solution is chosen. In such cases the extraction of the correct NMR parameters can still be achieved using an iterative computer program such as WIN-DAISY.

The main application for WIN-DAISY is to avoid all this time-consuming analysis of spectra. It has been shown in the sequence calculations (refer to Fig. 2.32) and explained using the disable option of spins (refer to Fig. 2.33) that while all 8 transitions in the AB part are visible in the spectrum the A and B parts of the spectrum may be interpreted as a heavily distorted 'doublet of doublet' - even in cases where both parts overlap.

Check it in WIN-NMR and WIN-DAISY:

Open... the "pseudo-experimental" ABX ...\999999.1R spectrum with WIN-NMR. From the **Analysis** pull-down menu chose the **Multiplets** option and analyze the spectrum signals as doublets of doublets. In the X region of the spectrum ignore the weak outer, combination lines. Connect the coupling constants and **Export...** the data to **WIN-DAISY**. Execute a **Simulation** and display the "experimental" and calculated spectrum in the WIN-NMR **Dual Display** (refer to Fig. 2.41).

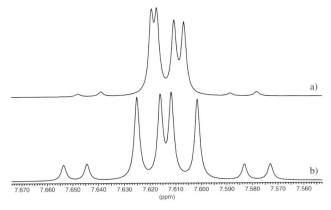

Fig. 2.41: a) 'Pseudo-experimental' AB part of the ABX spectrum.
b) Simulation based on the first order analysis data of a).

The calculated spectrum based on first order analysis (Fig. 2.41b) differs significantly from the original data (Fig. 2.41a), especially in the AB part of the spin system. Nevertheless, the approximate values obtained by this simple first order analysis are good enough as input parameters for a WIN-DAISY iteration. The default control parameters have to be modified because the correct lines in the experimental and calculated spectrum do not necessarily overlap.

This extended discussion of the different method available for the analysis of ABX spectra (or a simple first order analysis followed by a simple WIN-DAISY iteration) should indicate how much easier it is to extract the NMR parameters from a 1D data set using computer programs. However, it must be stressed that for a thorough understanding of the ABX spin system, particularly in teaching establishments, a rigorous manual analysis should be attempted initially and the manual results checked using WIN-DAISY.

Check it in WIN-DAISY:

Open the **Control Parameters** dialog box and select the **Advanced Iteration** mode and the option **Starting Frequencies poor**. Execute the parameter adjustment (**IT**) and check the calculation result in WIN-NMR **Dual Display**.

Using a first order analysis the correct values for J_{AB} and the resonance frequency v_X may be obtained but the values for v_A and v_B will always be wrong.

Check it in WIN-NMR and WIN-DAISY:

Open... again the "pseudo-experimental" spectrum based on the WIN-DAISY input document ABX_2.MGS with WIN-NMR. **Export...** again the **Linewidth**, interval and **Multiplets** data to WIN-DAISY. Execute a spectrum **Simulation** and display both spectra in the **Dual Display** mode of WIN-NMR. Switch to the frequency scale into Hz using <Ctrl + X>. **Zoom** the AB part of the spectrum. Click the **Move Trace** button; holding down the <Shift> key drag the second trace across the first trace until the A part of both spectra fit. The frequency shift is displayed in the header status line. Repeat this process for the B part of the spectrum. Toggle back to **WIN-DAISY** and modify the **Frequencies** of Spin 2 and Spin 3 by the appropriate amount (move both frequencies towards each other by about 1.7 Hz). Execute an **Iteration** using the *default* **Control Parameters** and compare the results using the **Dual Display** mode of WIN-NMR.

Check it in WIN-NMR:

Analyze the aromatic region of the proton spectrum ...\ABX\001001.1R. The benzene and pyrrole give two separated spin systems in the WIN-NMR **Multiplets** mode. Determine a **Linewidth** of one of the lines in the X part of the ABX spin system. In the spectrum the AX type system of the pyrrol frames the ABX type system of the benzene fragment.

4-(2,3-dichlorophenyl)-
1H-pyrrole carbonitrile

The coupling tree of the AB part of the ABX subsystem is shown in Fig. 2.42. It is obvious that in the 'B' multiplet (shift 4) not all four lines are resolved, because the two coupling constants involved are quite similar.

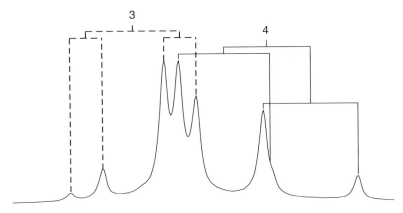

Fig. 2.42: Multiplet analysis of AB part of ABX subsystem.

Check it in WIN-DAISY:

Export... the **Multiplets** including the linewidth information to **WIN-DAISY**. Select the fragmentation option with **Yes** and **Simulate** the spectrum. Compare the simulated and experimental data in the **Dual Display** mode of WIN-NMR. As expected, the AB part of the ABX spin system is incorrect and the outer (combination) lines in the X part are shifted further away from v_X. However the lines of experimental and calculated spectrum already overlap, so that the default values are good enough as input values for WIN-DAISY.

Execute an **Iteration** and look at the result in WIN-NMR **Dual Display**.

The present ABX spin system is already very close to an ABC spectrum, visible by the roof effect in the X part of the spectrum (refer to Fig. 2.43a). In true ABX spectra the X part is totally symmetrical. Due to this reason the calculated artificial conjugated ABX spectrum (Fig. 2.43b) does not show exactly the same AB spectrum intensities as in the normal solution. However the NMR parameters obtained in the first order analysis are sufficient to achieve the correct solution.

The question is how to obtain the correct solution when analyzing the artificial ABX spectrum referring to the conjugated solution which is very far from the first order analysis.

2.3 Second Order Spin Systems 91

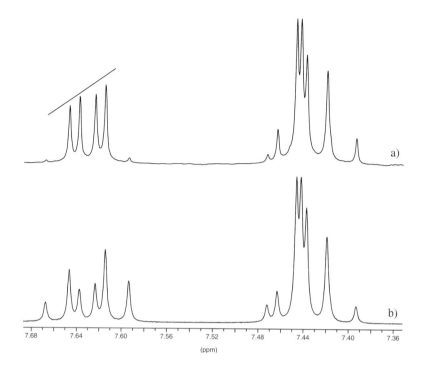

Fig. 2.43: a) ...\ABX\001001.1R (the line indicates the roof effect in the X part) representing the 'normal' ABX solution.
b) ...\ABX\002001.1R representing the 'conjugated' ABX solution.

Check it in WIN-NMR and WIN-DAISY:

Open... the pseudo experimental spectrum ...\ABX\002001.1R (Fig. 2.43b) with WIN-NMR. Determine the **Linewidth** and analyze the **Multiplets** in the *same* way as done previously for the real experimental spectrum. In the X multiplet of the ABX spin system ignore again the "weak" outer lines, so that an identical first order analysis is obtained. **Export...** the data to **WIN-DAISY**.

Execute a spectrum **Iteration** using the default values. Inspect the result in WIN-NMR **Dual Display** and examine the statistical information in the **Iteration Protocol** list.

As expected the *wrong* solution is achieved in this way, because the iteration control parameters are only sufficient to obtain the nearest solution, which is the 'normal' ABX solution, not the 'conjugated'. As discussed earlier the Control Parameters for the iteration have to be modified in order to achieve the correct solution.

Check it in WIN-DAISY:

Open the **Control Parameters** dialog box and select the **Advanced Iteration Mode**. Choose **Broadening high** and **No. of Cycles high**. Check in addition the option **Starting Frequencies poor**. Click on the **Scalar Coupl.** button and change manually the iteration limits for the coupling constants J_{12} and J_{13}. To cover the conjugated solution enter <–15> Hz for the **lower** limit and <+15> Hz for the **upper** limit. Quit the dialog boxes with **Ok**.

Execute an **Iteration** and inspect the result.

Check it in WIN-NMR and WIN-DAISY:

Measure a **Linewidth** and analyze the aromatic region of the proton spectrum ...\ABX\003001.1R in the **Multiplets** mode. Only one of the two separated spin systems shows a second order effect. Mark a peak of the spin system with the **maximum cursor**.

5-(2,4-dichloro-phenoxy)-2-nitrobenzoic acid methyl ester

Export... both the multiplet and linewidth information together with the iteration regions to **WIN-DAISY**.

Select the fragmentation option (**Yes**) and **Simulate** the spectrum. Compare the simulated and experimental data in the **Dual Display** mode of WIN-NMR. Although the intensities and AB pattern at high field is not quite correct, but there is overlap between the appropriate lines in the experimental and calculated spectrum.

Open the **Scalar Coupling Constants** dialog box of the second fragment and manually select the para coupling constant (J_{13}) for iteration: Click with the mouse pointer in the edit box of the desired coupling constant and check the **Iterate** option in the bottom line. Execute a spectrum **Iteration** with the default values. Check the result in WIN-NMR **Dual Display** and the **Iteration Protocol**. **Export** the iteration result to **WIN-NMR exp**. Display the **Report...** box while the **Windaisy** button is depressed. Note the values of the para coupling constants. Then *deselect* the **Windaisy** option and inspect the splitting in the **Report...** box determined for the para coupling from the spectrum, which is bigger.

This examples illustrates that in the second order spin system the para coupling constant could be analyzed directly, although its *value* is smaller than the linewidth parameter in the **Lineshape Parameters**.

Due to the second order effect the spin system data result in a signal *splitting* which is bigger than the parameter value. In the first order three-spin system the para coupling constant is also present, but only contributes to the line broadening.

2.3.4 Extended Spin Systems

Check it in WIN-NMR and WIN-DAISY:

Measure a **Linewidth** and analyze the proton spectrum ...\ABMX\001001.1R of *9-fluorenone*, assuming that there is no inter-ring proton coupling, using the **Multiplets** mode of WIN-NMR. In practice the system should be labeled as AMNX, because the two middle signals show a second order effect. It may be analyzed as a modified ABX type system using the first order approximation that a fourth proton gives an additional doublet splitting to each of the other signals.

9-fluorenone

Thus, each signal requires three couplings; signal A (low field) seems to have one ortho, one meta and one para coupling constant (two couplings seem to be quite similar leading to a *triplet* appearance) while signal X seems to contain two ortho and one meta coupling (ignore the weak outer lines for analysis). For the strongly coupled spins M and N the coupling patterns are not obvious, but by using the **Coupled Grid** facility in the Multiplet analysis it should be possible to supply one signal with two ortho and one meta coupling and the other with one ortho, one meta and one para coupling. Always start with the weak lines at the border of the signal. With the first coupled grid match the next appropriate doublet. Ignore the fact that the second coupled grid will not match on both sides peaks in the multiplet. If you like, you may use the possibility to **select** and **shift** the **individual** stick **lines** of the multiplet tree to match the peak tops. Connect the coupling constants in the **Report...** box. This has to be done manually because the values are quite different in this rough analysis. **Auto Connect** the coupling and manually connect the remaining values.

Mark one outer line of the high-field signal with the spectrum cursor. **Export...** the multiplet, linewidth and interval data to **WIN-DAISY**. Execute a spectrum **Simulation** and display both spectra using the **Dual Display** mode of WIN-NMR. **Change** the calculated spectrum name and perform an **Iteration** using the default **Control Parameters**.

Compare experimental and iterated spectra in the **Dual Display**. If the correct result is not achieved with the default parameters, open the **Control Parameters** dialog box and check the parameter limits for iteration, especially the **Scalar Coupl.** by clicking once the check boxes and compare the limit values with the parameter value. It is possible that the default limits for some of the coupling constants are not large enough to include the correct values (then the parameter value is identical are very close to one limit value) - if this is so extend the limit. Continue optimizing the parameters via **Iteration** until a satisfactory result is obtained. Then **Export...** the iteration result back to **WIN-NMR** experimental spectrum (refer to Fig. 2.44, zoomed MN region). Toggle back to **WIN-DAISY**.

Display the **Calculated Spectrum** in WIN-NMR. **Change** the calculated spectrum name and open the **Frequencies** dialog box. Check the **Disable** option for spin **3**. Run a spectrum **Simulation** with suppression of the transitions belonging to spin 3. Toggle to **WIN-NMR** and load manually the

94   2 General Characteristics of Spin Systems

new calculation as the second file into the **Dual Display**. Zoom in the MN signal region. Use the **Move Trace** option together with the <Ctrl> button to align the Y-scale and press the **Show Difference** button (Fig. 2.45) to display both signals of M and N spin separately.

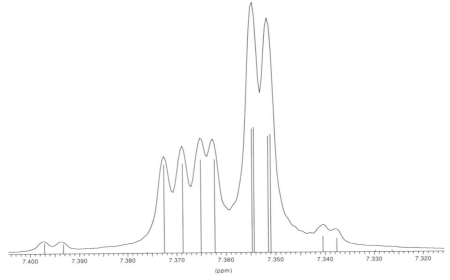

Fig. 2.44: WIN-DAISY calculated transition lines of the MN Multiplets.

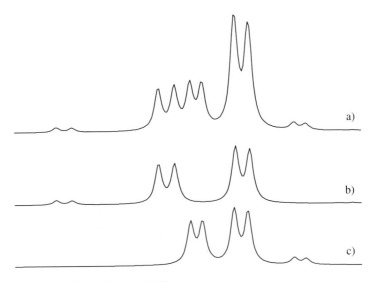

Fig. 2.45: a) Total simulation of MN spectral part.
         b) Simulation with suppression of N signals, showing signal M.
         c) Difference spectrum a) – b) showing signal N.

2.3 Second Order Spin Systems 95

2.3.4.1 Taking Advantage of Field Effects

The purpose of the following "Check its" is to illustrate that the NMR parameters extracted from spectra at one magnetic field strength can be compared directly with experimental data recorded at a different field strength using WIN-DAISY. In addition, the "Check its" also illustrate the increasing second order effects that can occur when changing to a lower value magnetic field strength.

Check it in WIN-NMR and WIN-DAISY:

Open... the 250 MHz proton spectrum of *ortho* acetyl-salicylic acid ("Aspirin") ...\AMRX \003001.1R with WIN-NMR and analyze it in the **Multiplets** Mode. **Export** the results to **WIN-DAISY**. In the **Scalar Couplings** dialog box select the para-coupling constant, currently 0 Hz, for **iteration**. Perform a standard **Iteration** and examine the results using the **Dual Display** mode of WIN-NMR.

o-acetyl salicylic acid

Do not close the WIN-DAISY document representing the results of the 250 MHz analysis. **Open...** the 80 MHz spectrum of the same compound ...\ABCD\001001.1R with WIN-NMR.

This spectrum now displays strong second order effects such that with the exception of the low field signals no separate multiplets can be identified. Because the 250 MHz spectrum of the same compound has just been analyzed we can check if the extracted NMR parameters also fit the 80 MHz spectrum. For this purpose it is useful to connect the spectral data of the 80 MHz experimental spectrum with the WIN-DAISY input document containing the 250 MHz result.

Check it in WIN-NMR and WIN-DAISY:

With the spectrum ...\ABCD\001001.1R displayed in WIN-NMR select the **Interval** option from the **Analysis** pull-down menu. Click on the **Regions...** button and **Load...** the predefined region file into WIN-NMR. Then mark the left out-most peak with the spectrum cursor and **Export...** the data from the Interval mode to **WIN-DAISY**. A number of dialog boxes will now appear. In case multiplet data are also available for the present spectrum answer **No** for the request to use the multiplet file and select **Ok** for the message to connect the WIN-NMR data to an existing document. Click on the **Connect** button and answer **Yes** to keep the ppm values of the chemical shifts and *adjust* the frequency values to the new spectrometer frequency (Change the order of the spin system!). Check the **Spectrum Parameters**, they belong to the 80 MHz experimental spectrum. Automatically the status line for experimental and calculated spectra name are updated. Run a spectrum **Simulation** and display both spectra using the **Dual Display** mode of WIN-NMR. There is only a slight frequency shift and some impurity signals present. Toggle back to **WIN-DAISY**, run an **Iteration**, **Dual Display** the result, toggle back to **WIN-DAISY** and **Export** the optimized parameter from WIN-DAISY back to **WIN-NMR exp.** spectrum.

Check it in WIN-NMR and WIN-DAISY:

Using only the 80 MHz proton spectrum of *ortho-acetyl salicylic acid* ...\ABCD\001001.1R, perform a full spectrum analysis. Start with the **Linewidth** determination and then the **Multiplets Analysis.** You will see how much more difficult the analysis is compared to the using of the high field 250 MHz spectrum as experimental input due to strong signal overlap and similar coupling constant values.

If you do not succeed in analyzing this spectrum in the Multiplet mode interactively you may use as assistance the previously stored WIN-DAISY result by depressing the **Windaisy** button.

Export your manual analysis of the spin system data to **WIN-DAISY** to optimize the NMR parameters.

A third way to analyze the spectrum is to leave the main part to the iteration program WIN-DAISY. Merely define the approximate location of four chemical shifts in the spectrum.

Check it in WIN-NMR:

Open... again the 80 MHz proton spectrum ...\ABCD\001001.1R. Measure a **Linewidth** and switch into the **Multiplets** mode. The left somehow separate signal can be identified, but the three others represent three overlapped signals. For left-hand and right-hand signal define a *doublet* of which only the shift value will be used later on (use the out-most lines of the signal to get the midpoint as chemical shift). In the remaining spectral part **Define** two **Singlet**s (use the context menu via right-hand mouse button). **Export...** the data to **WIN-DAISY** including the interval data (**Yes**) and answer the questions as follows.

Check it in WIN-DAISY:

Answer the fragmentation question with **No** in order to combine all four shifts in one fragment. Also deny to include the non-assigned coupling constant from the doublet - answer **No**. Open the **Options** dialog and define the **Parameter limits for iteration** for **Scalar Coupling Constants** to ± <10> Hz and do *not* **Allow sign changes**. Open the **Control Parameters** dialog box and smoothing parameters, e.g. **medium Broadening** and **medium No. of Cycles**. Increase the **Convergence Criterion** to <0.05>. You may either keep the **Standard Iteration Mode** or change to the **Advanced Mode** with **Starting Frequencies poor**. Click on the **Scalar Coupl.** button and **Select All** constants for iteration. Open the **Scalar Coupling Constants** dialog box and enter values, e.g. <1>, <2>, <3>, <4>, <5> and <6>. In the **Lineshape Parameters** *deselect* the **iterat**ion flag from the **global linewidth** parameter. Run an **Iteration** and inspect the result in WIN-NMR **Dual Display**.

The calculated spectrum shows distinct discrepancies from the experimental lineshape. An eyecatching difference is visible at the high-field signal, where the calculation tries to cover the impurity signals as well by inclusion of a to large coupling constant.

2.3 Second Order Spin Systems

Check it in WIN-DAISY:

Open the **Scalar Coupling Constants** dialog box and reset one of the large coupling constants (e.g. J_{34}) to <0> Hz and run another **Iteration** with identical **Control Parameters**. Check the result in the WIN-NMR **Dual Display** and the *R-factor* in the **Iteration Protocol** (about 0.5 % for fixed linewidth about 1 Hz).

The resulting calculated lineshape shows the correct pattern but the linewidth is too broad. The iteration of the linewidth parameter in this kind of spectrum will fail due to impurity signals and base line complications. You may see the splitting in the low-field signal by simulating with a lower global linewidth parameter.

Check it in WIN-DAISY and WIN-NMR:

Open the **Lineshape Parameters** dialog box and enter a lower **global linewidth** parameter (e.g. <0.8> Hz) and run a **Simulation**. Display both spectra in the WIN-NMR **Dual Display**. Now the small splitting in the low-field signal is visible. Toggle back to **WIN-DAISY** and **Export** the result **to WIN-NMR exp.**

The multiplet trees with the WIN-DAISY iteration result above the experimental lineshape show in direct comparison the difference between first order and second order spin system. The stick spectrum below the experimental lineshape represent the second order transitions calculated by WIN-DAISY.

2.4 Magnetic Equivalence

So far, all the first and second order spin systems that have been considered have only contained one nucleus in each particular environment represented by different letters for each nucleus, e.g. AX, AB, AMX etc. In a number of instances, limiting cases have been discussed which have arisen from the chemical shifts of particular nuclei becoming identical. NMR therefore has to distinguish between two basic types of equivalence:

Isochronicity* and magnetic equivalence.

This chapter deals with *magnetic* equivalence. The word equivalence already implies that one feature of this property is the *isochronicity* of nuclei under consideration:

The chemical shifts of the nuclei are identical.

A second condition to distinguish chemical from *magnetic* equivalence concerns the relationship of the isochronous spins to other members of the spin system which they may be coupled to:

The nuclei i and j are magnetically equivalent, if they show the same chemical shift and if the nuclei i and j exhibit the same coupling constants to all other nuclei k of the same spin system.

Therefore, the two conditions for *magnetic* equivalence are as follows:

1. $\nu_i = \nu_j$
2. $J_{ik} = J_{jk}$ for $k \neq i,j$

The reason *why* the coupling constants to other spins of the spin system are identical is irrelevant. In case where no spins other than the isochronous nuclei under consideration define the total spin system they are by definition magnetically equivalent, because no other nucleus which may couple to these spins is present.

This fact has already been briefly introduced in section 2.1.2.1 on non-coupled spin systems. In second order spin systems (section 1.3) the isochronicity of nuclei always occurred as the limiting case.

---

* Commonly used in literature is the synonymous expression *chemical equivalence*. This term suggests, that the chemical environment is connected which is not necessarily the case - chemical shifts may be also accidentally identical in different chemical environment. To avoid such ambiguous expressions only isochronicity is used [2.21].

Apart from the situation just described the two conditions for magnetic equivalence can be met either by structural symmetry or due to molecular mobility. An accidental coincidence of chemical shifts and coupling constants is most improbable.

In this section only magnetic equivalence of nuclei with I = ½ will be discussed, other spin values will be discussed in section 4.3.

2.4.1 Notational Aspects

As already mentioned in section 2.1.3, isochronous spins are labeled with the same capital letters of the alphabet. To distinguish between the isochronous spins the prime notation can be used. For isochronous spins that also satisfy the condition for magnetic equivalence the number of equivalent nuclei is written in subscript Arabic numbers behind the alphabetical letter.

For example a spin system consisting of only two isochronous will always be magnetic equivalent because of the lack of additional spins, so that the correct notation is A_2. To be able to distinguish between individual isochronous spins the prime labeling is used (i.e. to name the coupling constant between the two A spins in an A_2 spin system, the coupling constant is labeled $J_{AA'}$).

To understand the effect of magnetic equivalence on spectra the coupling interaction between two isochronous spins will be considered first.

2.4.2 The Coupled A_2 Spin System

In section 2.1.2.1 the non-coupled A_2 spin system has been discussed in detail with the construction of energy levels and the extraction of the transition frequencies and intensities. In fact, the present spin system can be regarded as one of the limiting cases of a two-spin second order spin system (AB) where the perturbation parameter has a vanishing chemical shift difference:

$$\lambda_{AA'} = \left|\frac{J_{AA'}}{\Delta_{AA'}}\right| = \left|\frac{J_{AA'}}{0}\right|$$

The limit of the perturbation parameter becomes infinite as follows:

$$\lim (\lambda_{AA'}) = \infty$$

This has been mentioned at various places in the discussion of second order spin systems in chapter 2.3.

Check it in WIN-DAISY and WIN-NMR:

Open... the WIN-DAISY input document AB.MGS, equalize both chemical shift values and **Simulate** the spectrum. **Change** the calculated spectrum name, enter <0> for the **Scalar Coupling Constant**, recalculate the spectrum (**SIM**) and compare both calculated spectra in **WIN-NMR** (load the latter calculation manually as the second trace into the **Dual Display**).

2 General Characteristics of Spin Systems

In the case of magnetically equivalent spins the coupling constant between both spins has no effect on the spectrum. Of course the coupling constants influences the HAMILTONian and in consequence the energy levels. In order to understand this circumstances the two energy levels for $m_T = 0$ are calculated. The A_2 spin system can be regarded as a limiting case of the AB spin system with $v_A = v_B$ and consequently $\Delta_{AB} = 0$ (refer to Table 2.8).

$$E_{2,3} = -\frac{1}{4} J_{AA'} \pm \frac{1}{2} D$$

where $D = J_{AA'}$

Table 2.16: Energy level values for an A_2 spin system.

Level Index	m_T	Energy Level
E_1	1	$-v_A + \frac{1}{4} J_{AA'}$
E_2	0	$+\frac{1}{4} J_{AA'}$
E_3	0	$-\frac{3}{4} J_{AA'}$
E_4	-1	$+v_A + \frac{1}{4} J_{AA'}$

In comparison to the energy levels of the non-coupled A_2 spin system in Fig. 2.5 the non-zero coupling constant removes the degeneracy of the energy levels E_2 and E_3.

Check it in WIN-DAISY:

In the **Main Parameters** dialog box select the **Output Option EIGENvalues**, enter in the Scalar Coupling Constants dialog any value and run a spectrum **Simulation**. Display the WIN-DAISY **Simulation Protocol** file and based on the *Resonance Frequencies* and the *Scalar Coupling* **calculate** the *EIGENvalues* (energy levels) using the above equations and compare the results with the values in given in the protocol file.

The ordering of the EIGENvalues within the same submatrix (same total spin value m_T) corresponds to increasing energy.

Table 2.17: Transition frequencies of an A_2 spin system.

Transition index	Energy level	Transition frequency
T_{A1}	$E_3 - E_1$	$v_A - J_{AA'}$
T_{A2}	$E_4 - E_2$	v_A
$T_{A'1}$	$E_2 - E_1$	v_A
$T_{A'2}$	$E_4 - E_3$	$v_A + J_{AA'}$

Check it in WIN-DAISY:

Open the **Main Parameters** dialog box of the A_2 spin system and change the **Minimum Intensity** to <0>, so that all transition frequencies - even transitions of vanishing intensity - are calculated. In the same dialog box select the Output Option **Transitions and energy levels** and *deselect* the **EIGENvalues**. Execute a **Simulation** and examine the **Simulation Protocol** file. Transfer

2.4 Magnetic Equivalence   101

the simulated spectrum to WIN-NMR via the red peak button. From the **Analysis** pull-down menu select the **Peak Picking** option. Perform a peak picking on the simulated spectrum.

The calculation protocol shows that the transitions T_{A1} and $T_{A'2}$ occur with a vanishing transition probability.

Check it in WIN-DAISY and WIN-NMR:

In the WIN-DAISY **Frequencies** dialog box change one of the chemical shifts slightly, e.g. by a value of the magnitude of the **Scalar Coupling Constant** in **Hz**. **Change** the calculated spectrum name and **Simulate** the spectrum. Again, examine the **Simulation Protocol** file before transferring the data to WIN-NMR and performing a **Peak Picking**.

By changing one of the chemical shifts the conditions for magnetic equivalence are no longer fulfilled. Consequently the AB spin system characteristics occur in the spectrum and the outer lines of the four-transition pattern appear. To understand the absence of the outer lines for nuclei being magnetically equivalent the EIGENfunctions describing the EIGENstates of $m_T = 0$ have to be determined.

In order to do this the equations for the AB spin system transition intensities of Table 2.10 using the limiting condition for the A_2 case $\Theta = 45°$ given in Table 2.13 are used:

Table 2.18: Transition intensities for an A_2 spin system.

Transition index	Transition frequency	Transition intensity
T_{A1}	$v_A - J_{AA'}$	$1 - \sin 90° = 0$
T_{A2}	v_A	$1 - \sin 90° = 2$
$T_{A'1}$	v_A	$1 - \sin 90° = 2$
$T_{A'2}$	$v_A + J_{AA'}$	$1 - \sin 90° = 0$

The corresponding EIGENfunctions shown in Table 2.10 result in:

Table 2.19: EIGENfunctions for an A_2 spin system.

EIGENfunction	Linear combinations
$|\Psi_1\rangle$	$|\alpha\alpha\rangle$
$|\Psi_2\rangle$	$+\frac{1}{\sqrt{2}}|\alpha\beta\rangle + \frac{1}{\sqrt{2}}|\beta\alpha\rangle$
$|\Psi_3\rangle$	$-\frac{1}{\sqrt{2}}|\alpha\beta\rangle + \frac{1}{\sqrt{2}}|\beta\alpha\rangle$
$|\Psi_4\rangle$	$|\beta\beta\rangle$

As already mentioned in the discussion about the limiting cases of AB spin systems, the two EIGENfunctions for total spin value $m_T = 0$ are very similar. The contribution of each bpf is equal to $2^{-½}$, only the signs are different. In such cases the EIGENfunction with all positive signs (here $|\Psi_2\rangle$) is called the symmetric linear combination of $|\alpha\beta\rangle$

and |βα⟩ and the other function (|Ψ₃⟩) the corresponding antisymmetrical combination. However it is not possible to distinguish between spins A and A' in the EIGENfunctions. The construction of the total spin diagram with the four EIGENvalues labeled makes it clear that the two transitions with vanishing intensity involve the antisymmatrical EIGENfunction. Transitions between wave functions of different symmetry type are forbidden.

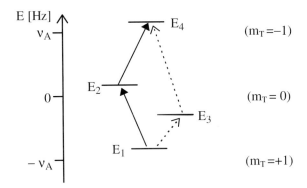

Fig. 2.46: Energy level diagram for a coupled A_2 system, $J_{AA'} > 0$
→ symmetry allowed transitions, ‑‑> symmetry forbidden transitions

Comparing the A_2 spin system with coupling (present section) and without coupling (section 2.1.2.1) it is obvious that the coupling between magnetically equivalent spins has no effect on the spectrum and consequently cannot be determined from the spectrum. For the non-coupled case four transitions with frequency v_A and intensity 1.0 result. By introducing the coupling constant the calculation is different and yields two transitions at frequency v_A both showing the intensity '2.0'; the remaining transitions are symmetry forbidden (refer to Fig. 2.46). In fact the coupling constant between magnetically equivalent spins cannot be determined because the corresponding transition have no intensity and do not appear in the spectrum.

2.4.3 The Composite Particle Approach

Because the inner-group scalar coupling constants of magnetically equivalent spins have no effect on the resulting spectrum it is not necessary to consider this type of coupling when simulating NMR spectra. By definition, the coupling to all other nuclei is identical and the calculation of spectra involving such type of spins can be simplified. A more detailed theory description is found in the appendix, section 6.2.

Because it is not possible to distinguish between the nuclei showing magnetic equivalence, they can be treated as a group or a *particle composed* of the single spins ("composite particle", abbr. "cp"). Such a particle is then not represented by the independent single spin quantum numbers *I* of the members of the spin group, but by a *particle spin value F* composed of the sum of the single spin values. Such a particle can be in different *states* according to the relative orientation of the particle members.

2.4 Magnetic Equivalence

For example, the A_2 particle - consisting of two single spins $I = ½$ - can be composed of two spins with parallel spin orientation ($m_T = 1$) or of two spins having anitparallel spin orientation ($m_T = 0$). For particles these values are called *actual particle spins* F_{act} and the *maximum Particle spin value* F_{max}. By analogy with atomic spectroscopy, these different *states* are called *terms* and labeled by their multiplicity:

$$\text{multiplicity } g = 2 \cdot F_{act} + 1 = 2 \cdot n \cdot I + 1$$

where n = number of nuclei in the particle
I = angular momentum of the single spins
$F_{act} = n \cdot I$ = actual particle spin

To distinguish between single spins and particles the latter type is printed in braces, with the multiplicity written as a superscript in front of the parentheses. The treatment of the A_2 particle using these rules is given in Table 2.20.

Table 2.20: Possible states of an A_2 particle.

A_2 particle	State 1	State 2
single spin orientations	parallel	antiparallel
spin symbols	↑↑	↓↑
F_{act}	1 ($\equiv F_{max}$)	0
multiplicity g	3	1
state	triplet	singlet
term symbol	$^3\{A_2\}$	$^1\{A_2\}$

Different terms - the so-called irreducible components - do not interfere and interact, so that they describe independent systems. In consequence they result in independent subspectra which can be calculated separately. The superposition of all subspectra results in the total spectrum of the spin system.

For the particle spin A_2, the triplet term represents the maximum particle spin value $F_{max}(\ ^3\{A_2\}\ ) = 1$ and has one additional actual particle spin value $F_{act}(\ ^1\{A_2\}\ ) = 0$. This means that the A_2 particle represents one term in which it behaves as a spin F = 1 particle, and another term where it is not spin active at all (F = 0 particle). The spin inactive term cannot cause any NMR signals, so the spectrum consists only of the resonances caused by the triplet term. The singlet term will become important when the spin system is extended to include other spin that are not members of the same particle (section 2.4.5).

At the beginning of chapter 2 the concept of chemical shift, spin quantum number and multiplicities were introduced (refer to Fig. 2.2). Thus a multiplicity of '3' - a triplet term - is connected with three different orientations of the spin - here particle - in the magnetic field. The singlet term has only one possible orientation in the spin system and is macroscopic spin inactive.

All the single spins of the spin systems discussed in the previous sections can be described by one term and because the discussion has been limited so far to $I = ½$ nuclei

104   2 General Characteristics of Spin Systems

this term is a doublet. The combination of all particle states results in a single spin system term, which is in any case identical with the quantum mechanical systems dealt with in the preceding sections.

Table 2.21: Possible states of single spin systems discussed in section 2.2.

Single spin system (I = ½)	Term description
A	$^2\{A\}$
AX	$^2\{A\}^2\{X\}$
AMX	$^2\{A\}^2\{M\}^2\{X\}$
AMRX	$^2\{A\}^2\{M\}^2\{R\}^2\{X\}$

What is actually done by using the composite particle approach is an exchange of the functional basis. By using the particles instead of the spins a *basis transformation* is done. The spin functions $|\alpha\rangle$ and $|\beta\rangle$ are no longer sufficient for the description of all possible terms, they are only adequate for doublet terms. For the representation of other particle multiplicities other symbols must be used. To keep the symbols for the particle functions descriptive the particle spin values are represented by symbols (e.g. $|+1\rangle$ for a particle with $F_{act} = +1$). The construction of the energy diagram for both irreducible components of the A_2 spin system is shown in Fig. 2.47.

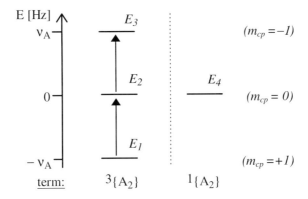

Fig. 2.47: Energy level scheme for the complete A_2 spin system.

Because no inter-term interactions are allowed only two transitions in the triplet term are present. The singlet term appears externally as spin inactive with only a single energy level; in the absence of other EIGENstates in the same term no singlet transition is possible.

Table 2.22: Transition frequencies and intensities for A_2 using the cp approach.

Transition	Frequency	Intensity
$E_2 \rightarrow E_1$	v_A	2.0
$E_3 \rightarrow E_2$	v_A	2.0

2.4 Magnetic Equivalence

Check it in WIN-DAISY and WIN-NMR:

Open... the input file A.MGS (or reduce in the previously used input file in the **Main Parameters** the **Number of spins or groups with magnetical equivalent nuclei** to <1>) and extend in the **Frequencies** dialog box the **spins in group** to <2>. In the **Main Parameters** select the **Output Options EIGENvalues** and **Transitions and energy levels** and perform a **Simulation**. View the **Simulation Protocol** list to check the previous discussion.

Open the **Main Parameters** dialog box and select the Output Option **Output of Subspectra** and **Simulate** the spectrum again. Another button is now available on the toolbar: **Generate Subspectra**. Click this button. WIN-DAISY will generate individual spectra containing the transitions for each of the irreducible components. The original simulated spectrum will not be overwritten by the calculated subspectra as WIN-DAISY automatically reduces the filename by 1 for each subspectrum calculated (here ...\999998.1R, ...\999997.1R). Toggle to **WIN-NMR** and **Open...** the first subspectrum. Switch to the **Dual Display** mode and load the second subspectrum as the second trace. Click the **Separate** button to show the difference between the two subspectran (the second subspectrum does not contain any peak). From the **Output** pull-down menu select the **Title...** option. The Text window that opens up contains information relating to the current subspectrum (first trace) such as filename and F_{act} value.

2.4.4 The A_n Spin Systems

As stated previously, any spin system consisting solely of a single group of magnetically equivalent spins without any coupling partner gives rise to a singlet in the NMR spectrum when measured in isotropic solution. The anisotropic case will not be discussed in this book. The A_2 case just discussed consists of a triplet and a singlet term but the latter has no transitions associated with it. This matter changes when considering systems where n > 2. General rules for the number and the characteristics of possible states (terms), energy levels and transitions can be derived as shown in Table 2.23.

Table 2.23: Formulae for general description of particle spins A_n with I = ½

Characteristic	Formula
number of terms	$t\ (n\ even) = F_{max} + 1$ $t\ (n\ odd) = \dfrac{2F_{max}+1}{2}$
actual particle spin	$F_{act} = F_{max} - j$ $j = 0, 1, ...t$
multiplicity per state i = number of energy levels	$g = 2 \cdot F_{act} + 1$
energy levels	$E = \dfrac{g-2i-1}{2} \cdot v_A$ $i = 0, ...g-1$
number of transitions	$g-1$
transition frequency	v_A

106 2 General Characteristics of Spin Systems

Although the isotropic case is not particularly interesting from the point of view of the transition frequency we will extend it to any number of spins n in the group A_n because it is important to know the general rules for generating all possible terms of a particle. In addition the A_n systems exhibit an important fact: the weighting of the different subspectra arising from the individual terms. This weighting factor depends on the so-called *particle multiplicities* of each composite particle state: $G(F_{act})$.

They are not only dependent on the actual particle spin F_{act}, but also on the number of nuclei in the particle n and the spin of the nuclei (in this chapter $I = \frac{1}{2}$). The particle multiplicity is determined based on the difference between binomial coefficients (refer to the appendix, section 6.2).

Check it in WIN-DAISY and WIN-NMR:

Open... the input document A.MGS and in the **Frequencies** dialog box extend the number of **spins per group** to <3>. In the **Main Parameters** dialog box select the **Output Options Output of Subspectra**, **EIGENvalues** and **Transitions and energy levels**. Execute a **Simulation** and examine the **Simulation Protocol** file. Check the data determined by the program and compare it with the values given in Table 2.24 and in Fig. 2.48.

Click the **Generate Subspectra** button. Toggle to **WIN-NMR** and **Open...** the first subspectrum. Switch to the **Dual Display** mode and load the second subspectrum as the second file. Click the **Separate** button to show the difference between the two subspectra. Both subspectra now contain intensity. Again, selecting the **Title...** from the **Output** pull-down menu displays information relating to the current subspectrum loaded first.

Table 2.24: Composite Particle data of an A_3 spin system.

A_3 particle	state 1	state 2
spin symbols	↑↑↑	↑↑↓
F_{act}	$\frac{3}{2}$ (=F_{max})	$\frac{1}{2}$
multiplicity g	4	2
state	quartet	doublet
term symbol	$^4\{A_3\}$	$^2\{A_3\}$
particle multiplicity G	1	2

The particle multiplicity of 2 for the doublet term (Table 2.24) means a double intensity weight of the second subspectrum. How this value is determined refer to appendix, section 6.1.2. The energy level diagram is given in Fig. 2.48.

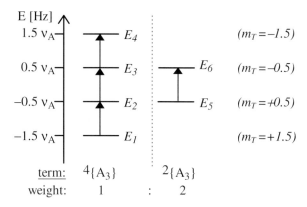

Fig. 2.48: Energy level scheme for the complete A_3 spin system.

Check it in WIN-DAISY:

Carry on extending the number of **spins per group** (e.g. 4,5,6 and so on) in the **Frequencies** dialog box and check the data reported in the **Simulation Protocol** with manually determined values.

When the **Main Parameters Output Option Output of Subspectra** is selected, **Generate** the independent **Subspectra** and inspect them in the WIN-NMR **Multiple Display**. The number of Subspectra generated is reported in the **Simulation Status** line of the WIN-DAISY document main window.

A spin system consisting of four magnetically equivalent spins is the smallest composite particle where there are three different actual particle spins F_{act}.

Table 2.25: Possible states of an A_4 particle.

A_4 particle	state 1	state 2	state 3
spin symbols	↑↑↑↑	↑↑↑↓	↑↑↓↓
F_{act}	2 ($\equiv F_{max}$)	1	0
multiplicity g	5	3	1
state	quintet	triplet	singlet
term symbol	$^5\{A_4\}$	$^3\{A_4\}$	$^1\{A_4\}$
particle multiplicity G	1	3	2

2.4.5 First Order Spin Systems

2.4.5.1 The A_2X Spin System

The problem becomes more complicated when another chemical shift becomes involved. The easiest first order example is a composite particle with two magnetically equivalent spins (A_2) and one X spin or vice versa: AX_2.

Initially the two types of spins are treated independently:

A_2 states: $^3\{A_2\}$ and $^1\{A_2\}$
X state: ^2X

All the possible combinations of particle states result in the total number of subspectra:

$$^3\{A_2\}^2X \qquad ^1\{A_2\}^2X$$

In all the following spin systems, because more than one frequency is present, it is clearer if m_T rather than energy level diagrams are used.

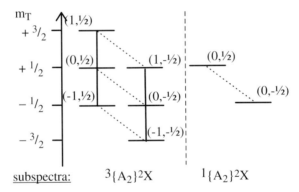

Fig. 2.49: Total spin value scheme for the complete A_2X spin system, with subspectra states in brackets ; (1,½): $F_{act}\{A_2\}=1$, $F_{act}\{X\}=½$.
——— A transitions (vertical) ,   ······ X transitions (diagonal)

Check it in WIN-DAISY:

Open... the input file A2X.MGS with WIN-DAISY and execute a **Simulation**. Display the **Calculated Spectrum** in WIN-NMR. Look at the **Simulation Protocol** list and compare the number of different *transition frequencies* with the m_T scheme given in Fig. 2.49.

Obviously the A_2 particle gives rise to a doublet in the spectrum (the two vertical transition paths in Fig. 2.49) and this appears twice, because we have two identical - non-distinguishable - A spins present in the spin system:

2.4 Magnetic Equivalence   109

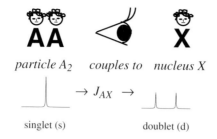

Fig. 2.50: Signal of particle A_2 due to interaction with single spin X.

The X nucleus appears as a three line pattern, somehow similar to the situation that is found in cases where the coupling constant to two other nuclei is quite similar (refer to section 1.2.3, example …\AMX\004001.1R in Fig. 2.22):

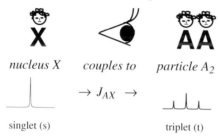

Fig. 2.51: Signal of nucleus X due to interaction with particle A_2.

For first order spin systems the so-called *"n+1"* rule can be used to determine the multiplicity of the signals due to scalar coupling constant interactions:

*The signal of the spin coupled to the group of **n** magnetically equivalent spins is split into **g = n+1** lines.*

The intensity of the lines is then given by the binomial coefficients (Pascal Triangle, refer to appendix A, section 6.2). The two non-distinguishable neighboring spins A_2 cause the X signal to be split into three lines with the intensity ratio 1:2:1. This pattern is called a *triplet* as shown in Fig. 2.25.

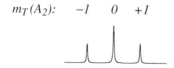

Fig. 2.52: X signal pattern labeling due to spin orientation of the coupled particle.

The double intensity of the middle transition of the triplet pattern can be explained using the m_T scheme shown in Fig. 2.49. The left hand side irreducible component includes one transition for each spin orientation of particle A_2 (the dotted diagonal lines).

But the right hand side term represents only the total vanishing spin value for the A_2 particle, so that the corresponding transition is identical with the middle transition of the left hand side subspectrum (compare the corresponding subspectra states mentioned in brackets).

The reasoning behind these observation can be worked out exactly when the HAMILTONian matrix is generated using the X-approximation due to the fact that the difference between resonance frequencies v_A and v_X is much bigger than the value of the coupling constant J_{AX}. Based on this rule the HAMILTONian off-diagonal elements can be neglected and the diagonal matrix elements correspond directly to the EIGENvalues. In this case the X-approximation can be used to determine the entries in the HAMILTONian as listed in Table 2.26.

Table 2.26: EIGENvalues for an A_2X spin system including X-approximation.

m_T	sub-spectrum	basis function	element number	matrix element = EIGENvalue
+1.5	1	$\|+1,+\frac{1}{2}>$	$H_{11}=E_1$	$-\frac{1}{2}v_X - v_A + \frac{1}{2}J_{AX}$
+0.5	1	$\| 0,+\frac{1}{2}>$	$H_{22}=E_2$	$-\frac{1}{2}v_X$
+0.5	1	$\|+1,-\frac{1}{2}>$	$H_{33}=E_3$	$+\frac{1}{2}v_X - v_A - \frac{1}{2}J_{AX}$
−0.5	1	$\|-1,+\frac{1}{2}>$	$H_{44}=E_4$	$-\frac{1}{2}v_X + v_A - \frac{1}{2}J_{AX}$
−0.5	1	$\| 0,-\frac{1}{2}>$	$H_{55}=E_5$	$+\frac{1}{2}v_X$
−1.5	1	$\|-1,-\frac{1}{2}>$	$H_{66}=E_6$	$+\frac{1}{2}v_X + v_A + \frac{1}{2}J_{AX}$
+0.5	2	$\| 0,+\frac{1}{2}>$	$H_{77}=E_7$	$-\frac{1}{2}v_X$
−0.5	2	$\| 0,-\frac{1}{2}>$	$H_{88}=E_8$	$+\frac{1}{2}v_X$

Check it in WIN-DAISY:

Open... the input file A2X.MGS. Using the NMR parameters in this file and the formulae given in Table 2.26, manually **calculate** the transition frequencies. (Don't forget to take into account the selection rule). To enable all transitions, including forbidden transitions, to be listed in the protocol file set the **Minimum intensity** in the **Main Parameters** dialog box to <0>. Perform a **Simulation** and compare the manually calculated transition frequencies with those listed in the **Simulation Protocol** file.

The nature of the only forbidden transition is the following:

$$|-1, \tfrac{1}{2}> \leftarrow |1, -\tfrac{1}{2}>, \text{ with } \Delta m_T = -1, \text{ but } \Sigma|\Delta m_{z,i}| \neq 1$$

which means, that the net change in total spin value satisfies the selection rule, but more than one spin alters their orientation at the same time. As discussed earlier in section 2.2.2.4 about first order three spin systems, such transitions are forbidden in first order applications.

Check it in WIN-DAISY:

Open... the input document A2X.MGS with WIN-DAISY and execute a **Simulation**. Examine the **Simulation Protocol** file. This file will contain the

2.4 Magnetic Equivalence

subspectra with their corresponding *transition frequencies* and *intensities*. Determine which transitions arise from the second subspectrum.

Click the **Generate Subspectra** button. Toggle to **WIN-NMR** and **Open...** the first subspectrum. Switch to the **Dual Display** mode and load the second subspectrum as the second file. Using the **Zoom** mode expand the high-field part of the spectrum showing a triplet in the total simulation (refer to Fig. 2.53). The composition of the 1:2:1 triplet pattern arises from a 1:1:1 triplet plus a singlet at the center frequency.

Zoom in on the low-field doublet. As expected only the first subspectrum contains all the transitions for this signal.

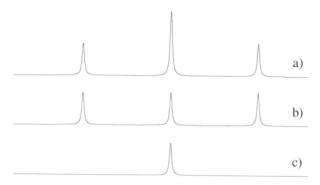

Fig. 2.53: a) Total signal pattern of X particle transitions.
b) Subspectrum of irreducible component $^3\{A_2\}^2X$
c) Subspectrum of irreducible component $^1\{A_2\}^2X$

The weighting of each subspectrum is determined as the product of involved particle multiplicities (refer to appendix, section 6.2).

Check it in WIN-NMR:

Open... the spectrum ...\AX2\001001.1R into WIN-NMR. The spectrum is of the aromatic region of a dichloroanisole compound. From the Analysis pull-down menu select the **Linewidth** option. Measure the linewidth and then switch to the **Multiplets** mode. Analyze the spin system. With this information and the signal intensities (Analysis **Integration** mode) determine if the spin system is best described as either A_2X or AX_2.

Note: This spectrum is the first example where one of the signals is not based on a doublet pattern. In the Multiplets analysis mode the doublet can be defined as usual. To analyze the triplet: with the cursor in **maximum cursor mode** select the middle peak of the triplet. Click the right hand mouse button and in the new pop-up menu select **Free Grid**: An new pop-up menu appears - select the appropriate number of distance lines, in this case **3 Distance Lines**. Click the left hand mouse button again, and release it to match the outer lines of the pattern. Connect the two couplings with each other. mark one peak with the spectrum cursor. In the WIN-NMR **Report...** box the different **Multiplicity** of both signals are mentioned.

Check it in WIN-NMR and WIN-DAISY:

Export... the data to **WIN-DAISY** and examine the data in the **Frequencies** dialog box. Using the established coupling connections WIN-DAISY can determine the number of magnetically equivalent **spins in** the coupled **group** as illustrated in Fig. 2.51 and expressed in the *'n+1' rule*. Execute an **Iteration** and compare the experiment and iterated spectrum using the **Dual Display** mode of WIN-NMR.

Return to **WIN-DAISY** and **Export** the calculated transitions, and adjusted NMR parameters to the **WIN-NMR exp.** spectrum. Using the **Zoom** mode expand the individual resonances. Because of the slight second order nature of the spectrum, the two center lines of the triplet and the two transitions for each of the doublet lines are not fully degenerate. Switch back to **WIN-DAISY** and examine the **Simulation Protocol** file to confirm these observations.

For this type of spin system only symmetrical substituted anisole (methoxy benzene) structures are possible, otherwise no magnetically equivalence could be present. From the coupling constant value deduce which of the two shown in Fig. 2.54 belongs to the spectrum ...\AX2\001001.1R.

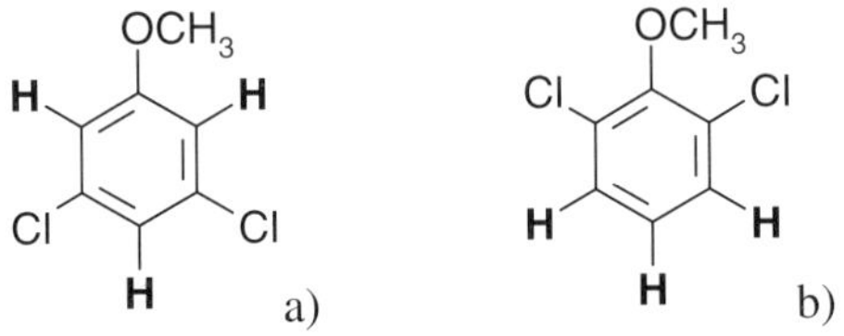

Fig. 2.54: a) Structure of *3,5-dichloroanisole*.
b) Structure of *2,6-dichloroanisole*.

Check it in WIN-NMR and WIN-DAISY:

Open... the spectrum ...\AX2\002001.1R relating to the other structure in Fig. 2.54 with WIN-NMR. This spectrum also contains the methoxy resonance. Analyse the aromatic region of the spectrum in the same way as before using the WIN-NMR **Linewidth** and **Multiplets** mode. **Export...** the data to **WIN-DAISY**. Open the **Spectrum Parameters** dialog box and click the **Iter. Regions** button. This will open a new dialog box that contains the two specral regions used for iteration. (You may toggle this display between **ppm, Hz** and **Points** units.) Two regions for the total spin system including the methoxy signal are defined. Deselect the low-frequency part by clicking on the corresponding line of the list box, because this part of the spin system is not yet included in the data of the multiplet analysis. Run a **Simulation** and use the **Dual Display**. In WIN-NMR you need the <Ctrl+Y> command to switch the y-gain from absolute to relative, because of the methoxy signal. Use the **Move Trace** option with depressed <Ctrl> key to overlay the peaks. Carry on with an iteration of the aromatic region. Label the spin system and check if the assignment of the shifts are consistent for both ansiole spectra.

You may also generate the subspectra. In the **Main Parameters** dialog box select the **Output Option Output of subspectra**. Perform a **Simulation** and use the **Generate Subspectra** button. Toggle to **WIN-NMR** and **Open...** the first subspectrum. Switch to the **Dual Display** mode and load the second subspectrum as the **second file**.

The treatment of the singlet which corresponds to the methoxy group used to calculate the total spin system will be discussed in chapter 3.x. As mentioned already, magnetic equivalence can result from different reasons. The magnetic equivalence of the examples discussed in the latter section arose from molecular symmetry. The anisole structures in Fig. 2.54 have a twofold symmetry. The symmetry operation (a C_2 axis or a mirror plane going through the methoxy group and the para proton) moves the protons and chlorines from one side of the symmetry element to the other. They are therefore *isochronous* because of their identical chemical environment. The magnetic equivalence is controlled by the coupling relationship of the proton para to the methoxy group. Both isochronous protons have got the same coupling path to the para proton and are therefore magnetically equivalent.

2.4.5.2 The AX$_3$ Spin System

Another way achieving magnetic equivalence is molecular mobility. So far all the structures discussed in section 2.2, 2.3 and 2.4 have been aromatic systems where the aromatic protons have been rigidly fixed. The scalar coupling mechanism works through chemical bonds and the magnitude and sign of the coupling constant is mainly determined by bonding geometry and environment. Consequently it is easy in many cases to distinguish between ortho, meta and para coupling pattern because of the *molecular configuration*.

The biggest source of composite particles is found in aliphatic mobile structures. The time scale of NMR measurements is such that the method is not capable of registering every individual rotamer. NMR spectra only represent a time-average picture over all *conformations*. In detail this fact will be discussed in section 3.2 and 4.3, here only the facts are stated.

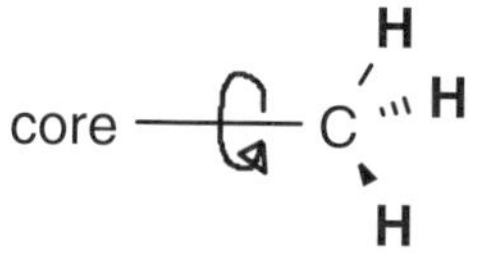

Fig. 2.55: Scheme of a rotating methyl group

Consider a methyl group bound to an extended structure that is free to rotate with respect to the core-carbon bond in Fig. 2.55. As the group rotates the position of all the protons in the methyl group is averaged so that as a function of time they all have an identical chemical environment and are therefore isochronous. Because the protons in the methyl group are isochronous, their coupling relations to other spin active nuclei of the core structure are identical. In fact the spins contained in the core structure are not able to distinguish between the three methyl protons. Consequently the methyl group can be described as a composite particle which consists of three protons. The geminal coupling constant $^2J_{HH}$ between the members of the methyl-particle is *not zero*, but it cannot be determined (as described earlier, refer to section 2.3.2.3, 2.3.3.3 and especially 2.4.2).

114   2 General Characteristics of Spin Systems

Methods to determine such "hidden" coupling constants are described in section 4.2.1.2. For an AX$_3$ spin system only two chemical shifts and one scalar coupling constant, J_{AX}, is of interest. Application of the 'n+1' rule results in a (3+1=) 4-line pattern for the A-signal a so-called *quartet*. The connected intensity ratio will be worked out in the following section.

Combination of the two different states for the X$_3$ particle (Fig. 2.48) with the single spin A leads to two irreducible components (subspectra):

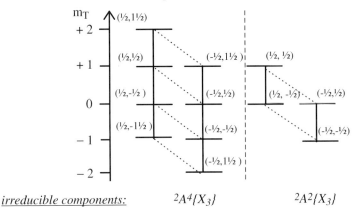

Fig. 2.56: Total spin value scheme for the complete AX$_3$ spin system,
—— X transitions (vertical), ······ A transitions (diagonal)

The diagonal dotted transitions in Fig. 2.56 of the first subspectrum give the 1:1:1:1 quartet pattern of the A signal pattern. The second irreducible component consists of two doublet states, in which the particles A and X$_3$ both act as a single spin with I=½, referring to a simple first order two-spin system as known from section 2.2.1 leading to doublets in A and X spectrum respectively. Spin A contributes only the two inner transitions of the four line pattern:

$$(½,½) \to (-½,½) \quad \text{and} \quad (½,-½) \to (-½,-½).$$

The determination of the weight of each subspectrum is based on the particle multiplicities (for the composite particle data of a three-spin I=½ particle refer to Table 2.24, the determination of the particle multiplicities is given in the appendix, section 6.2).

A: $G(½) = 1$

X$_3$: $G(1½) = 1$
            $G(½) = 2$

The product of the particle multiplicities leads to the statistical weight for each subspectrum:

$$^2A^4\{X_3\}: \quad S(\tfrac{1}{2}, 1\tfrac{1}{2}) = 1 \cdot 1 = 1$$
$$^2A^2\{X_3\}: \quad S(\tfrac{1}{2}, \tfrac{1}{2}) = 1 \cdot 2 = 2$$

Because of the double weight of the second subspectrum the intensity ratios of the transitions are as follows:

$$^2A^4\{X_3\}: \quad A: \quad 1:1:1:1 \quad X: \quad 8:8$$
$$^2A^4\{X_3\}: \quad A: \quad 2:2 \quad\quad\quad X: \quad 4:4$$

Check it in WIN-NMR and WIN-DAISY:

Open... the spectrum ...\AX3\001001.1R with WIN-NMR. This spectrum only contains the aliphtic region. From the Analysis pull-down menu select the **Multiplets** option and analyze the spectrum. To define the quartet pattern mark one of the two inner peaks with the arrow cursor before selecting the 4-distance lines **Free Grid**.

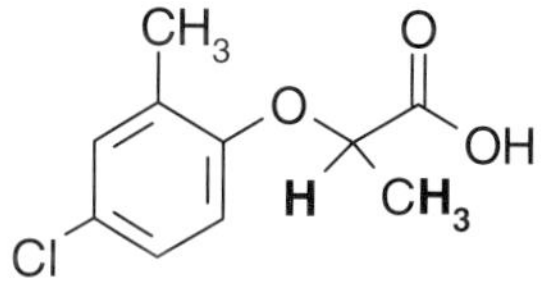

2-(4-chloro-2-methyl-phenoxy) propionic acid

Connect the couplings, **Export...** the data to **WIN-DAISY** and optimize the parameters via **Iteration**. **Export** the data back to **WIN-NMR experimental** spectrum. You may compare the intensity ratio with the data reported in the **Simulation Protocol** file.

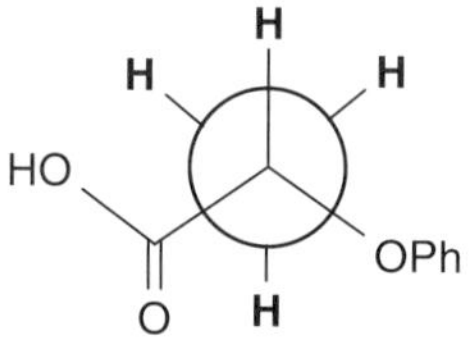

Fig. 2.57: NEWMAN projection of the propionic acid

The NEWMAN projection of propionic acid (Fig. 2.57) shows that each proton of the methyl group is in a different chemical environment and the coupling relation to the vicinal proton is different. Free rotation of the methyl group around the projection axis rotates each methyl proton into all the different orientations so that the NMR picture gives for all three protons a time averaged magnetic equivalence. In the present case of the methyl group there will be no preferred orientation, because all substituents are equal (protons). For methyl protons a free rotation is always possible. To decide if the magnetic equivalence conditions are still met, the forming of all possible conformers in NEWMAN projection is quite helpful. In cases where the rotation could be slowed down or stopped, the NMR parameter values change, because of different weighting of individual rotamers, what is a question of the "NMR time scale" and "life time" of the species in a dynamic exchange process, what is subject of section 5.3. The discussions in chapters before that deal always with fast processes resulting in averaged NMR parameters.

The situation that one substituent is different will lead always to initial non-isochronicity of the remaining two identical substituents in Fig. 2.57 in spite of fast rotation is present. This will be discussed in detail in chapter 3. From simple NMR spectra it is possible to evaluate important information about the molecular properties based on magnetic equivalence, isochronicity caused by symmetry (refer to section 3.1) or even non-equivalence (refer to section 3.2).

2.4.5.3 A_nX_m Spin Systems

This concept of signal multiplicity based on the '$n+1$ rule' can be extended without limits even to systems where both the A and X type nuclei consist of magnetically equivalent spins. All these spin systems are described by only one scalar coupling constant and two chemical shifts.

Check it in WIN-NMR and WIN-DAISY:

Open... the proton spectrum ...\AX3\002001.1R with WIN-NMR. It shows the aliphatic part of the given structure. Determine a **Linewidth**. Name the two independent spin systems and determine the multiplicity of the signals in the **Multiplets** mode.

2-[4-(6-chloro-2-benzoxazolyloxy)-phenoxy] propionic acid ethyl ester

The assignment of the two quartet pattern is easily by evaluation of the coupling constant connections. As well you can use the **Analysis Integration** mode to determine the corresponding relative number of spin responsible for the signals.

Export... the data to **WIN-DAISY** with fragmentation of the spin system into two independent parts (**Yes**). Optimize the parameters via **Iteration** and display the spectra using the WIN-NMR **Dual Display**.

Check it in WIN-DAISY:

Try to determine all of the irreducible components for the ethyl-group spin system. To check your results open the **Main Parameters** dialog box of the corresponding 5-spin system fragment in WIN-DAISY and select the Output Option **Output of subspectra**. Examine the terms and weightings listed in the **Simulation Protocol** file after **Simulation**. Construct a m_T diagram similar to Fig. 2.56. Compare your result with the scheme given in Fig. 2.58.

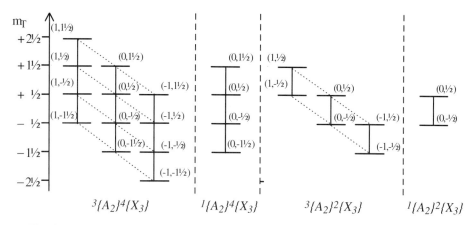

Fig. 2.58: Total spin value scheme for the complete A_2X_3 spin system,
——— X transitions (vertical), ······ A transitions (diagonal)

Check it in WIN-NMR and WIN-DAISY:

Toggle to WIN-NMR and **Export...** the multiplet data again to WIN-DAISY but this time do not fragment the total spin system (**No**). In the **Main Parameters** dialog box select the Output Option **Output of subspectra** and deselect the **Frequencies before degeneracy** flag. Execute a **Simulation** and compare the calculation times for this and previous simulation subdivided into two fragments. Check also the number of irreducible components and the list of transition frequencies in the **Simulation Protocol** file.

From the comparison of both calculation protocols it can be deduced that fragmentation simplifies the calculation significantly - without fragmentation the number of irreducible components increases from six (2 for the first and 4 for the second fragment to eight with a corresponding increase in calculation time (at least doubled). In spin systems where there is definitely no coupling interaction between subsystems, fragmentation should always be used. In the present example the number of transitions decreases from 672 for the non-fragmented spin system to 42 (14+28) transitions when the spin system is subdivided into two independent fragments. The un-fragmented system contains a large number of totally degenerate lines. Thus if there is no interaction between the separate spin systems no errors are introduced into the analysis using fragmentation. Even in cases where some interaction between the subsystems exists, fragmentation should always be chosen as the first step in any analysis. The inclusion of these long range interactions can be introduced in a second step by merging (Edit Merge fragments). This will be discussed in detail in section 3.3. For the present rather small spin systems the time difference is not too serious but the calculation time requirement increases not linear, so that it is very important for larger spin systems.

Check it in WIN-DAISY:

Select the WIN-DAISY document consisting of two fragments. Execute the command **Edit I Merge selected fragment with...** or press the button **Merge fragment**. Accept the appearing dialog with **Ok** (there is only one destination fragment) and answer the question with **Yes** in order to delete the superfluous fragment after merging. Then only one fragment remains, contain both spin systems. Open the **Frequencies** and **Scalar Coupling Constants** dialog box to examine the parameters. Run a spectrum **Simulation**. As expected the calculation time is now about doubled because of the larger matrices to be diagonalized.

Check it in WIN-NMR:

Try to analyze the isopropyl part of the proton spectrum ...\AX6\001001.1R in the WIN-NMR **Multiplets** mode. For high multiplicities it might be necessary to decrease in the **Options...** dialog the value of **Min. Intensity**, e.g. to <2> % to match all lines of the multiplet. What conclusions can you draw from the pattern arising from the single proton regarding the methyl group protons?

isopropyl-4,4-dibromo benzilate

The septet signal pattern of the single isopropyl-proton indicates that the two methyl groups can be regarded as magnetically equivalent. The scalar coupling path from one methyl group to the single proton is identical with the path from the other methyl group; thus both methyl groups can be combined into one composite particle consisting of 6 protons leading to a septet. This is only possible in case there is no coupling between the two methyl groups present. This would be a $^4J_{HH}$ coupling and can be regarded to be zero in this kind of compounds. First of all we are going to verify this statement while defining the septet as a quartet of quartets and the doublet for the methyl groups twice at the same position.

Check it in WIN-NMR:

Mark the second transition of the septet with the **maximum cursor**, click the right-hand mouse button and select **Free Grid I 4 Distance Lines**. Use the left-hand mouse button while moving the mouse cursor to match the four left-outmost lines of the signal. **Define** the **Multiplet**. Select the **Coupled Grid** button, press the left-hand mouse button and match the whole signal pattern. Mark one of the doublet lines with the **maximum spectrum cursor**, press the right-hand mouse button, select **Free Grid I 2 Distance Lines** and define the doublet. Redo this action for the same doublet to obtain two identical doublets. Open the **Report...** box, **Auto Connect** the coupling constants and **Export...** the data to **WIN-DAISY**.

Check it in WIN-DAISY:

Run a **Simulation** and display both spectra in WIN-NMR **Dual Display**. Toggle back to **WIN-DAISY** and open the **Frequencies** dialog box and change one of the identical resonance frequencies [Hz] slightly and switch into the **Scalar Coupling Constants**. Check the **iterate** option for the long-range coupling constant J_{23}. Open the **Spectrum Parameters** and press the **Iter. Regions** button. Change the display to ppm and *deselect* the aromatic region and the OH-proton region (at about 4.4 ppm) from iteration. Click on the corresponding line in the list box, that it is no longer highlighted. Quit the dialogs with **Ok**. Run an **Iteration** and inspect the **Iteration Protocol**. In case the *R-factor* obtained is about 50% and the linewidth parameter increased up to the upper limit you forgot to deselect the other spectral regions. Otherwise a very high *standard deviation* for the long-range coupling J_{23} much larger than the corresponding parameter value implies, that the coupling constant J_{23} cannot be determined from the spectrum and thus the methyl groups are magnetically equivalent. **Close** the input document and toggle to **WIN-NMR**.

Check it in WIN-NMR and WIN-DAISY:

Measure a **Linewidth** and define a septet and a doublet in the WIN-NMR **Multiplets** Mode, connect the coupling constant, mark the left outmost transition of the septet with the maximum cursor and **Export...** the isopropyl spin system to **WIN-DAISY**. *Deselect* again the first and third **Iter. Region** in the **Spectrum Parameters** and look at the data in the **Frequencies** and **Scalar Coupling Constants** dialog boxes. Run an **Iteration**. Inspect the result in the **Iteration Protocol** and WIN-NMR **Dual Display**. Toggle back to WIN-DAISY and **Export** the result to **WIN-NMR exp**.

2.4.5.4 First Order Spin Systems with more than one Coupling

The pattern multiplicities generated by one group containing a variable number of magnetically equivalent neighbors are given in the following Table 2.27 according to the first order PASCAL triangle (binomial coefficient):

Table 2.27: Signal multiplicities for spins I= ½ due to coupling interaction with n magnetically equivalent neighbors:

No. of neighbors	Multiplicity g		Intensities
0	singlet (s)	1	1
1	doublet (d)	2	1 : 1
2	triplet (t)	3	1 : 2 : 1
3	quartet (q)	4	1 : 3 : 3 : 1
4	quintet (qi)	5	1 : 4 : 6 : 4 : 1
5	sextet (sx)	6	1 : 5 : 10 : 10 : 5 : 1
6	septet (sp)	7	1 : 6 : 15 : 20 : 15 : 6 : 1
n		n+1	$B_I(n,i) = \binom{n}{i}, i = 0...n$

120   *2 General Characteristics of Spin Systems*

There has already been detailed discussion of the multiple coupling of single spins; thus if a single spin is coupled to two different coupling partners the first coupling constant gives rise to a doublet and the second coupling partner duplicates the doublet pattern. This multiple coupling scheme can also be extended to systems where either one or both of the different coupling partners are composite particles; the original multiple pattern caused by the first coupling partner is replicated in the multiplet pattern of the second. However, unlike the scheme involving all single spins where the coupling pattern is easily recognized, the patterns observed when composite particles are involved are often more complicated.

The first step in understanding this replicated coupling scheme will be the examination of the spectrum of pure ethanol - without any solvent. The whole compound consists of 6 protons arranged in three groups as shown in Fig. 2.59:

$$CH_3-CH_2-OH$$

Fig. 2.59: Structure of ethanol.

Check it in WIN-NMR and WIN-DAISY:

Open... the ethanol spectrum ...\A2MX3\001001.1R with WIN-NMR. Identify the signal by integrating the shift regions. Because the sample is neat liquid without any solvent the coupling constant over the oxygen bridge is visible. The linewidth of the methyl group signal is smaller than of the others indicating some dynamic process which will not be considered at the moment.

Measure a **Linewidth** and analyze the signal pattern in the WIN-NMR **Multiplets** mode. The CH_2-signal is coupled with one coupling constant to the CH_3-group (quartet) and with another coupling constant to the OH-proton (doublet). You may start the analysis of the doublet of quartet either with the doublet splitting or with the quartet pattern. It doesn't make any difference. Try out both ways (refer to Fig. 2.60). The magnitude of the two coupling constants for both multiplicities is known already from the analysis of the other signals due to the rule of repeated spacings.

Connect the coupling constants and **Export...** the data to **WIN-DAISY**. In WIN-DAISY check the entries in the **Frequencies** dialog box. The multiplicity of the signals is evaluated by the program to form the number of spins contained in the composite particle. Optimize the parameters via **Iteration** and display the result in WIN-NMR **Dual Display**. Toggle back to **WIN-DAISY**.

Remark:
It is obvious, that the linewidth is different between the methyl group signal and OH- and CH_2-signals. This problem will be discussed in detail in section 5.3 on molecular dynamics.

Check it in WIN-DAISY:

Open the **Lineshape Parameters** and check the **Nuclei specific linewidth** option and click the **Iterate All** button. Redo an **Iteration**.

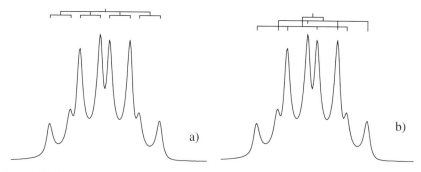

Fig. 2.60: CH_2-signal pattern a) quartet of doublets, b) doublet of quartets.

More complicated spin systems relating NMR parameters to molecular structure will be covered in chapter 3 section 3.2 and 3.3. The following examples does not relate directly to molecular structure, is has been chosen to illustrate the different multiplicities that can be analyzed using WIN-NMR.

Check it in WIN-DAISY and WIN-NMR:

Open... the input file 4CP.MGS in WIN-DAISY. You may include some noise: Open the **Spectrum Parameters** dialog box and click the box named **include noise**. **Simulate** the spectrum and display it with WIN-NMR. Define the spectral regions under consideration in the **Analysis | Interval** mode and **save...** the **Regions...** under the default file name. Measure the **linewidth** and switch over to the **Analysis Multiplets** mode. Try to identify the splitting pattern, connect the couplings and **Export...** the data to WIN-DAISY back.

2.4.6 Second Order Spin Systems

In a similar manner to single spins changing from first order to second order spin systems, composite particles can also be extended to second order systems. In an analogous way to single spins, the development of second order effects on composite particles will be based on the modification of simple first order particle systems.

2.4.6.1 The A_2B Spin System

The simplest way to illustrate the second order effect in this three spin system is to start with the previous A_2X example and to calculate a simulation sequence starting with the values analyzed earlier.

Check it in WIN-DAISY:

Open... the input document A2X_SEQ.MGS in WIN-DAISY. This input file contains the solution for example ...\AX2\002001.1R, but with a smaller linewidth (0.5 instead of 1.5 Hz). The predefined simulation sequence moves the low-frequency shift of spin X towards and beyond spins A. Execute the **Simulation Sequence** and **Display** first the reference **Calculated Spectrum** with the red peak button in WIN-NMR.

Check it in WIN-NMR:

Use the **Multiple Display** mode to display the 21 calculated spectra. Try to work out the maximum number of transitions for this three spin system. E.g. **Open...** manually the spectrum ...\999983.1R and perform a **Peak Picking**.

It is already obvious in the reference spectrum ...\999999.1R that the low-frequency signal is not a triplet but a four-line pattern. Up to now this information has been obscured by the broad linewidth and has not been of interest. Closer examination of the sequence calculation shows that the maximum number of lines is 9, yet the m_T diagram for the A_2X spin system Fig. 1.54 shows only 8 transitions. The sequence simulation makes it clear that not only are the degenerate transitions turned into very different transition frequencies going from a first to second order system but a very weak, combination line appears as well. As already mention in section 2.2 on single spin second order spin systems, this is one basic feature of real second order spin systems. This combination line refers to a type of transition that does not comply with the original selection rule. The basic selection rule states that only one spin is allowed to change its spin value while all other spin values must remain the same. Combination lines do not break this rule but comply in a different way. Thus if more than one spin changes its spin orientation there must be other concerted spin flips in the opposite direction to compensate the changes in m_T, so that the net difference in m_T remains -1. For the A_2B system there is only one transition that can satisfy this condition:

$$(1,-½) \rightarrow (-1, ½)$$

Fig. 2.61 shows the corresponding m_T diagram with the combination line illustrated. All three spins change their spin orientation at the same time, but the net m_T difference strikes the original selection rule. It is convincing that such a transition is quite seldom compared with 'normal' single quantum transitions and therefore the lines are very weak.

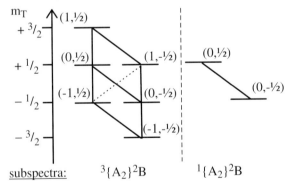

Fig. 2.61: m_T diagram for A_2B spin system; ······· combination line.

To summarize what has been discussed about the first order A_2X spin system, there were four A-transitions and four X transitions, the combination line was not observed. The A-transitions are degenerate in pairs to give a doublet and the X-transitions form a

2.4 Magnetic Equivalence

triplet whose center transition is degenerate, one part arising from the main subspectrum and the second part caused by the single transition in the second subspectrum. Applying these facts to the second order extension of the spin system to A_2B the degeneracy of the transitions forming the doublet and triplet is abolished and the maximum number of transitions including the combination line are observed in the spectrum.

The two 2 x 2 matrices appearing for the first irreducible component for $m_{cp} = +\frac{1}{2}$ and $m_{cp} = -\frac{1}{2}$ can be solved algebraically as shown for the 2 x 2 matrix of the AB spin system using the p-q formula given in section 2.3. The result can be simplified by using the abbreviation $D_\pm$ to obtain the EIGENvalues given in Table 2.28.

$$D_\pm = \sqrt{\Delta_{AB}^2 \pm \Delta_{AB} \cdot J_{AB} + \frac{9}{4} J_{AB}^2}$$

Table 2.28: EIGENvalues for A_2B spin systems

m_T	Sub spectrum	Element number	EIGENvalue
+1.5	1	E_1	$-\frac{1}{2}\nu_B - \nu_A + \frac{1}{2}J_{AB}$
+0.5	1	$E_{2,3}$	$-\frac{1}{2}\nu_A - \frac{1}{4}J_{AB} \pm D_-$
−0.5	1	$E_{4,5}$	$+\frac{1}{2}\nu_A - \frac{1}{4}J_{AB} \pm D_+$
−1.5	1	E_6	$+\frac{1}{2}\nu_B + \nu_A + \frac{1}{2}J_{AB}$
+0.5	2	E_7	$-\frac{1}{2}\nu_B$
−0.5	2	E_8	$+\frac{1}{2}\nu_B$

All possible transitions, including the combination line, are listed in Table 2.29.

Table 2.29: Transition frequencies for A_2B spin systems, *combination line

peak label (origin)	transition	transition frequency
T_8 (A)	$E_2 - E_1$	$\frac{1}{2}(\nu_A + \nu_B) - \frac{3}{4} J_{AB} + D_-$
T_4 (B)	$E_3 - E_1$	$\frac{1}{2}(\nu_A + \nu_B) - \frac{3}{4} J_{AB} - D_-$
T_7 (A)	$E_4 - E_2$	$\nu_A + D_+ - D_-$
T_9 (*)	$E_4 - E_3$	$\nu_A + D_+ + D_-$
T_2 (B)	$E_5 - E_2$	$\nu_B - D_+ + D_-$
T_5 (A)	$E_5 - E_3$	$\nu_B + D_+ - D_-$
T_1 (B)	$E_6 - E_4$	$\frac{1}{2}(\nu_A + \nu_B) + \frac{3}{4} J_{AB} - D_+$
T_6 (A)	$E_6 - E_5$	$\frac{1}{2}(\nu_A + \nu_B) + \frac{3}{4} J_{AB} + D_+$
T_3 (B)	$E_8 - E_7$	ν_B

Check it in WIN-NMR and WIN-DAISY:

Open... the last spectrum, the 21st ...\999979.1R, of the simulation sequence in WIN-NMR and compare this spectrum with the experimental spectrum ...\AB2\001001.1R using the **Dual Display** mode. Normalize the display <Ctrl + Y> and click the **All** button. Click the Move Trace button and holding down the <Shift> key slide the original spectrum to the right until both spectra overlap exactly. What can be said about the structure of the compound used to measure the original experimental spectrum?

From the Output pull-down menu select the **Title...** option. The text window that opens up contains information about the actual chemical shift value for the sequence simulation in Hertz: Place the mouse cursor on the number and double click to mark the number - type <Ctrl + C> to paste this number into the clipboard. Switch back to **WIN-DAISY** and in the **Frequencies** dialog box double click the second frequency entry and past the number in from the clipboard <Ctrl + V>.

This example illustrates the fact, that the appearance of the signal pattern is only dependent on the *difference* between chemical shifts and the size of the coupling constant, independent of the absolute chemical shift values. In the simulation sequence the coupling constant was not changed, so that it might be possible that the compound refers to an 1,2,3 trisubstituted benzene (substituents 1 and 3 identical) as well.

Check it in WIN-NMR and WIN-DAISY:

Change the WIN-NMR X-scale into Hertz <Ctrl + X> and in the Dual Display status line note the value of **XInc** (ca. 50 Hz). Toggle back to **WIN-DAISY** and open the **Spectrum Parameters**. Enter the determined XInc value into the field **Offset [Hz]** and press the button **Apply!**. The dialog box closes immediately and the entered offset value will be applied to all the frequency values in the input document. Run a spectrum **Simulation** and **Display** the **Calculated Spectrum** in WIN-NMR. Load the spectrum ...\AB2\001001.1R as the **second file** into the **Dual Display** and compare. The parameters in the WIN-DAISY input document almost correspond to the experimental spectrum.

Open... the experimental spectrum ...\AB2\001001.1R alone with WIN-NMR. From the **Analysis** pull-down menu choose the **Interval** option and load the predefined spectral regions (**Regions... | Load...**). Click the **Export...** button. In case there are multiplet data available a message box appears which has to be answered with **No** in order to export the Interval data exclusively. When WIN-DAISY is moved to the foreground, accept the message with **Yes**, that the experimental data have to be connected with an input document. Click the **Connect** button to connect the experimental data. In the **Control Parameters** dialog box select the **Standard Iteration** mode with **none Broadening**. The shifts, the coupling and the linewidth are already selected for iteration in the loaded input document. You may check the Iteration limits in the **Frequencies** and **Scalar Coupl.** sub dialog boxes, but they are adapted together with the offset correction. Execute an **Iteration** and look at the result in WIN-NMR **Dual Display**. Keep the WIN-DAISY document open!

It has to be kept in mind that the nine transition frequencies refer to only three NMR parameter: two resonance frequencies v_A and v_B and one coupling constant J_{AB}. These values can be determined directly from the spectrum provided at least the eight main transitions can be identified as belonging to either the A or B part of the spectrum. This identification is possible based on the ninth combination line which is situated on the composite particle part (A-part in case of A_2B spin system) or merely on the integral values of left-hand four lines compared to the right-hand four lines.

2.4 Magnetic Equivalence

Check it in WIN-NMR:

Open... the experimental spectrum ...\AB2\001001.1R with WIN-NMR. From the Analysis pull-down menu choose the **Integration** option. Integrate the region of the eight main transitions and then split the integral into two parts, each containing four lines, by clicking the **Split Int.** button. Does the spectrum belong to an A_2B or an AB_2 spin system?

It is possible to extract the NMR parameters directly from the spectrum using analytical equations, that can be derived from Table 2.29. In order to do this, first the A and B part of the spectrum have to be identified as done in the previous "Check it". For an A_2B spin system the following rules can be applied (For AB_2 systems the labeling of the resonance frequencies has to be inverted if the transition numbering is kept):

$$\nu_B = T_3$$

$$\nu_A = \frac{T_5 + T_7}{2}$$

$$J_{AB} = \frac{(T_1 - T_4) + (T_6 - T_8)}{3}$$

Check it in WIN-NMR, WIN-DAISY and the Calculator:

Using the same spectrum ...\AB2\001001.1R select the **Peak Picking** option from the **Analysis** pull-down menu and perform a **Mouse PP.** with **Interpolation** and peak labels in **Hz**. If necessary use the **Edit Cursor** mode to remove any unwanted peaks (place the spectrum cursor on the peak you want to remove from the list and press the right mouse button). Click the **Report** button to display the picked peaks.

Toggle to **WIN-DAISY** and run the **Calculator**. Using the three equations above, calculate the NMR parameters and compare these values with those in the WIN-DAISY **Iteration Protocol** file. Then use the **Edit | Disconnect experimental data** command to recover the original WIN-DAISY document A2X_SEQ.MGS.

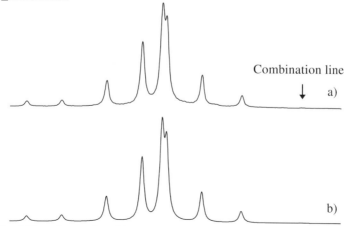

Fig. 2.62: a) Experimental and b) calculated spectrum of example AB2\001001.1R.

126 2 General Characteristics of Spin Systems

It seems extraordinary that one transitions should correspond exactly to the resonance frequency of the single spin in a real second order spectrum. But inspection of Table 2.29 and Fig. 2.61makes clear that the second subspectrum includes only one transition which is caused by a spin system state in which the composite particle is spin inactive. Therefore is the transition frequency is unperturbed by any second order effects and represents purely the single-spin resonance frequency.

The evaluation of the irreducible representations and the analysis of the Peak Picking values make clear, that this spin system cannot be analyzed using first order approximations as done before. What has been discovered is, that one resonance frequency can be found in the spectrum identified by one transition and the other resonance frequency can be determined by the arithmetic mean of two distinct transitions. Consequently there is a possibility to extract at least these NMR parameters from WIN-NMR using the Multiplet Analysis mode.

Check it in WIN-NMR and WIN-DAISY:

Open... the experimental spectrum ...\AB2\001001.1R in WIN-NMR. If not done so already, use integration to determine whether the spectrum is a A_2B or AB_2 spin system. Use a suitable signal and measure the **Linewidth**. From the **Analysis** pull-down menu choose the **Multiplets** option. With the cursor in **maximum cursor** mode, position the cursor on top of the T_3 transition. Click the right hand mouse button and in the pop-up menu click the **Define Singlet** option to store the single spin resonance frequency. Using T_5 and T_7, define a doublet - the arithmetic mean of these two signals corresponding to the second resonance frequency.

2,6-dichlorobenzonitrile

Export... the data to **WIN-DAISY**: choose the interval file, do *not* check for fragmentation (**No**) and ignore non-assigned coupling constants (**No**). In this way the measured frequencies are taken as single spins into one fragment. In order to define the complete spin system, in the **Frequencies** dialog box change the number of **spins in group** for the composite particle to <2>. In the **Scalar Coupling Constant** dialog box enter a suitable value for an ortho coupling e.g. <8> Hz. Execute a **Simulation**. Use the **Dual Display** mode of WIN-NMR to compare the spectra.

Return to **WIN-DAISY** and check the assignment of your spin system by changing the composite particle from spin 1 to spin 2 or vice versa - in the **Frequencies** dialog box exchange the number in the **spins in group** boxes. Execute a **Simulation** and compare the calculation results.

Comparing the two methods available for the analysis of A_2B / AB_2 spin systems it is obvious that computer simulation using WIN-DAISY is the by for the easiest and fastest method provided that some information about the spin system is available. If no information about the coupling constant is available, the following method is possible: Either run a simulation sequence with variation of the coupling constant or run an iteration with smoothing parameters.

2.4 Magnetic Equivalence 127

Check it in WIN-DAISY:

Proceed in a similar manner to the previous "Check it" but do not enter any Coupling constant. Open the **Control Parameters** dialog box and select **medium broadening** and **medium No. of Cycles** for the default **Standard Iteration** mode. Click the group parameters **Scalar Coupl.** button and select the scalar coupling constant for **iteration** and change the lower limit to <0> Hz and the upper to <10> Hz. Perform an **Iteration** and check the results in the **Dual Display** mode of WIN-NMR.

This example shows how powerful computer programs such as WIN-DAISY are - to produce a satisfactory result it was only necessary to change the limits of the coupling constant to cover the range of expected values.

Finally the generation of subspectra of the AB_2 spin system should lead to the second subspectrum containing the transition at v_A:

Check it in WIN-DAISY:

Open the **Main Parameters** dialog box and select the **Output Option Output of Subspectra** and **Simulate** the spectrum again. Click the button **Generate Subspectra**. Toggle to WIN-NMR and load both simulated spectra into the Dual Display (**Open...** ...\AB2\001998.1R first and load ...\001997.1R as the **second file**, Fig. 2.63).

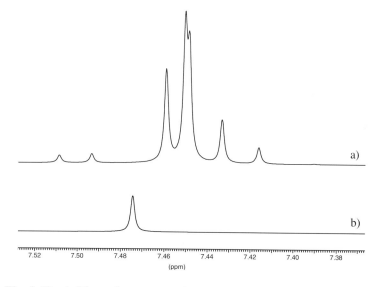

Fig. 2.63: a) First subspectrum of AB_2 spin system.
b) Second subspectrum of AB_2 spin system.

Note: Strong second order effects can prevent the exact identification of transitions from the EIGENvectors, so that this method is strictly applicable only to first order systems.

Check it in WIN-NMR and WIN-DAISY:

Open... the experimental spectrum ...\AB2\ 002001.1R. Using Integration identify the spin system as either AB$_2$ or A$_2$B. As in a previous "Check it", define in the **Multiplets** mode a **Singlet** and doublet to obtain the chemical shift values.

2,6-dichlorobenzoic acid

Export the data to **WIN-DAISY** choosing interval file (**yes**), **No** fragmentation and ignore non-assigned coupling constants (**No**). In the **Frequencies** dialog box change the number of **Spins per group** for the composite particle to <2>. Select **Smoothing parameters** in the **Control Parameters**, e.g. medium/medium, select the **Scalar Coupl.** for Iteration and ensure that the range of possible values is correct. Perform an **Iteration** and check the results in the **Dual Display** mode of WIN-NMR.

Compare the chemical structures of the two compounds used in these last examples and the corresponding spin systems?

2.5 References

[2.1] Purcell, E.M., Torrey, H.C., Pound, R.V., *Phys. Rev.*, 1946, *69*, 37.
[2.2] Bloch, F., Hansen, W.W., Packard, M.E., *Phys. Rev.*, 1946, *69*, 127.
[2.3] Proctor, W.G., Yu, F.C., *Phys. Rev.*, 1950, *77*, 717.
[2.4] Dickinson, W.C., *Phys. Rev.*, 1950, *77*, 736.
[2.5] Ramsey, N.F., *Phys. Rev.*, 1950, *78*, 699.
[2.6] Taft, R.W., *J. Amer. Chem. Soc.*, 1957, *79*, 1045.
[2.7] Hammett, L.P., *Trans. Faraday Soc.*, 1938, *34*, 156.
[2.8] Kalinowski, H.-O., Berger, S., Braun, S., *Carbon-13 NMR Spectroscopy*, Chichester: Wiley, 1994.
[2.9] Kutzelnigg, W., Fleischer, U., Schindler, M., in: *NMR Basic Principles and Progress*, Diehl, P., Fluck, E., Kosfeld, R (Eds.). Berlin: Springer, 1991, *23*, 165-262.
[2.10] Gauss, J., *Chem. Phys. Lett,*. 1992, *191*, 614 and
Gauss J., *Ber. Bunsenges. Phys. Chem.*, 1995, *99*, 1001.
[2.11] Harris, R.K., *Nuclear Magnetic Resonance Spectroscopy*, Harlow: Longman, 1986.
[2.12] Pople, J.A., Schneider, W.G., Bernstein, H.J., *High-resolution Nuclear Magntic Resonance*, McGraw-Hill, New York, 1959.
[2.13] Akitt, J.W., *NMR and Chemistry- An introduction to modern NMR spectroscopy*, 3rd ed., London: Chapman & Hall, 1992.
[2.14] Günther, H., *NMR Spectroscopy - Basic Principles, Concepts and Applications in Chemistry*, 2nd ed., Chichester: Wiley, 1995.
[2.15] Friebolin, H., *Basic One- and Two-Dimensional NMR Spectroscopy*, 2nd ed., Weinheim: VCH, 1993.
[2.16] Bernstein, H.J., Pople, J.A., Schneider, W.G., *Can. J. Chem.*, 1957, *35*, 1060.
[2.17] Bigler, P., *NMR Spectroscopy: Processing Strategies*, Weinheim: Wiley-VCH, 1997.
[2.18] Garbisch, E.W., *J. Educ. Chem.,* 1968, *45*, 311.
[2.19] Pople, J.A., Schaefer, T., *Mol Phys.*, 1961, *3*, 547.
[2.20] Diehl, P., Pople, J.A., *Mol. Phys.*, 1961, *3*, 557.
[2.21] Pretsch, E., Clerc, J.T., *Spectra Interpretation of Organic Compounds*, Weinheim: Wiley-VCH, 1997.
[2.22] Brügel, W., *Nuclear Magnetic Resonance Spectra and Chemical Structure*, Darmstadt: Steinkopff, 1967.
[2.23] Stephenson, S., *Encyclopedia of Nuclear Magntic Resonance,* Grant, D.M., Harris, R.K. (Eds.), Chichester: Wiley, 1996, *1*, 816-821.
[2.24] Stephenson, S., Binsch, G., *J. Magn. Reson.*, 1980, *37*, 395 and 409.

3 Structure and Spin System Parameters

In the previous chapter the main aim was to explain the basic principles about spin systems in high resolution NMR spectra. The majority of examples were aromatic structures with the exception of the section about magnetic equivalence where it was necessary to introduce some aliphatic molecules to discuss higher multiplicities. The main point of this chapter is to establish the strong connection between the NMR parameters and the chemical structure. It will focus on the huge amount of information contained in simple spin systems. The careful interpretation of NMR parameters may lead directly to the chemical structure or at least prove or disprove a proposed structure.

The first section 3.1 will deal with symmetry properties of rigid compounds. The second section will introduce the variety of structural information that can be extracted from spin systems using the techniques discussed in chapter 2 and expanded in the first section of this chapter. The second part will also discuss the possibility of using NMR to identify configurational (3.2) and conformational (3.3) isomers. Both these types of isomers are referred to as stereo isomers and arise because molecules are a three dimensional arrangement of atoms. Configurational isomers can only be converted into one another by breaking a chemical bond and can usually be chemically isolated. Conformational isomers can be converted into one another by rotation about single bonds and usually cannot be isolated.

3.1 Symmetry Effects

In chapter 2 it was shown that NMR could be used to distinguish between structural (constitutional) isomers in aromatic systems based on the magnitude and connectivity's of the coupling constants and on the chemical shifts. Only at the end of the chapter in the discussion on A_2B and AB_2 spin systems was a symmetrical compound considered. This 1,2,6-trisubstituted benzene compound featured symmetry elements because the substituents at position 2 and 6 were identical. This type of molecule posses a mirror or twofold axis of rotation which exchanges the positions of the identical substituents as illustrated in Fig. 3.1.

The symmetry operation exchanges the R_2 substituents into each other, the protons H_A with each other and leaves R_1 and H_B unchanged because they are located within the symmetry plane or axis. Thus the protons H_A are isochronous.

132   3 Structure and Spin System Parameters

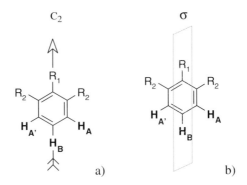

Fig. 3.1: a) Application of a twofold rotation to the 1,2,3-substituted benzene
b) Application of a mirror reflection to the 1,2,3-substituted benzene

The structural relation between the isochronous protons H_A is homotopic, because the substitution of either one or the other leads to the same compound. But the coupling paths to proton H_B are related by symmetry (reflection or rotation), and therefore only enantiotopic and *not* homotopic. In fact the coupling constants J_{AB} between H_A and H_B and $J_{A'B}$, the coupling constant between $H_{A'}$ and H_B, are identical, so the protons H_A and $H_{A'}$ are not only isochronous, they are magnetically equivalent because of the spin system symmetry.

In the following sections we will mainly discuss the spectra of symmetrical compounds that do *not* feature magnetic equivalence.

3.1.1 Structural and Notational Aspects

In this section we will deal with the remaining types of aromatic spin systems: 1,2- and 1,4-substituted aromatic spin systems. In the first type of compounds both substituents must be the same (Fig. 3.2a) while in the second type of compound they must be different (Fig. 3.2b). In case that R = Cl in Fig. 3.2b, and no other type of spin active nucleus coupled to the protons, all protons are magnetically equivalent, and only a singlet obtained in the proton spectrum.

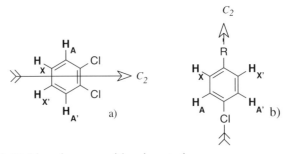

Fig. 3.2: a) *1,2-Dichloro benzene* with spin notation,
b) 1,4-Chloro phenyl-compound with spin notation.

3.1 Symmetry Effects

The assignment of the proton labels A and X depend on the nature of the substituents. Instead of the rotation symmetry C_2 the reflection C_s point group is as well a correct description of the symmetry of the molecules. Using the substitution criterion the pairs of protons H_A, $H_{A'}$ and H_X, $H_{X'}$ are homotopic, because the substitution of either the one or the other in each pair results in the same molecule.

Looking at the symmetry relationship of the coupling constants the case is different; the couplings J_{AX} and $J_{AX'}$ are *not* identical and *not* related by symmetry! The same is true for the coupling constants J_{AX} and $J_{A'X}$. Thus the protons H_A, $H_{A'}$ and H_X, $H_{X'}$ are only *isochronous* and *not* magnetically equivalent. As discussed before isochronicity can arise either by chance or because of chemical structure. In NMR literature isochronous spins that are not magnetically equivalent are called *chemically equivalent*. But this notation does not distinguish between accidentally isochronous nuclei and isochronicity that occurs due to molecular structure and hence are related by symmetry. Both expressions (isochronicity and chemical equivalence) describe the same circumstances but the latter is ambiguous in its meaning. Consequently, in all further discussions only the expression isochronous will be used.

The spin systems in Fig. 3.2 can be called AA'XX' using the prime notation for spins that are isochronous but are not magnetically equivalent. The prime notation indicates that the spin system has the following properties:

- $v_A = v_{A'}$ and $v_X = v_{X'}$: *Both pairs of spins are isochronous.*

- $J_{AX} \neq J_{A'X}$ and $J_{AX} \neq J_{AX'}$ and $J_{A'X'} \neq J_{A'X}$ and $J_{A'X'} \neq J_{AX'}$
  Both v_A and $v_{A'}$ as well as v_X and $v_{X'}$ are only isochronous and not magnetic equivalent.

This notation does not express anything about the *point group symmetry* of the spin system. In fact the spin system features also has the following characteristics:

- $J_{AX} = J_{A'X'}$ and $J_{AX'} = J_{A'X}$
  The isochronicity is not accidental but is related by symmetry.

To highlight this important characteristic in the spin system notation, the HAIGH notation should be used [3.1]. In this notation the symmetry elements are symbolized with square brackets and the order is appended as a subscript outside the square brackets - no primes are used in the description of the spin system. The point group can be added in brackets and italic letters after the spin notation. Thus the spin systems shown in Fig. 3.2a and Fig. 3.2b, can be labeled [AX]$_2$ because the symmetry operation (either C_2 or σ) transfers the structure fragment containing spins AX into A'X'. To distinguish between the isochronous spins in the parameters - e.g. in coupling constants: J_{AX}, $J_{A'X}$ and $J_{AA'}$ - the primes are still used.

For such a simple spin system it is not necessary to add the point group - C_2 or C_s. The HAIGH notation was introduced in 1970; up to that time the notation used in the literature often did not discriminate between the different types of equivalence. Consequently this often led to confusion because no notational difference was made between magnetic equivalence and isochronicity or even accidental and non-accidental isochronicity.

HAIGH'S [3.1] original paper contained recommendations for spin notation and was intended as a basis for further discussion, but more than 25 years after his original proposals this notation is still not in common use. The most commonly used notation is still the general prime notation even though it does not *define* the symmetry properties it only *implies* them. The main reason for the popularity of the prime system is the popular fallacy that the prime notation defines the symmetry characteristics. Another reason is that for small spin systems with low symmetry the HAIGH notation does not add greatly to the understanding of the spin system. Furthermore, since most organic molecules generally only have twofold symmetry, the point group label would normally not be used in the HAIGH notation. In this book however, the HAIGH nomenclature of spin systems will be used consequently, for completeness the old prime notation is mentioned.

Another thing to be considered is that the spin system point group symmetry is always a subgroup of the main molecular point group [3.2]. This means that the spin system can be of a lower order than the molecular symmetry, displaying only some of the main symmetry elements. Both the structures in Fig. 3.2 belong to the *point group* C_{2v} [3.3] and have the symmetry operations E, C_2, σ_v and $\sigma_{v'}$. The unity operation E leaves the molecule unchanged and the twofold rotation axis C_2 turns the isochronous spins into each other. The mirror planes σ_v and $\sigma_{v'}$ also include the rotation axis (v for vertical and not h for horizontal) One mirror plane is represented by the molecular plane which, like the unity operation E, leaves the molecule unchanged while the other, perpendicular to the molecular plane, has the same effect on the spin system as rotation. Because the point groups C_2 and C_s are subgroups of the main point group C_{2v}, the molecular point group contains both the symmetry groups used to define the spin system in this example. Using the higher C_{2v} symmetry would not simplify the spin system in any way and consequently, the spin system is completely defined as either C_2 and C_s. Both these point groups are equivalent as discussed at the beginning of this section. Full use of symmetry, to simplify spectral analysis, will be made in the following examples. In chapter 4 examples of inorganic compounds are given exhibiting higher symmetries.

3.1.2 The [AX]₂ Spin System

These symmetrical spin systems usually involve strong second order effects. Although the two non-isochronous spins can have a first order relationship (spins A and X) the isochronous spins are always second order and the coupling constants $J_{AA'}$ and $J_{XX'}$ influence the spectrum. Because A and A' are isochronous, they have no difference in resonance frequency and the spectral perturbation parameter $\lim (\lambda_{AA'}) = \infty$. The same is true for the spins X and X'. As an introduction into this, the simple AX spin system will be extended to build up an [AX]₂ spin system.

Check it in WIN-NMR:

Open... the experimental spectrum of *2-chloro-6-nitro-3-phenoxy-aniline* ...\AX\002001.1R in WIN-NMR. Switch to the **Analysis I Multiplets** mode and press the **Windaisy** button to show the iteration result of the AX spin system of two ortho-connected aromatic protons performed in chapter 2. Open the

3.1 Symmetry Effects

Report... box and **Export...** the spin system data to **WIN-DAISY**. Accept all appearing dialogs with **Yes**.

Check it in WIN-DAISY:

Run a spectrum **Simulation** and check that the expected pattern of two doublets for an AX spin system is obtained (**Display Calculated Spectrum**). We now need to extend this spin system so that it represents the 1,4-disubsituted benzene compound shown in the figure beside. Open the **Main Parameters** dialog box and alter the **Number of spins or groups with magnetical equivalent nuclei** from "2" into <4>. Click the **Next** button in the bottom line to of the window to open the **Symmetry Group** dialog box. The default symmetry group is always **C1** which only posses the identity operation and implies *no* symmetry.

Whenever a symmetry group other than C_1 is selected, clicking the **Ok** button opens the **Symmetry Description** dialog box where the permutation operators are to be defined. The **Symmetry Description** dialog box will then always contain a minimum of two entries. The first entry is always the identity operation E which defines the initial spin labeling and cannot be altered. The remaining line(s) always start with the symmetry descriptor (operation) and show how the labels in the identity row are changed by the symmetry operation.

Check it in WIN-DAISY:

Click any radio button in the **Symmetry Group** dialog box and then the **Ok** (or **Next**) button. Examine the symmetry operations in the **Symmetry Description** dialog box. Click on the **Previous** button to return to the **Symmetry Group** dialog box and repeat this process with a different point group.

The symmetry descriptors (operations) must be in the given order, because the WIN-DAISY calculation uses a character table of the corresponding point group symmetry based on the given order of symmetry operations. The Greek letter σ is symbolized by "sig", "S3^2" means S^2_3 and so on. For the symmetry operations refer to [3.3] or [3.4].

Check it in WIN-DAISY:

Finally return to the **Symmetry Group** dialog box and click the **C2** radio button and then the **Ok** button.

The initial AX spin system is referred to by the spin label number "1" and "2"; the nuclei added to the system are given the labels "3" and "4". You are free to assign these last numbers to the remaining protons either as shown in the figure or the other way round. The following discussion is based on the labels shown in the figure above.

Check it in WIN-DAISY:

In the **Symmetry Description** edit fields of the second row identified by the symmetry operation **C2**, enter the permutation operators. In the first edit column below the index 1 enter the destination spin label into which nucleus 1 is transferred after execution of the twofold rotation. Looking at the figure, the twofold rotation transfers spin **1** into spin <4>, spin **2** into <3> and at the same time spin **3** into <2> and spin **4** into <1>. Enter these values into the edit fields using the <TAB> key to switch between fields. As reference the required symmetry description is given in Fig. 3.3.

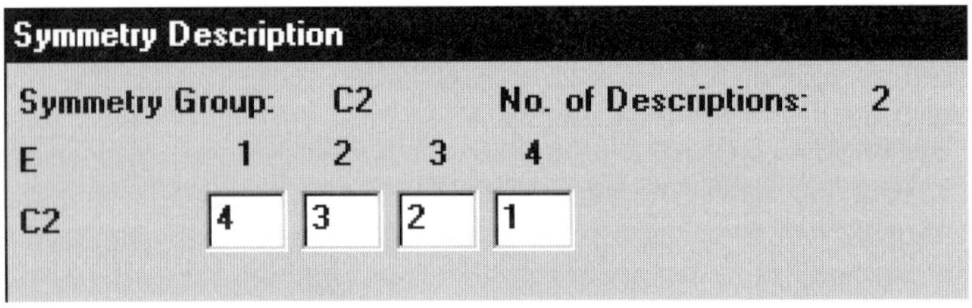

Symmetry Description				
Symmetry Group: C2			No. of Descriptions:	2
E	1	2	3	4
C2	4	3	2	1

Fig. 3.3: Symmetry Description for [AX]$_2$ spin system

Check it in WIN-DAISY:

Click the **Next** button to open the **Frequencies** dialog box. According to the symmetry properties of the spin system just defined, only the independent resonance frequencies are shown. Thus only the frequency v_A (index **1**) is displayed and the identical frequency $v_{A'}$ (index 4) is hidden. The same is true for v_X (index **2**) and $v_{X'}$ (index 3, hidden). Click on the **Next** button to open the **Scalar Coupling Constants** dialog box. This dialog box contains the minimum number of independent coupling constants needed for the spin system: $^3J_{AX}$ (J_{12}, ortho; = $^3J_{A'X'}$, J_{34}), $^5J_{AX'}$ (J_{13}, para; = $^5J_{A'X}$, J_{24}), $^5J_{AA'}$ (J_{14}, meta) and $^3J_{XX'}$ (J_{23}, meta). The coupling constant J_{AX} is filled in already because the original two-spin system already contained this ortho coupling constant. Without entering any values into the three remaining edit fields the original AX spin system will be reproduced. To verify this run a spectrum **Simulation** and display the experimental and simulated spectrum using the **Dual Display** mode of WIN-NMR. Toggle back to **WIN-DAISY**.

To define the complete spin system parameters open the **Scalar Coupling Constants** dialog box and enter appropriate values for the para coupling <0.5> Hz and for the meta couplings, e.g. <2.0> and <1.5> Hz. The meta coupling constants involve different coupling pathways and can have different values. **Change** the spectrum name to keep the original calculations, e.g. to ...\002799.1R. **Simulate** the spectrum and display the classic pattern, typical of 1,4-disubstituted benzene compounds in WIN-NMR **Dual Display**. Use the **Separate** button to display both spectra separately (Fig. 3.4b). Toggle back to **WIN-DAISY**.

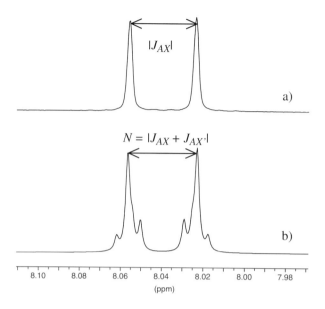

Fig. 3.4: a) Experimental X spectrum part of AX spin system.
b) Calculated X spectrum part for p-substituted phenyl fragment.

The AX lines (Fig. 3.4a) seem to be almost reproduced in the [AX]$_2$ spectrum (Fig. 3.4b), but the distance between the intense doublet lines, the *N-lines* is not J_{AX} but:

$$N = |J_{AX} + J_{AX'}|$$

The reason the line positions look similar is that the ortho coupling constant, J_{AX}, is considerably larger than the para coupling constant, $J_{AX'}$, so that the parameter N is in 1,4 disubstituted benzene compounds essentially controlled by J_{AX}.

Check it in WIN-DAISY and WIN-NMR:

Display the **Calculated Spectrum** [AX]$_2$ in WIN-NMR. In the **Analysis | Multiplets** mode, mark one of the N-lines with the spectrum cursor and use the **Free Grid** to measure the separation in Hz between the N-lines of a multiplet. Toggle to **WIN-DAISY** and run the **Calculator**. Open the **Scalar Coupling Constants** dialog box; add the ortho (J_{12}) and para (J_{13}) coupling constant together and compare this value with the determined value.

The spectrum of an [AX]$_2$ spin system consists of two halves, the A and the X resonances, which are mirror images of each other. Each half spectrum is composed of two (a$_2$) type subspectra and two (ab) type subspectra. Both (a$_2$) subspectra and one of the (ab) subspectra are called symmetrical, the other (ab) subspectrum is called anti symsymmetrical (ab)_. Separate plots of symmetrical and antisymmetrical subsystems are shown in Fig. 3.5.

138   3 Structure and Spin System Parameters

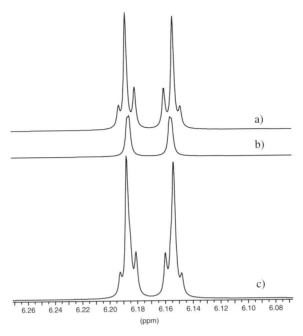

Fig. 3.5: a) Symmetrical subspectrum of [AX]$_2$ half spectrum.
b) Antisymmetrical subspectrum of [AX]$_2$ half spectrum.
c) Sum of a) and b) = [AX]$_2$ half spectrum.

Check it in WIN-DAISY:

Open the **Main Parameters** dialog box and check the **Output Option Output of subspectra** and **Simulate** the spectrum again. Click the **Generate Subspectra** button. As usual, the process number of the first subspectrum will be one less than the original simulated spectrum process number. Toggle to **WIN-NMR**.

Check it in WIN-NMR:

Open... the first subspectrum ...\AX\002798.1R. From the **Process** pull-down menu select the **File Algebra** option and load the second subspectrum ...\AX\002797.1R as the **second file**. Click the **Add./Sub.** button to *add* both subspectra together. The upper trace represents the symmetrical subspectrum, the second trace the antisymmetrical subspectrum and the lower trace the sum of both spectra (Fig. 3.5).

Quite often in asymmetrical 1,4-disubstituted compounds of [AX]$_2$ the antisymmetrical (ab)_ subspectrum is not resolved and is almost hidden under the two (a$_2$) subspectra (representing the N-lines). The repeated distance between the (ab)$_+$ lines refer to the parameter K which is the sum of the two meta coupling constants:

$$K = |J_{AA'} + J_{XX'}| = J_{(ab)+}$$

and the repeated distance between the (ab)_ subspectrum corresponds to the difference between both meta coupling constants:

$$M = |J_{AA'} - J_{XX'}| = J_{(ab)-}$$

Often M cannot be determined. The parameter L representing the difference between the ortho and para coupling constants can be extracted from the square root of the product of S_i and S_o of either the (ab)$_+$ or (ab)$_-$ subspectrum. In the present example of 1,4-disubstituted benzenes, the parameter values are such that only the (ab)$_+$ subspectrum lines are resolved:

$$L = |J_{AX} - J_{AX'}| = \sqrt{S_i \cdot S_o}$$

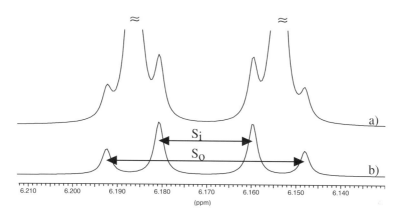

Fig. 3.6: a) [AX]$_2$ half spectrum, (a$_2$) transitions intensity cut.
b) (ab)$_+$ subspectrum.

Since the meta coupling constants do not differ by very much, the (ab)_ subspectrum parameter M is usually very nearly zero. Because the ortho coupling is much larger than the para coupling (), the relative signs of J_{AX} and $J_{AX'}$ can be determined: when $N>L$ the signs are the same, and when $N<L$ they are opposite.

If the parameters N,L,K and M can be determined the four coupling constants can be calculated.

$$J_{AX} = \tfrac{1}{2}(N+L) \quad J_{AX'} = \tfrac{1}{2}(N-L)$$
$$J_{AA'} = \tfrac{1}{2}(K+M) \quad J_{XX'} = \tfrac{1}{2}(K-M)$$

Check it in WIN-DAISY, WIN-NMR:

Run the **Calculator** from the WIN-DAISY toolbar. **Display** the **Calculated [AX]$_2$ Spectrum** in the spectrum window of WIN-NMR. From the **Analysis** pull-down menu choose the **Peak Picking** option; switch the **Peak Labels** to **Hz** and **Execute** a peak picking with **Interpolation**. Zoom in either the A or X resonance half spectrum and type <Alt + TAB + TAB> to switch to the WINDOWS Calculator to determine the values of N, L and K. Determine the relative signs of the ortho and para coupling constants and calculate the coupling constants using the equations above.

Compare the values with the parameters defined in the WIN-DAISY **Scalar Coupling Constants**. **Change** the calculated spectrum name and examine the effect of the signs of J_{12} and J_{13}, e.g. both negative or one positive and one negative, and compare the new **Simulations** with the original simulation using the Dual Display mode of WIN-NMR (**Display** the **Calculated Spectrum** with WIN-NMR and load the original calculation …\002799.1R as the second trace into the **Dual Display**). Toggle back to **WIN-DAISY**.

Exchange the values of the meta coupling constants J_{14} and J_{23}. Check the effect on the spectrum. Repeat this for the ortho and para coupling constant J_{12} and J_{13}. **Display** the **Calculated Spectra** in comparison with the original simulation …\002799.1R in WIN-NMR Dual Display.

Since the parameter M cannot be determined from the spectrum it implies that both the meta coupling constants are quite similar, and for all practical purpose they can be considered almost equal. The simulations with exchange of the meta coupling constants with each other and the ortho and para values with each other, result in identical spectra showing that the spectrum is independent of the assignment of these coupling constants. The reason for this is clear when the parameters N, L, K and M and the spin labels used in the original structure are examined (Fig. 3.7a). N, L, K and M are absolute values while exchanging the spin labels 2 and 3 interchanges the indices of the coupling constants such that J_{12} now corresponds to the para coupling $J_{AX'}$ and J_{13} to the ortho coupling J_{AX} - the meta couplings $J_{AA'}$ and $J_{XX'}$ are of course left unaltered (Fig. 3.7b). Finally, while the ortho and para couplings can be assigned, it is not possible to assign the two meta couplings unambiguously and this should always be kept in mind when analyzing [AX]$_2$ spectra.

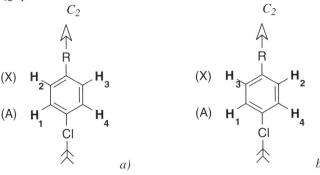

Fig. 3.7:   a) Original labels   b) Equivalent labeling

So far, the point group C_2 has been used in the discussion of the [AX]$_2$ spin system but the point group C_S is equally as well.

Check it in WIN-DAISY and WIN-NMR:

Toggle back to **WIN-DAISY**. Change the spectrum name, e.g. to ...\002699.1R. Open the **Symmetry Group** dialog box and change the point group from **C2** to **Cs**. Press the **Ok** or **Next** button. The **Symmetry Description** dialog box appears automatically on the screen. This point group contains two entries: the identity operation **E** that defines the spin labeling and the reflection plane **SigH** (σ_H). The effect of either rotation or reflection on the spin system is the same. Check this by inspection of the **Frequencies** and **Scalar Coupling Constants** dialog boxes. Verify that the C_2 and C_s point group give the same result by running a **Simulation** and **Display**ing the **Calculated Spectrum**. Then load in WIN-NMR the original calculated spectrum as the second trace into the **Dual Display**. Toggle back to **WIN-DAISY**.

Finally the molecular symmetry C_{2v} will be used to demonstrate that there is no benefit in using the highest molecular symmetry, but the spin symmetry is sufficient.

Check it in WIN-DAISY:

Open the **Symmetry Group** dialog box and select the point group **C2v**. Press the **Ok** or **Next** button. The displayed **Symmetry Description** dialog box contains four rows according to the four symmetry operations. The first two lines are kept: the identity: **E** representing the spin labeling and the twofold rotation axis **C2** with the permutations "4", "3", "2" and "1". In the next lines permutations for the first vertical reflection plane **SigV** (σ_v) and second vertical reflection plane **SigV'** ($\sigma_{v'}$) must be entered. "Vertical" means that they include the axis C2 (and are not perpendicular to it). The order which reflection is first or second is ambiguous. You may give first the reflection representing the molecule plane: <1>, <2>, <3> and <4> and then the reflection being perpendicular to the molecule plane: <4>, <3>, <2> and <1> (Fig. 3.8) or vice versa and quit the dialog with **Ok** to accept the values.

Open the **Frequencies** and **Scalar Coupling Constants** dialog boxes and prove that the parameter structure remains identical. Open the **Main Parameters** dialog box and make sure that the **Output Option Output of subspectra** is selected. Run a **Simulation** and **Generate Subspectra**. The simulation status informs that four subspectra were calculated. **Display** the **Calculated Spectrum** with the total simulation ...\AX\002998.1R by pressing the red peak button.

Check it in WIN-NMR:

Load the original calculation (with C_2 or C_s symmetry) ...\AX\002799.1R as the **second file** into WIN-NMR **Dual Display**. Then load the four calculated subspectra ...\AX\002698.1R up to ...\AX\002695.1R into the **Multiple Display** and use the **Separate** option. Press the **Lock X** and **Lock Y** buttons and **zoom** in one half spectrum.

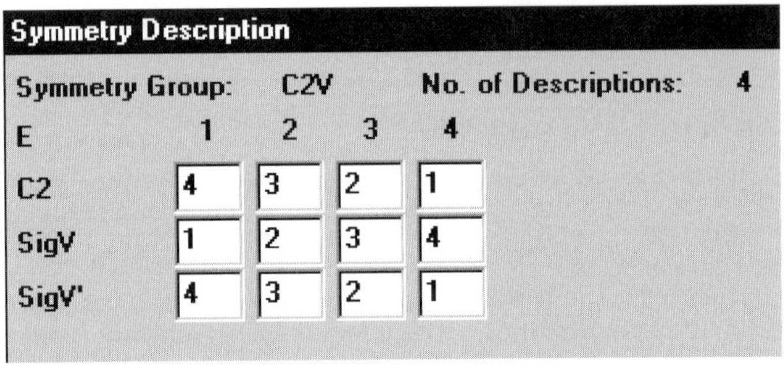

142   3 Structure and Spin System Parameters

Symmetry Group:	C2V				No. of Descriptions:	4
E	1	2	3	4		
C2	4	3	2	1		
SigV	1	2	3	4		
SigV'	4	3	2	1		

Fig. 3.8: Symmetry Description dialog box for C_{2v} point group symmetry.

Obviously two of the subspectra do not contain any transition and the other two spectra are identical with the subspectra obtained when using the spin system point group symmetries either C_2 or C_s. Here it is shown that the high molecular symmetry has no advantage over the spin system symmetry. This can be seen in the **Symmetry Description** dialog box where the permutations **E** and **SigV** as well as **C2** and **SigV'** are identical and also in the **Frequencies** and **Scalar Coupling Constants** which already show the minimum number of parameters when using C_2 or C_s symmetry.

Check it in WIN-NMR:

Open... the spectrum ...\AX6\001001.1R in WIN-NMR. The aliphatic region of this compound has already been analyzed, but we now want to focus on the aromatic region. Measure a suitable linewidth in the **Analysis | Linewidth** mode and then switch to the **Analysis | Multiplet** mode. Click the **Delete All** button to remove the aliphatic multiplets and then zoom in on the aromatic region. Define for each [AX]$_2$ half spectrum a doublet based on the N-lines and connect the two distances in order to define a pseudo AX-spectrum. **Return** to normal display under saving the new Multiplets data (**Yes**).

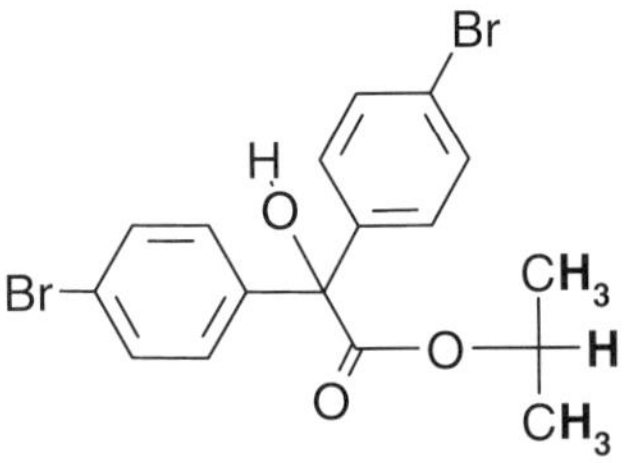

4,4-dibromo benzilic acid isopropyl ester

Switch to the **Analysis | Interval** mode and use the **perpendicular cursor** mode to define a spectral region for the aromatic signals. **Export...** the interval and multiplets data (**Yes**) to **WIN-DAISY**.

Check it in WIN-DAISY:

Open the **Main Parameters** dialog box and extend the **Number of spins or groups of magnetical equivalent nuclei** in the from "2" to <4>. Click the **Next** button and select the **C2** radio button in the **Symmetry Group** dialog box. Click the **Ok** or **Next** and in the **Symmetry Description** dialog box enter the permutations in the second row identified by the symmetry operation **C2** (<4>, <3>, <2>, <1>).

In the **Frequencies** dialog box the two independent resonance frequencies are shown. The **Scalar Coupling Constants** dialog box contains the four independent coupling constant edit fields in which only the first number is non-zero, showing the *N* parameter determined in the multiplets mode. Enter approximate values for the para (e.g. <0.5> Hz) and both meta coupling constants (e.g. <2.0> Hz) and press the **Iterate all** button to select all coupling constants for iteration. Call the **Control Parameters** and select the **Output Option** to report the **Correlation matrix**. Execute a spectrum **Iteration**.

The calculation is soon finished and a comparison of calculated and experimental spectrum in the WIN-NMR **Dual Display** shows no improvement. Load the **Iteration Protocol** file and scroll to the output of the *Parameter Correlation Matrix*.

The parameter correlation matrix describes the inter-dependency of the optimized parameters (consecutively numbered) in a triangular form. Large off-diagonal elements show a strong inter-dependency of the related parameters. Inspection of the matrix entries shows a value of "−1.0" for elements 5-6, the correlation between the two meta couplings, indicating a perfect negative correlation such that as one meta coupling increases the other decreases. This negative correlation causes linear dependent lines in the matrix of the second derivatives which becomes singular. (When this occurs, a warning is printed in the iteration protocol file.) The program tries to overcome this problem mathematically but in the in the present case it does not succeed.

Check it in WIN-DAISY:

To help the program to overcome the singularity problem modify one of the meta coupling constants in the **Scalar Coupling Constants** dialog box e.g. change either J_{14} or J_{23} to <1.5> Hz and rerun an **Iteration**. Examine the **Iteration Protocol** file to determine whether the spin labels have been exchanged (visible in the size of the coupling constants: J_{12} refers to as ortho coupling and J_{13} as para or vice versa), before inspecting the result using the **Dual Display** mode of WIN-NMR.

By permutating the spin labels with initially different meta couplings the program is able to find one of the correct solutions (refer to Fig. 3.7).

Check it in WIN-DAISY:

Another way to perform the analysis is the iteration using smoothing parameters. Open the **Scalar Coupling Constants** dialog box and change all coupling constant values to <0.0>. In the **Control Parameters** dialog box leave the **Mode** to **Standard Iteration** and select the smoothing parameters, **Broadening** and **No. of Cycles**, any value except none will be suitable for these options. Click the **Scalar Coupl.** button and set the iteration limits for the coupling constants, e.g. for each coupling enter the values <0> for **Lower limit** and <12> for **Upper limit**. Click the **Ok** button and run a spectrum **Iteration**. Check the standard deviation values for the meta couplings in the **Iteration Protocol** file. If these values are high, return to the **Scalar Coupling Constants** dialog box and modify the value of one of the meta couplings. Rerun the **Iteration** and examine the results.

To store aliphatic and aromatic iteration results with the spectrum do as follows:

Check it in WIN-DAISY and WIN-NMR:

Close all other documents *except* the aromatic spin system of ...\AX6\001001.1R, if there are any present. **Display** the **Experimental Spectrum** in WIN-NMR. Enter the **Analysis | Multiplets** mode and select the **Windaisy** option. Open the **Report...** box and **Export...** the aliphatic optimized spin system parameters back to **WIN-DAISY** (answer all questions with **Yes**).

The aliphatic multiplets are now present in a separate WIN-DAISY document. *Two* documents shall be present now in WIN-DAISY (the aliphatic and aromatic spin system of spectrum ...\AX6\001001.1R). Use the **Edit Copy fragment to...** command (the forth button from the left in the toolbar) and accept the appearing dialog with **Ok**. **Close** the actual document. The remaining document contains now both fragments. Run a **Simulation** and display both spectra in WIN-NMR **Dual Display**. Compare the signal intensities.

The simulation differs in the integral of the aromatic spectral region. If you remember the molecular structure of the compound under investigation, it becomes clear, that the aromatic spin system is contained twice in the structure.

Check it in WIN-NMR and WIN-DAISY:

Toggle back to **WIN-DAISY** and open the **Main Parameters** of the aromatic fragment and enter a **statistical weight** of <2>. Run a **Simulation** and use the **Dual Display** command. Toggle to **WIN-DAISY**. Execute the command **Export WIN-NMR exp.** to store the result with the experimental spectrum.

The present $[AX]_2$ example already shows a second order effect between A and X, because the N-lines do not have equal intensity but show a distinct roof effect towards the other half spectrum (refer to Fig. 3.9).

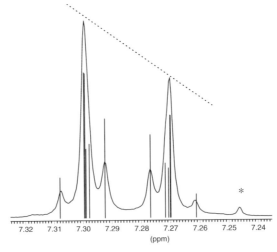

Fig. 3.9: Experimental high-field $[AX]_2$ half spectrum with stick spectrum below and visualized roof effect, *$CHCl_3$ impurity.

Check it in WIN-NMR and WIN-DAISY:

To derive the symmetrical 1,2-disubstituted [AX]$_2$ spectrum reload again the experimental spectrum of 2-chloro-6-nitro-3-phenoxy-aniline named ...\AX\002001.1R, into the spectrum window of WIN-NMR. Switch to the **Analysis | Multiplet** mode and click on the **Windaisy** button. **Export...** the optimized AX spin system to **WIN-DAISY**. Proceed in a similar manner to the 1,4-disubstituted spin system discussed at the beginning of this section.

In the **Main Parameters** dialog box, extend the **Number of spins or groups with magnetical equivalent nuclei** from "2" to <4>, select twofold **C2** symmetry in the **Symmetry Group** dialog box and set the **Symmetry Description**. Open the **Scalar Coupling Constants** dialog box and enter a meta coupling constant for J_{13} (e.g. <2.0> Hz, a para constant for J_{14} (e.g. <0.5> Hz) and another ortho coupling for J_{23} (e.g <8.0> Hz).

Change the spectrum name, e.g. to ...\002599.1R. Run a spectrum **Simulation** and display the experimental AX spectrum together with the calculated symmetrical ortho-disubstituted benzene type spectrum in **Dual Display** of WIN-NMR.

The only difference between the simulation of both types of [AX]$_2$ spectra is found in the *values* of the coupling constants. The asymmetric para-substituted phenyl type features one ortho, two meta and one para coupling constant, the present symmetrical ortho-disubstituted type shows two ortho, one meta and one para coupling constant. The signal patterns for 1,2-disubstituted and 1,4 disbustituted systems is so distinctive, it is very easy to distinguish between them.

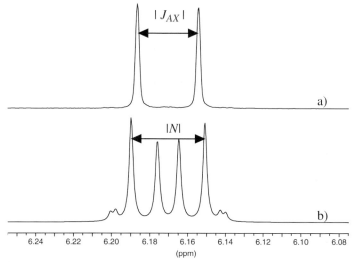

Fig. 3.10: a) Experimental AX half spectrum.
b) Calculated [AX]$_2$ half spectrum including both ortho couplings.

Check it in WIN-DAISY:

Open the **Main Parameters** dialog box and select the **Output Option Output of subspectra**. Run a spectrum **Simulation** to generate the frequency list ordered according to subspectra. Click the **Generate Subspectra** button. Again the process number of the first subspectrum will be one less than the original simulated spectrum process number. Toggle to **WIN-NMR** and load the first subspectrum into the spectrum window (e.g. if original spectrum number was ...\AX\002599.1R, load the spectrum ...\AX\002598.1R). From the **Process** pull-down menu select the **File Algebra** option and load the second file ...\AX\002597.1R. Click the **Add./Sub.** button to add both subspectra together. The upper trace represents the symmetrical (ab)$_+$ and two (a$_2$) subspectra and the second trace the antisymmetrical (ab)$_-$ subspectrum and the lower the sum of both spectra. (Fig. 3.11).

As can be seen in Fig. 3.11b, the inner transitions of both subspectra are almost degenerate but unlike the 1,4-disubstituted [AX]$_2$ systems, the outer transitions of the (ab)$_-$ subspectrum are clearly resolved. Each of the two N-lines are again composed of the two degenerate transitions of the (a$_2$) subsystems. Each half spectrum consists altogether of 12 transitions. Due to the degeneracy of the (a$_2$) and almost degeneracy of the inner lines of the (ab) subspectra, normally 8 resolved lines are detected in this type of spectra.

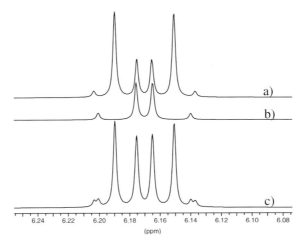

Fig. 3.11:   a) Symmetrical subspectrum (half) of [AX]$_2$.
             b) Antisymmetrical subspectrum (half) of [AX]$_2$.
             c) Sum of a) + b) = total spectrum (half) of [AX]$_2$.

Check it in WIN-DAISY and WIN-NMR:

Toggle back to **WIN-DAISY**. Run the **Calculator** and press the red **Display Calculated Spectrum** button to transfer the complete calculated spectrum to WIN-NMR. Switch into the **Analysis | Peak Picking** mode and perform a peak picking. Using the values in the **Report...** box calculate the parameters N, L, K and M (use <Alt + TAB + TAB> to move the Calculator to the foreground). Using these parameters, determine the four coupling constants J_{AX}, $J_{AX'}$, $J_{AA'}$ and $J_{XX'}$ and compare them with the entries in the **Scalar Coupling Constants** dialog box in WIN-DAISY.

The parameter relation N > L and K ≈ M describes this type of disubsituted benzene compound. N is the sum of one ortho and one meta coupling constant (L is their difference) and K the sum of the other ortho and the para coupling constant. Because the para coupling is small compared to the ortho coupling the parameters K and M are very similar in size. All the parameters are determined as absolute values.

Check it in WIN-NMR and WIN-DAISY:

Open... the proton spectrum ...\SYMMETRY \001001.1R of the naphtoquinone derivative in WIN-NMR. Measure a suitable **Linewidth** and then in the **Analysis I Multiplet** mode define two doublets to represent the parameter N. **Export...** the data to **WIN-DAISY** and build up the four-spin system as described in the previous "Check its".

2,3-dichloro-1,4-naphtho quinone

Open the program **Options** and give **Iteration limits** for **Scalar Coupling Constants** to ± <12> Hz. In the **Control Parameters** dialog box, select smoothing parameters (**Broadening** and **No. of Cycles**) and click the **Scalar Coupl.** and **Select all** four coupling constants for iteration and check the iteration limits. Run a spectrum **Iteration** and examine the results.

The $[AX]_2$ spin system just discussed includes first order relation between spins A and X and second order relation between A and A' as well as between X and X'. The spectrum appearance is independent of the difference between the resonance frequencies of spin A and X, Δ_{AX}. It has already been mentioned, that a roof effect in the N-lines indicates a second order influence in the spin system.

Check it in WIN-NMR and WIN-DAISY:

Open... the experimental 300 MHz proton spectrum ...\SYMMETRY\002001.1R of 1,2-dichlorobenzene with WIN-NMR. Measure a **Linewidth** and then define two doublets in the **Analysis I Multiplet** mode. Open the **Report...** box and connect the "coupling constants".

1,2-dichlorobenzene

The question marks **?** that appear beside the connected couplings indicates a second order effect (although the distance measured in the Multiplets mode is not a coupling constant but the parameter N, the sum of an ortho and meta coupling). Mark one peak of the spin system with the spectrum cursor.

Export... the data to **WIN-DAISY** and build up the four-spin system with twofold symmetry. Without entering any other coupling constant values, run a spectrum **Simulation** without entering any additional coupling constant and display both spectra in the **Dual Display** mode of WIN-NMR.

The result is shown in Fig. 3.12. Because no other coupling constant has been entered, the calculated spectrum (Fig. 3.12b) only contains the N-lines. Because the parameter N represents the sum of J_{AX} (ca. 9.5 Hz) and $J_{AX'}$ (ca. 0 Hz) the lines almost exactly reproduce the roof effect. By zooming in one half spectrum a slight shift in the line positions becomes visible indicating the start of second order effects.

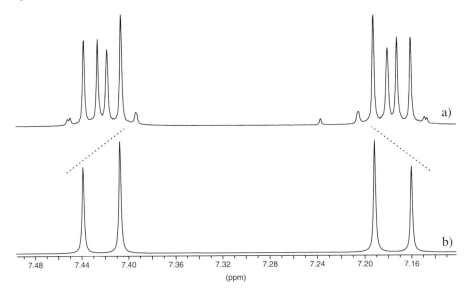

Fig. 3.12: a) Experimental *1,2-dichlorobenzene* spectrum.
b) Calculated spectrum representing solely the N-lines with roof effect.

Check it in WIN-DAISY:

Call the **Options** from the toolbar and give default iteration limits for Scalar Coupling constants of e.g. ± <10> Hz. In the **Control Parameters** dialog box select smoothing parameters. Click the **Scalar Coupl.** button and **Select all** four coupling constants for iteration and check the iteration limits. Run an **Iteration** and inspect the result in the **Dual Display** mode of WIN-NMR. Toggle back to **WIN-DAISY**.

Export the iteration result from WIN-DAISY back into **WIN-NMR exp**. **Zoom** in one half spectrum (Fig. 3.13). A close examination shows that the lines are almost degenerate but this degeneracy will be removed the stronger the second order effects become.

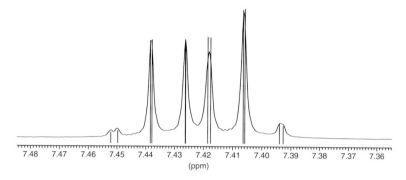

Fig. 3.13: Half spectrum with discrete transitions lines as stick spectrum

3.1.3 The [AB]$_2$ spin system

There are two ways to modify an [AX]$_2$ spin system to obtain an [AB]$_2$ spectrum: a simulation sequence that moves the frequency of X towards the frequency of A or the reduction of the spectrometer basic frequency to modify the A and X shifts at the same time. First we are going to consider asymmetric 1,4-disubstituted phenyl compound systems.

Check it in WIN-DAISY and WIN-NMR:

Open... the experimental spectrum ...\AX6\001001.1R in WIN-NMR, call the **Analysis | Multiplet** mode and press the **Windaisy** button. **Export...** the data to **WIN-DAISY**. From the Edit pull-down menu use the **Disconnect** option to disconnect the experimental spectrum from the spin system data and **Delete** the aliphatic fragment. To restore the symmetry properties open the **Symmetry Group** dialog box and select **C2**. And enter a proper **Symmetry Description**. Open the **Spectrum Parameters** dialog box and press the **Auto Limits** button. Use the **Next** button to open the **Control Parameters** and select the **Sequence Simulation** mode. Change the **Default number of steps** to <16> and the **Default Parameter Increment [Hz]** to <3.0>. Press the group button **Frequencies** button and select **Spin 2** for variation (only two shifts are displayed here, otherwise no symmetry has been defined. Accept the dialog boxes with **Ok**. Change the spectrum name, e.g. to ...\001899.1R and run the **Simulation Sequence**. Display the reference **Calculated Spectrum** with WIN-NMR and switch into the **Multiple Display** mode. If necessary, click the **Delete All** button before selecting all 17 simulation traces (...\001899.1R to ...\001883.1R) for display. Click the **Stack Plot** button and zoom in on the left hand side of the three spectra nearest the bottom of the display window. Click the ***2** button until a suitable y-gain is obtained. Use the vertical slider to scroll up the multiple display. In this way the stepwise removal of the degeneracy in lines becomes visible. The uppermost spectrum shows only an intense singlet peak with very weak outer lines in a deceptively simple spectrum type.

Toggle back to **WIN-DAISY**. Open the **Main Parameters** dialog box and select the **Output Option Output of subspectra**. Call the **Spectrum Parameters** and change the **Frequencies adapted to Spectrometer**

Frequency to **80 MHz**. Answer **Yes** in the message box that appears to adapt the spin system to the new spectrometer frequency. Call again the Spectrum Parameters and select the **Auto Limits** option. Change the spectrum name, e.g. to ...\001799.1R. Run a **Simulation** and then click the **Generate Subspectra** button. **Display** the 80 MHz **Calculated Spectrum** in WIN-NMR.

Switch into the **Dual Display** mode and load the first subspectrum (...\001798.1R) as the **second file**. Press the **Show Diff.** button to display the antisymmetrical (ab)_ subspectrum.

In comparison with the [AX]$_2$ spectrum type the overall appearance does not change dramatically. The antisymmetrical subspectrum does move slightly away from the N-lines position and is better resolved. Besides the proximity of the resonance frequencies, the most significant change in the spectrum is the removal of the degeneracy of the "inner N-lines" (the A N-line and the X N-line, that finally overlap in the sequence simulation) by second order effects.

Based on these observations, the analysis of this type of [AB]$_2$ spectra is relatively straight forward. Start the analysis in the same way as for the first order [AX]$_2$ case by defining a doublet in each half spectrum to represent the N-lines separation.

Check it in WIN-NMR and WIN-DAISY:

Open... the spectrum ...\SYMMETRY\003001.1R in WIN-NMR. Measure a **Linewidth** and define two doublets in the **Multiplet** to represent the N-lines. Open the **Report...** box and connect the coupling constants. Note that a question mark is displayed.

4-bromobenzonitrile

Export... the data to **WIN-DAISY**. Extend the **Number of spins or groups with magnetical equivalent nuclei** to <4> and select twofold symmetry (**C2** or **Cs**) and define the permutations. Call the **Options** dialog and define the **Scalar Coupling Constants parameter limits for Iteration** to ± <10> Hz. Open the **Scalar Coupling Constants** dialog box and press the **Iterate all** button. Call the **Control Parameters** and select fairly strong smoothing parameters, e.g. **high Broadening** and **high No. of Cycles** Run a spectrum **Iteration**. If the *R-Factor* in the **Iteration Protocol** is larger than *1.25%*, change the value of one of the meta coupling constants (J_{14} or J_{23}) back to <0> and run the **Iteration** again.

Slight differences between the experimental and optimized spectrum will remain, because of the distorted lineshape of the experimental signals caused by shimming problems.

The conversion of a [AX]$_2$ spin system into a [AB]$_2$ spin system for symmetrically 1,2-disubstituted benzene compounds will now be considered.

Check it in WIN-NMR and WIN-DAISY:

Open... the 300 MHz experimental spectrum of *1,2-dichlorobenzene*, ...\SYMMETRY\002001.1R with WIN-NMR. Switch to the **Analysis | Multiplet** mode and press the **Windaisy** button. **Export...** the data to **WIN-DAISY** and apply either **C2** or **Cs** symmetry and enter the permutations. Open the **Lineshape Parameters** and enter a small **Global Linewidth**, e.g. <0.3> Hz. Call the **Control Parameters** and select the **Sequence Simulation** mode. Change the **Default number of steps** to <15> and the **Default parameter increment** to <5> Hz. Press the **Frequencies Group Parameter** button and select the **Spin 2**. Accept the dialog boxes with **Ok** and run a **Sequence Simulation**. Toggle to **WIN-NMR** and call the **Multiple Display** mode and, if necessary, click the **Deselect All** button before selecting all 16 simulation traces for display. Click the **Stack Plot** button. Again, zoom in on the left hand side of the three spectra nearest the bottom of the display window, and after adjusting the y-gain (<Ctrl + Y>, *2), scroll up to get an impression of how the signals change during the simulation sequence. The removal of the degeneracy is clearly visible.

Inspection of the simulations leads to the observation that the maximum number of transitions for a [AB]$_2$ spectrum is 24 lines - 12 lines for each half spectrum. The stronger the second order effect, the more the half spectra are shifted towards each other until they overlap completely giving a deceptively simple spectrum.

Check it in WIN-NMR and WIN-DAISY:

Open... the 13th simulation (...\SYMMETRY\002987.1R) with WIN-NMR. From the **Output** pull-down menu select the **Title...** option. The title string contains the frequency of the second spin, "2213.9872" Hz. Paste this value into the edit window for spin **2** in the WIN-DAISY **Frequencies** dialog box. **Disable** either spin **1** or **2**. Change the spectrum name, e.g. to ...\SYMMETRY\002899.1R. Run another **Simulation**. Toggle back to **WIN-NMR** *without* loading any new spectrum (to keep the previously loaded) and in the **Dual Display** mode load the second simulation ...\SYMMETRY\002899.1R, with suppression of one half spectrum, as the **second file**. The display (Fig. 3.14) shows quite clearly that the inner lines of both half spectra overlap. Toggle back to **WIN-DAISY**.

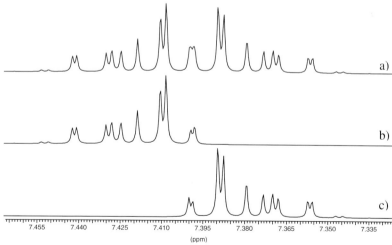

Fig. 3.14: a) Total simulation of [AB]$_2$ spectrum.
b) Simulation of one half spectrum by suppression of spin 2.
c) Difference spectrum, showing the other half spectrum.

Check it in WIN-DAISY:

Open the **Main Parameters** dialog box. Select the **Output Option Output of subspectra**. Open the **Frequencies** dialog box and remove the **disable** flag set in the previous "Check it". Run a spectrum **Simulation** and then click the **Generate Subspectra** button. Toggle to **WIN-NMR**.

Check it in WIN-NMR:

Open... the first subspectrum ...\SYMMETRY\002898.1R and select in the **Dual Display** mode the second subspectrum ...\SYMMETRY\002897.1R as the second file (Fig. 3.15). The simulation of the subspectra shows that the degeneracy is removed for the inner lines of the (ab) subspectra and that the antisymmetrical (ab)_ subspectra of both half spectra are shifted toward each other that they are nested.

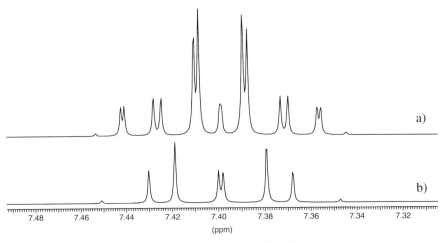

Fig. 3.15: a) Simulation of symmetrical part of [AB]$_2$ spectrum.
b) Simulation of antisymmetrical part of [AB]$_2$ spectrum, representing the two anti symmetrical (ab) subspectra.

All the simulations in the simulation sequence were based on the modification of the chemical shift difference - all coupling constants were kept constant. A strong second order effect results in the overlapping of the signals belonging to the different spins A and B. This signal overlap should be kept in mind when analyzing [AB]$_2$ spin systems.

Check it in WIN-NMR:

Open... the experimental proton spectrum ...\SYMMETRY\004001.1R with WIN-NMR. Measure a suitable **Linewidth** and then switch into the **Analysis I Multiplets** mode.

The spectrum displays a strong second order effect so that an analysis based on the N-lines would make little sense.

1,2-dihydroxybenzene

Define two **Singlets** (context menu when clicking the right hand mouse button), one in each half spectrum to estimate the A and B shifts. The selected lines for the singlets should *not be exactly* in the center of each half spectrum, but shifted slightly towards the other resonance. Remember that in a AB spectrum the resonance frequency is not the center of each pair of lines, but is shifted towards the other resonance frequency. **Export...** the data to **WIN-DAISY** *without* fragmentation (**No**). The spin system will be built up using only the chemical shifts defined by the singlets.

Expand the spin systems in the **Main Parameters** dialog box to <4> spins. Select **C2** or **Cs** symmetry in the **Symmetry Group** window and in the second row in the **Symmetry Description** dialog box enter <4>, <3>, <2> and <1>. In the **Scalar Coupling Constants** dialog box enter some approximate values for 1,2-disubstituted aromatic compounds, e.g. <8>, <1.5>, <0.5> and <7.5> Hz. Run a spectrum **Simulation** and compare both spectra in the **Dual Display**. Toggle back to **WIN-DAISY**.

Open the **Control Parameters** dialog box and select smoothing parameters, e.g. **low Broadening** and **low No. of Cycles**. Press the **Scalar Coupl.** button and click on the **Select All** button. Run a spectrum **Iteration**. The result is acceptable if the *R-Factor* in the **Iteration Protocol** is below 0.5%. Examine the result in the **Dual Display**.

After a satisfactory result has been obtained, click the red **Display calculated spectrum** button to display the spectrum in WIN-NMR. Toggle back to **WIN-DAISY** and **Change** the calculated spectrum name and **Disable** one of the spins in the **Frequencies** dialog box and **Simulate** the spectrum. Toggle to WIN-NMR and load the actual calculation as the **second trace** into the **Dual Display** and **Show Diff**. The overlap of A and B parts is now clearly visible (Fig. 3.16).

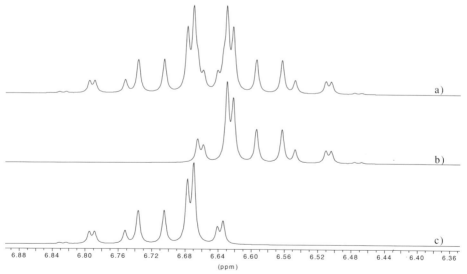

Fig. 3.16: a) Total simulation of [AB]$_2$ spectrum.
b) Simulation of one half spectrum by suppression of spin 1.
c) Difference spectrum representing the other half spectrum.

Check it in WIN-NMR and WIN-DAISY:

Open... the 250 MHz experimental proton spectrum ...\SYMMETRY\005001.1R in WIN-NMR. Measure a suitable **Linewidth** and then switch into the **Analysis | Multiplets** mode.

This molecule exhibits as well a twofold symmetry. How would you label the spin system?

1,2,3-trihydroxybenzene

Decide the most appropriate way to analyze the aromatic part of the spectrum. **Export...** the data from the **Report...** box to **WIN-DAISY** in an appropriate way. Prepare the spin system data so that an **iteration** can be performed. Display the result in the **Dual Display** mode of WIN-NMR.

What comments can you make about the meta coupling constant?

Check it in WIN-NMR and WIN-DAISY:

Then load the 80 MHz experimental spectrum of the same compound ...\SYMMETRY\006001.1R into the spectrum window of WIN-NMR. Measure a suitable **Linewidth** and analyze the aromatic region of the spectrum in the **Analysis | Multiplets** mode in the same way. Open the **Report...** box to **Export...** the data to **WIN-DAISY**. Modify the spin system and optimize the parameters via **Iteration**. Does the low field spectrum contain more information than the high field spectrum analyzed before?

Due to the twofold symmetry of the compound, the two protons next to the chlorines are magnetically equivalent, because the show the *same* coupling to the proton situated in between. Therefore the meta coupling constant cannot be determined in the spectrum! There is no advantage to record the spectrum at lower frequency, because only the second order effect is increased in the AB_2 spin system, the meta coupling constant remains non-determinable. Always decide first whether the spins are only isochronous or even magnetically equivalent before defining the spin system. Symmetry in spin systems can result in isochronicity or magnetic equivalence of nuclei.

Check it in WIN-NMR:

Open... the experimental spectrum of *1,2-dimethoxybenzene* showing the aromatic part ...\SYMMETRY\007001.1R in WIN-NMR. Measure a suitable **Linewidth** and switch to the **Analysis | Multiplets** mode.

Again **Define** two **Singlets**, one in each half spectrum to estimate the chmical shifts of A and B. **Export...** the data to **WIN-DAISY** without fragmentation. Build up a 4-spin twofold system with twofold symmetry.

1,2-dimethoxybenzene

Check it in WIN-DAISY:

3.1 Symmetry Effects

Using this spin system as a model, a different way to analyze the spectrum will be tested. Display the program **Options** and change the iteration limits for **Scalar Coupling Constants** from ± "2" Hz to ± <10> Hz. Leave the dialog box with **Ok**. Open the **Control Parameters** and press the **Scalar Coupl.** button and **Select all** for iteration. Click the **Ok** button. Change the **Mode** to **Advanced Iteration** and select smoothing parameters, e.g **high Broadening** and **medium No. of Cycles**. Quit the dialog with **Ok** and open the Scalar Coupling Constants. Enter random values for the couplings, e.g. <4>, <3>, <2> and <1> to give non-zero values which are all different. This is necessary for the Advanced Iteration mode.

Run a spectrum **Iteration**. The result is acceptable if the *R-Factor* in the **Iteration Protocol** list is below *0.54%*. Examine the result using the **Dual Display** mode of WIN-NMR. If a satisfactory result is not yet obtained, reset one of the large coupling constants to zero and run another **Iteration** until a reliable solution is obtained.

When a satisfactory result has been obtained, check out the overlapping of the signals by disabling either the A or the B spin as described in previous "Check its".

Check it in WIN-NMR:

Open... the experimental 80 MHz spectrum of 1,2-dichlorobenzene ...\SYMMETRY\008001.1R in WIN-NMR. The spectrum is referenced to be centered around zero, because it will be used later in an analytical solution.

Measure a suitable **Linewidth** and then switch into the **Analysis | Multiplets** mode.

1,2-dichlorobenzene

Define again two **Singlet**s as a starting point for the chemical shifts (e.g at about –7.5 and +7.5 Hz) and **Export...** the data to **WIN-DAISY**.

Check it in WIN-DAISY:

Build up the 4-spin system with twofold symmetry as described previously. Set up the Program **Options** and **Control Parameters** as in a previous "Check it". Because of the very sharp lines in the experimental spectrum strong smoothing parameters are necessary to optimize the parameters via total line-shape fitting. Thus select **Advanced Iteration Mode** and define **high Broadening** and **high No. of Cycles**. Check the option **Starting Frequencies poor**. For the selected Advanced iteration mode the starting values for the coupling constants should not be zero as discussed in the previous "Check it". Open the **Scalar Coupling Constants** dialog box and enter different values for all the couplings, e.g. <1>, <2>, <3> and <4>. Run a spectrum **Iteration**. The result is acceptable if the *R-Factor* in the **Iteration Protocol** file is below *1.74%*. Examine the result using the **Dual Display** mode of WIN-NMR.

If a satisfactory result is not obtained, change the value of one or more coupling constants in the **Scalar Coupling Constants** dialog box. It is possible that the calculation might fall into a local minimum defined by coupling constants with values of about 1, 5, 3 and 3 Hertz. In order to escape from this wrong solution alter one of the equal coupling constants to a different value. Make sure that the **Advanced Iteration Mode** with **Starting Frequencies Poor** is selected and run another **Iteration** until the correct solution is reached. Examine the result in WIN-NMR **Dual Display**. **Export** the iteration result from WIN-DAISY to **WIN-NMR exp.** spectrum and **Close** the WIN-DAISY document.

As well as this very fast iterative method for analyzing [AB]$_2$ spin systems, a manual approach using special equations and the transition frequencies is also available. The present spectrum is a good example to use to try out this type of analysis; the A and the B parts are well separated and all 24 of the theoretically possible transitions are observed. First of all we are going to restore and modify the WIN-DAISY input data.

Check it in WIN-NMR:

Open.. the experimental spectrum of the 80 MHz *1,2-dichlorobenzene* example ...\SYMMETRY\008001.1R into the spectrum window of WIN-NMR. Switch into the **Analysis | Multiplets** and press the **Windaisy** button to visualize the WIN-DAISY result. **Export...** the data to **WIN-DAISY**.

Check it in WIN-DAISY:

All spin system data except the symmetry are restored. Build up the symmetry as described in previous "Check its". Perform a spectrum **Simulation**.

Change the name of the calculated spectrum (e.g. ...\008998.1R) and open the **Spectrum Parameters**. Use the **Edit | Disconnect** command to remove the information about experimental data. Increase the **Frequencies adapted to Spectrometer Frequency** from "80" to **800** MHz. To ensure adequate digitization click the **Auto Limits** button. **Run** a spectrum **Simulation**. **Display** the actual **Calculated Spectrum** with WIN-NMR by using the red peak button.

Check it in WIN-NMR:

Switch to the **Dual Display** mode and load the original simulation at 80 MHz as the second trace ...\SYMMETRY\008999.1R. Change the x-axis units to Hertz <Ctrl + X>. Click the **Move Trace** button and holding down the <Shift> key drag the second trace over one half spectrum of the first trace. Using a combination of "dragging" and "zooming" adjust the second trace until it overlays with the first trace (Fig. 3.17). The comparison of both spectra using frequency units makes it easier to assign the transitions in the second order spectrum.

Due to the strong second order influence in this spin system, the analysis using the two (a$_2$) and two (ab) type subspectra as discussed for the [AX]$_2$ spin system is not applicable. The degeneracy of the lines in the (a$_2$) subsystem is totally removed and four lines are found in the spectrum, although they are not easy to identify. The (ab) subspectra are highly distorted. The assignment of the transitions in the spectra to these subsystems is essential if the solution of DISCHLER and ENGLERT [3.5] is to be used to analyze the spectrum. These authors use an alphabetical labeling of the lines forming the subspectra as shown in Fig. 3.17: The four lines of the symmetrical (ab)$_+$ subspectrum are labeled g, h, i and j, the antisymmetrical (ab)$_-$ subspectrum c, d, e and f. The former N-lines which previously displayed two degenerate lines in two (a_2) subspectra now split into four lines, labeled a,k and b,l. To determine which transition belongs to which line, the lines of the (ab) subspectra must first be assigned.

The highly distorted (ab) subspectra can be identified using repeated spacings of the J_{ab+} and J_{ab-} "coupling". The following rules are still valid:

$$J_{ab+} = c-d = e-f$$
$$J_{ab-} = g-h = i-j$$

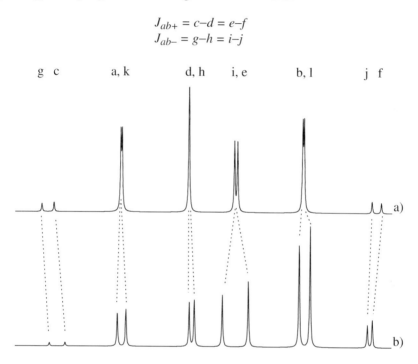

Fig. 3.17: Zoomed half spectra of both simulations (20 Hertz expansion).
a) Simulation at 800 MHz proton resonance.
b) Simulation at 80 MHz proton resonance.

Check it in WIN-NMR:

Open... again the 80 MHz experimental spectrum of *1,2-dichlorobenzene* ...\SYMMETRY\008001.1R with WIN-NMR. Use the **Analysis | Multiplet** mode to determine the repeated spacing by definition of two dd in one half spectrum. Determine the values of J_{ab+} and J_{ab-}, they are given in the **Report...** box. In this way the transitions c,d,e,f and g,h,l,j are identified.

158   3 Structure and Spin System Parameters

Unlike [AX]$_2$ spectra, the pseudo coupling constants of these subspectra do not refer to parameters K and M. When the (ab) subspectra are identified only four transitions remain: a,b and k, l. The assignment of these labels is based on the following rules:

$$g - a = b - j$$
$$a - i = h - b$$

Check it in WIN-NMR:

Return to the normal spectrum display in WIN-NMR. From the **Analysis** pull-down menu select the **Peak Picking** option. **Execute** a peak picking using **Interpolation** option the **peak labels** in **Hz**. Only the values from one half spectrum are needed (Fig. 3.18). Via **WIN-DAISY** you can call the **Calculator** to assign the transitions a, b and k, l.

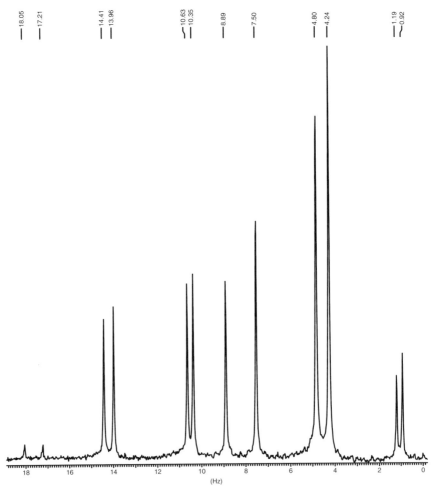

Fig. 3.18:   Peak Picking of one half spectrum of the [AB]$_2$ spectrum.

3.1 Symmetry Effects

The [AB]$_2$ spin system is described by the four auxiliary variables N,K,L and M and the chemical shift difference Δ_{AB}. These variables can be determined using the following equations. For each parameter one equation is sufficient, but by using the other equations, the correct assignments of the subspectra can be verified.

$$\Delta_{AB} = \sqrt{4 \cdot a \cdot b}$$
$$= \sqrt{2 \cdot (c \cdot e - d \cdot f)}$$

$$N = a - b$$

$$K = g + i + k - 2 \cdot a - l$$
$$= 2 \cdot b + k - h - j - l$$
$$= b + g + k - a - h - l$$
$$= b + i + k - a - j - l$$

$$M = \frac{(c \cdot e - d \cdot f)}{\Delta_{AB}}$$

$$L = \sqrt{(c - e)^2 - M^2}$$
$$\sqrt{(d - f)^2 - M^2}$$

Check it in WIN-DAISY:

Use the peak picking data and the assignments of the 12 frequencies (letters a to l) to calculate the five variables N, K, L, M and Δ_{AB}. Using these variables **Calculate** the four scalar coupling constants J_{AB}, $J_{AB'}$, $J_{AA'}$ and $J_{BB'}$ and compare these values with the numbers in the **Scalar Coupling Constants** dialog box in WIN-DAISY.

This type of analysis for [AB]$_2$ spin system is extremely tedious. Furthermore the assignments of the frequency labels a to l can be considerably more complicated when the A and B part overlap as is the case in the spectra of *1,2-dimethoxybenzene* and *1,2-dihydroxybenzene*. The analysis of such spectra using the iterative refinement of parameters is the fastest way to obtain a reliable result for the relevant NMR parameters.

3.2 Configuration Isomers in NMR

The main part of this section about configuration isomers will deal with cis and trans isomers. This type of configuration isomerism can occur about double bonds and in alicyclic compounds. The values and signs of coupling constants and chemical shifts in these systems will be discussed as well as the possibility of identifying individual isomers present in a spectrum containing a mixture of isomers.

3.2.1 Cis/trans Isomerism

Three different types of scalar coupling constants are present in vinylic spin systems (Fig. 3.19). A rough classification of coupling constants can be done based on the number of bonds in between the coupled protons.

H-H coupling constants over two bonds are called geminal: $^2J_{gem}$. H-H coupling constants over three bonds vicinal: $^3J_{vic}$. Two different stereochemical arrangements of vicinal protons are possible in the vinyl skeleton: $^3J_{cis}$ and $^3J_{trans}$. As a general rule and aid to assigning individual spins: $^3J_{trans} > \,^3J_{cis} > \,^2J_{gem}$.

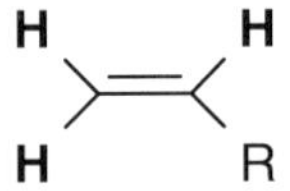

Fig. 3.19: Vinyl skeleton

The discussion of cis and trans isomerism about double bonds will start with a simple two spin system.

Check it in WIN-NMR and WIN-DAISY:

Open... the experimental proton spectrum ...\AX\003001.1R with WIN-NMR. Identify the vinylic protons AX spin system. Measure a suitable **Linewidth** and analyze the **Multiplets**. **Export...** the data to **WIN-DAISY** perform an iteration and **Export** the data back to **WIN-NMR exp.**.

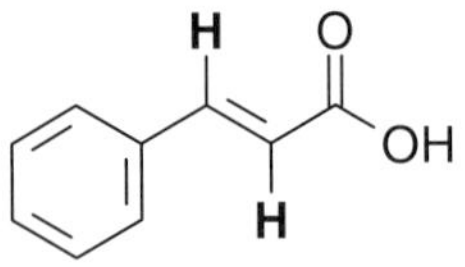

trans cinnamic acid

Compare the experimental spectrum and calculation result in the **Dual Display** mode. There is an obvious difference in the linewidth of the A and X signals which implies that the signal with the broader linewidth is displaying long-range coupling to the ortho protons of the benzene and so must be next to the benzene ring.

If the aromatic ring system is not included in the simulation, this influence can be taken into account by using different linewidths. Open the **Linewidth Parameters** dialog box and enable the **Nuclei specific linewidths** option. Select both linewidths for iteration and run a second **Iteration**.

Table 3.1: Range of H-H coupling constants for the vinyl skeleton.

Coupling Constant	Coupling range [Hz]
$^2J_{gem}$	-3 to $+7$
$^3J_{cis}$	$+3$ to $+18$
$^3J_{trans}$	$+12$ to $+24$

A vinylic three spin system including all three types of coupling constants is represented by the following example.

Check it in 1D WIN-NMR and WIN-DAISY:

Open... the experimental spectrum of *styrene*, ...\AMX\007001.1R in WIN-NMR. In the **Multiplets** mode analyze the vinylic signals and then **Export...** the data to **WIN-DAISY**.

styrene

Optimize the NMR parameters via **Iteration** and label the three spins of the vinyl fragment according to the PSB notation of spin systems based on the magnitude of the coupling constants. Keep the WIN-DAISY document open.

First order spin systems without any second order roof effect are independent of the signs of coupling constants. However, in practice many spin systems show a noticeable roof-effect.

3.2.1.1 Determination of the Signs of Coupling Constants

Check it in WIN-DAISY:

Open the **Iteration Protocol** file and note the *R-Factor*, which is about *4.46%*. Keep the file open. Display experimental and calculated spectra in the **Dual Display** mode of WIN-NMR. **Zoom** in the signals and inspect the intensities. The quite high error value results from a slight negative baseline in the experimental spectrum and different linewidths which shall be not taken into account in the present analysis. Open the program **Options** and check the option **Allow sign changes**, because the geminal coupling constant might be negative!

Change the sign of the geminal coupling constant (J_{23}) within the **Scalar Coupling Constants** dialog box. Perform a spectrum **Iteration**, accept to overwrite the old iteration list (**Yes**). Compare the experimental and simulated spectrum in the **Dual Display** mode of WIN-NMR. Examine again the signal intensities and display the actual **Iteration Protocol** file. The *R-Factor* obtained is higher (about *4.71%*) with negative geminal coupling constant.

The determination of the sign of the geminal coupling constant is possible due to the roof effect of the M signal, even though the spectral parameter is very low e.g. about $r_{gem} \approx 0.025$. Even though the parameter limits for the geminal coupling constant covers positive as well as negative values, the iteration cannot determine the correct solution starting with a negative geminal coupling constant without smoothing. In the process of iteration the program must be able to escape from the local minimum (negative coupling

162   3 Structure and Spin System Parameters

constant) and overcome higher error-values on the way to the global minimum, which is in the present case not much deeper than the local one.

Check it in WIN-DAISY:

Open the **Control Parameters** dialog box and select the **Advanced Iteration Mode** with **high Broadening** and **high No. of Cycles**. Enter the **maximal number of iterations** of <200>. Run an **Iteration** and examine the **Iteration Protocol**.

The program was able to find the correct solution starting from the wrong parameter set by application of strong smoothing.

Check it in WIN-DAISY:

Change the sign of one or more of the coupling constants in the **Scalar Couplings** dialog box. Perform an **Iteration** and compare the calculated and experimental spectrum in the **Dual Display** mode of WIN-NMR. Find out which sign combinations gives the best fit. At the end of each iteration open the **Iteration Protocol** file and from the **File** pull-down menu select the **Save As...** option and save the file under a different name e.g. <1.RST>, <2.RST> etc., for use in the next "Check it". Keep the document open.

By checking out all possible eight sign combinations you can discover two sets of parameters that give the best result. Both sets represent the same relative sign combinations and corresponds with the fact that the effect on a spin system of sign reversal is invariance provided that the signs of ALL the coupling constants is reversed. Comparing the evaluated coupling constants with the characteristic ranges found in vinyl systems listed in Table 3.1 the most likely sign combination is obvious.

Check it in WIN-DAISY:

Load the iteration protocol files stored during the previous "Check it" into the WINDOWS Notepad editor and compare the *statistical information* for all the calculations. The exact meaning of these values is reported in the appendix section 6.3, but concentrate on the *R-Factor (%)* as it reflects the percentage deviation between the experimental and calculated lineshape. Give positive signs to all **Scalar Coupling Constants** and keep the document open.

The differences in the parameter optimization statistics is not very large, even when the sign of $^3J_{trans}$ is inverted, so that it is not possible to determine the relative signs of coupling constants with any degree of certainty using this method and other techniques must be found. Increasing the magnetic field strength for the measurement will decrease the second order effect more, so that the determination of the coupling constant sign will be even more difficult.

Check it in WIN-NMR and WIN-DAISY:

Open... the experimental 250 MHz spectrum ...\STYRENE\001001.1R of styrene and call the **Analysis | Interval** mode. Click on the **Regions...** button and **Load...** the predefined spectral regions. **Export...** them to **WIN-DAISY** and accept the message box. Press the **Connect** button and answer the question about spectrometer frequency with **Yes**. Run a spectrum **Simulation** and display both spectra in WIN-NMR **Dual Display**. Use the Move Trace option with depressed <Shift> key to overlay the signals. With <Ctrl + X> switch the x-units into Hz and note the shift difference. Toggle to **WIN-DAISY**. Open the **Spectrum Parameters** and give the negative frequency shift (e.g. about <5> Hz) and **Apply!** the offset correction. Run an **Iteration** and display the result in WIN-NMR **Dual Display**. Toggle back to **WIN-DAISY** and look at the *R-Factor* in the **Iteration Protocol**. Change the sign of the geminal **Scalar Coupling Constant** J_{23} and run another **Iteration**. Inspect the *R-Factor* in the **Iteration Protocol**.

In the 250 MHz spectrum no decision which sign is the correct one can be made from the line-shape iteration. Additionally the line-shape is not symmetrical, so that it has to be made use of other methods for sign determination in such cases. One such technique is the double resonance spin tickling experiment. In addition to the static homogeneous field B_o, this method uses a second decoupler field B_2 which has a frequency ν_2.

The theoretical treatment of double resonance calculations will not be discussed here, the user is referred to the references [3.6]–[3.8]. The effect on the resulting HAMILTONian matrix is the appearing of additional off-diagonal elements even in first order spin systems.

The LARMOR frequencies of the nuclei ν_i appear relatively to the irradiated frequency ν_2. The conditions for the spin tickling experiment are:

1. *the strength of the field:* $\quad B_{2eff} = \frac{\gamma}{2\pi} B_2 \approx$ *line width (weak field)*

2. *the frequency of the field:* $\nu_2 = T_{Xi} \quad$ *(one of the transition frequencies)*

Check it in WIN-DAISY:

In WIN-DAISY click the **Export WIN-DR** button to export the data and to generate an input document for the double resonance simulation program which is automatically loaded into the program WIN-DR.

Check it in WIN-DR:

Using the original data, run a **Simulation** representing the unperturbed experiment and examine the **Display**ed **Calculated Spectrum** in WIN-NMR. Toggle back to **WIN-DR**. Open the **Perturbation Parameters...** dialog box and select the **Double Resonance** option. Click the **Tickling** experiment radio button and then the **Calc. Transitions: Unperturbed** button. In the list box the transition frequencies are listed along with the energy levels involved in the transition. Double click on one of the entries to select a transition, e.g. the very first one (labeled 8-7). Level 8 has $m_T = +1.5$ and level 1 $m_T = -1.5$. Using this information, design an m_T diagram using the energy level labeling of WIN-DR in such a way, that all transitions arising from one nucleus appear as parallel arrows in the representation (see section 2.2.3).

For the present three spin system the resulting diagram can be represented as a cube in which the eight corners represent the energy levels and all the parallel edges (three types referring to the number of different spins) correspond to the transitions of one distinct spin of the system as illustrated in Fig. 3.20. The numbering of the energy levels beside the transitions given in the list box within one m_T value might permute. But the important thing is, that according to the values and signs of the coupling constants, the *energy levels* are shifted. All the numberings refer to the unperturbed spin system.

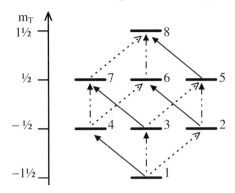

Fig. 3.20: m_T scheme for the AMX spin system
↗ A transitions
↑ M transitions
↖ X transitions

Two transitions having one energy level in common are named progressive, if the m_T values of the two remaining energy levels differ by two units (Fig. 3.21a).

Two transitions having one energy level in common are called regressive, if the m_T values of the two remaining energy levels are identical (Fig. 3.21b).

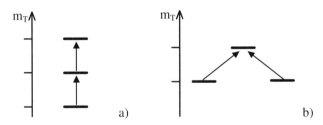

Fig. 3.21: a) Progressive connected transitions. b) Regressive connected transitions.

If we focus on the A transition at lowest field, which is labeled 8-7 in the case where all coupling constants possessing a positive sign, we may construct a sub m_T level diagram by extracting only the four directly connected transitions and total spin levels as shown in Fig. 3.22.

3.2 Configuration Isomers in NMR 165

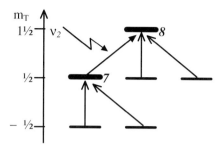

Fig. 3.22: m_T subscheme with lines directly connected to transition 8-7.

Two regressive and two progressive transitions are involved when irradiating line 8-7 as shown in Fig. 3.22. Each of the transitions will be split into two transitions with different relative linewidths dependent on if they are progressive or regressive connected with the irradiated transition.

The assignment of the four transitions to the line frequencies in the spectrum is dependent on the signs of the coupling constants. As mentioned already for the AX spin system the sign of the coupling constant interchanges the labeling of the transition lines in the spectrum by shifting the values of energy levels (refer to section 2.2.2). If it was possible to distinguish between progressive and regressive transitions in the spectrum, the relative signs of the coupling constants could be determined.

This can be achieved using the spin tickling experiment, which perturbs the energy levels in a well defined way by irradiating one distinct transition frequency with a weak magnetic field. Consequently, only the directly connected transitions are influenced and they split into two signals. How the signals behave depends on whether they are a progressive or regressive transition:

Progressive connected transitions show a BROADER linewidth in splitting.
Regressive connected transitions show a SHARPER linewidth in splitting.

Check it in WIN-DR:

Using the WIN-DR document set up in the previous "Check it", calculate the spectrum when the transition 7-8 in the spectrum is tickled. Open the **Perturbation Parameters** dialog box. Ensure that the **Tickling** and **Hz** radio buttons are set. Click the **Calc. Transitions: Unperturbed** button and double click in the list box on transition No. 1. Set **Field in B2** to the experimental linewidth e.g. <3.5>Hz. To ignore any frequency "spikes" set **Min. Absol. Linewidth [Hz]** to <0.1>Hz. **Simulate** the spectrum and **Display** the **Calculated Spectrum**. In WIN-NMR load one of the tickling experimental spectra (...\STYRENE\002001.1R up to ...\006001.1R) as the **second file** into the **Dual Display**. (When the second filename is selected, mark the full path with double mouse click and copy the name into the clipboard <Ctrl + C> to paste it for the next dual display <Ctrl + V> and simply change the number.) Align the Y-gain of the spectra with <Ctrl + Y> and press the **ALL** button. **Zoom** in the signals and with **Move Trace** overlay the signals. With a negative geminal coupling the simulated spectrum will look like Fig. 3.23c. Toggle back to **WIN-DR**. In the **Scalar Coupling Constants** dialog box change the sign of

J_{23}, the geminal coupling. Run a **Simulation** and examine the result in WIN-NMR (Fig. 3.23b).

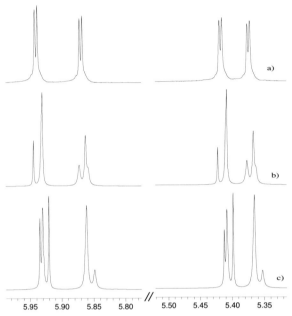

Fig. 3.23: a) Unperturbed reference spectrum zoomed MX spectrum part.
b) *Positive* geminal coupling constant when transition 8-7 tickled.
c) *Negative* geminal coupling constant when transition 8-7 tickled.

Check it in WIN-DR:

The experimental spectra (...\STYRENE\007001.1R up to ...\009001.1R) correspond to the tickling of the lowest field transition of signal M - transition no **5**. (6-8 for positive and 3-5 for negative geminal coupling constant).

Open the **Perturbation Parameters** press the **Calc. Transitions [Hz] unperturbed** button and double click on the transition **No. (5)** in the list box. Simulate as well the tickled spectra and check out if this is a suitable transition to determine the sign of the geminal coupling constant.

The comparison of the tickled 7-8 transition calculated spectrum with the experimental spectra shows, that the coupling constants is positive. The result of the tickling of the lowest field M-transition does not give different results according to the sign of the geminal coupling constant and is therefore not a suitable experiment.

Another feature available within WIN-DR is the simulation of the homonuclear decoupling experiment. In this experiment one spin is selected and its resonance frequency is irradiated. The power of the irradiating field is so high that the BOLTZMANN statistics are disturbed, the irradiated nucleus becomes saturated and is essentially removed from the spin system. The practical result is that the decoupled spin and all coupling constants associated with it vanish, simplifying the spectrum and allowing the identification of the decoupled spin's coupling partners.

3.2 Configuration Isomers in NMR

For the successful decoupling of a nucleus X the following conditions are required:

1. *the strength of the decoupler field:* $B_{2eff} = \frac{\gamma}{2\pi} B_2 \gg 2J_{iX}$ *(strong)*
2. *the frequency of the decoupler field:* $\nu_2 = \nu_X$ *(resonance frequency of X)*

Check it in WIN-DR:

To simulate the effects of a homonuclear decoupling experiment use again the present WIN-DR data of the ...\STYRENE\001001.1R data exported from WIN-DAISY to WIN-DR.

Open the **Perturbation Parameters** dialog box. Choose the **Decoupling experiment** and set the **Field in B2** to <1000>Hz. Click the **Effected Spin(s)** checkbox number **3** to select spin 3 for decoupling. Execute a **Simulation** and look at the result in WIN-NMR. By irradiating spin 3, $^3J_{cis}$ and $^2J_{gem}$ have been removed and the simulated spectrum corresponds to two protons exhibiting a trans coupling.

The unperturbed three-spin system can be calculated by either

- setting the decoupler field to zero
- or turning the perturbation option off (**Unperturbed** radio button).

3.2.1.2 Extended Spin Systems

This section will consider examples involving magnetic equivalence, different linewidths, fragmentation of large spin systems containing a vinyl group and the analysis of spectra containing mixtures of isomers.

Check it in WIN-NMR and WIN-DAISY:

Open... the spectrum ...\A2KMRX\001001.1R in WIN-NMR. Determine the spin system represented by the structure opposite and analyze the spectrum using the **Analysis | Multiplets** mode.

Export... the data to WIN-DAISY with fragmentation of the spin system. Optimize the NMR parameters and look at the result in **Dual Display** mode of WIN-NMR.

3-(3,5-dichlorphenyl)-5-methyl-5-vinyloxazolidine-2,4-dion

This present example contains two three spin systems of different types. A three spin vinylic system and a symmetrical aromatic system involving magnetical equivalence. The vinylic part is easily analyzed as an AMX type with the coupling constant J_{MX} zero. The aromatic system can be treated as an A_2X type spectrum for the first order analysis even though it already shows signs of an A_2B spin system type.

168   3 Structure and Spin System Parameters

Check it in WIN-NMR and WIN-DAISY:

Open... the experimental proton spectrum ...\AMNR2X\001001.1R in WIN-NMR. **Analyze** the spin system but keep in mind that the multiplicities present in the spectrum must be based on the 'n+1' rule. Include the OH proton in the analysis. **Export...** the data to **WIN-DAISY** with fragmentation of the spin system.

propene-3-ol

Check in the **Frequencies** dialog box the number of independent frequencies and decide whether the spin system has been correctly defined. If the number of independent frequencies is wrong, toggle back to **WIN-NMR** and look at the **Multiplets** analysis. Look carefully at the chemical structure and your analysis (in case there are still problems, refer to Fig. 3.24). To check that the spin system has been correctly defined perform a spectrum **Simulation** and compare spectra in the **Dual Display** mode of WIN-NMR.

The major difference between the simulated and experimental spectrum is in the linewidth for the OH-proton fragment. For optimization of the NMR parameters the **Upper** linewidth **limit** for the second fragment should be raised to about <15> Hz in the **Lineshape Parameters** dialog box. Perform an **Iteration** and examine the result in the **Dual Display** mode of WIN-NMR.

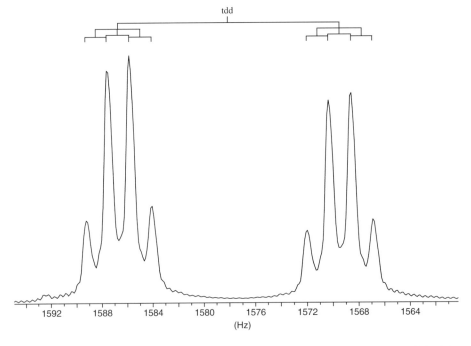

Fig. 3.24:   Multiplet pattern of a signal representing a triplet of doublets of doublets.

The purpose of this example is to illustrate that in any spectrum analysis reference should always be made to the chemical problem under investigation, otherwise the pattern shown in Fig. 3.24 might easily be misinterpreted as a doublet of quartets. Such problems arise in cases where different nuclei have similar coupling constants.

The protons of the allylic methylene group show magnetic equivalence and therefore the maximum multiplicity occurring in the spectrum is of order 3. The low-field signal shows the constellation, that one coupling is not of the *same* magnitude but rather an integral multiple of another coupling constant. This constellation makes the doublet of doublets of triplets even a more complicated outlook.

Quite often chemical synthesis can give products containing constitutional isomers. If these isomers cannot be chemically separated, the NMR spectrum will contain the signals of all the isomeric compounds. For the purpose of spectrum analysis, each isomer can be considered as an independent spin system, but the first and most important step in the analysis is to assign the signals to the individual species.

Check it in WIN-NMR:

Open... the experimental proton spectrum ...\MIX\001001.1R in WIN-NMR. Switch to the **Analysis | Integration** mode and integrate all the signals.

Ignore the overlapping multiplet at about 6 ppm and concentrate on the high and low-field signals. Is there any difference in the ratio of the individual species present?

trans and cis 1,3-dichloropropene

The ratio of the two isomers is not quite 1:1 and this information may help in assigning the signals to each species. The next step is to the perform a multiplet analysis for each isomer.

Check it in WIN-NMR:

Measure a suitable **Linewidth** and then switch to the **Analysis | Multiplets** mode. Keep the integrals displayed during the multiplet analysis as they might be helpful in the assignment procedure. If this is confusing, the integrals may be switch off and on using <Ctrl + I> or the **Display | Options...** command.

Before starting the analysis look at the chemical structure and using the 'n+1' rule decide what multiplicities should be present in the spectrum. Use the values in Table 3.1 to assign the signals to the cis and trans isomers. The overlapping signal region is shown in Fig. 3.25. The assignment of coupling constants should be done manually, do *not* use the **Auto Connect** Option.

Start with the trans isomer: The easiest coupling constant to assign is the large trans coupling. The two remaining triplet splitting are connected to the methylene group at high field. Because the coupling constants are similar in magnitude, consider the integral values when assigning the methylene signal of the trans isomer.

The coupling pattern of the cis isomer is not so trivial. Keep in mind the multiplicities when connecting the coupling constants or the wrong spin system will be obtained.

Check it in WIN-DAISY:

Export the multiplet data to WIN-DAISY with fragmentation of the spin system into two independent spin systems. Check the **Spins in group** displayed in the **Frequencies** dialog box. Run a spectrum **Simulation** and compare the calculated and experimental data in the **Dual Display** mode of WIN-NMR.

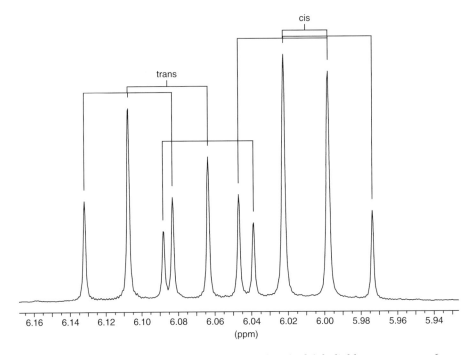

Fig. 3.25: Multiplet analysis of overlapped signal of *1,3-dichloropropene* at 6 ppm.

Check it in WIN-DAISY:

The slight different in the ratio of the isomers should be taken into account. Open the **Main Parameters** dialog box. The parameter **Statistical weight** is used in the normalization of the intensities. If there is only one fragment the **Statistical weight** has no real effect on the spectrum but if there is more than one fragment, this parameter may be set to reflect the *relative* proportion of each fragment, thus only one weighting factor need to be changed, e.g. enter for the trans fragment the **Statistical weight** of <0.9>. (Ensure that the value corresponds to the correct isomer, do not enter the cis value into the trans fragment and vice versa.) Perform a spectrum **Iteration**.

Export to **WIN-NMR exp.** the optimized NMR parameters. This command automatically transfer the parameters into the **Multiplet** mode of WIN-NMR. Note that the **Windaisy** button is also selected. Open the **Report...** box. The data is ordered, not by chemical shift, but according to the fragments generated by WIN-DAISY. In this way, the WIN-DAISY optimized NMR parameters can be stored with the WIN-NMR spectrum data (except the

3.2 Configuration Isomers in NMR

statistical weights of the **Main Parameters**) and if necessary exported back into WIN-DAISY to create an input document.

In the **Report...** box, click the **Export...** button to export the optimized data back into **WIN-DAISY** and generate a new input document. The spectrum of the cis isomer only can be simulated by deleting the trans isomer fragment. Select the second fragment in the list box of the WIN-DAISY document main window. Either press the **Delete fragment** button or use the pull-down menu **Edit | Delete fragment** option. Click the **Yes** button in the WIN-DAISY message window. After executing this command the list box will contain only one entry corresponding to the cis isomer fragment. Run a **Simulation** and display calculated and experimental spectra in the **Dual Display** mode of WIN-NMR.

Another example of the analysis of a mixture of cis/trans isomers is based on the cyclopropyl fragment shown in Fig. 3.26.

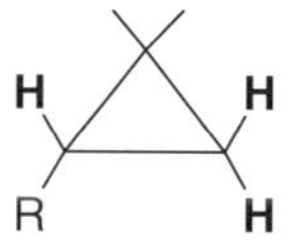

Fig. 3.26: Arrangement of protons for a cyclopropyl fragment.

In a similar manner to a vinyl group, the protons about the single bond of the three-membered ring shown in Fig. 3.26 have one geminal coupling and two three bond couplings, $^3J_{cis}$ and $^3J_{trans}$. Table 3.2 gives the range of values expected for the three possible types of coupling constants in a cyclopropyl fragment.

Table 3.2: Range of H-H coupling constants for the cyclopropyl skeleton.

Coupling Constant	Coupling range [Hz]
$^2J_{gem}$	-10 to -4
$^3J_{cis}$	$+5$ to $+12$
$^3J_{trans}$	$+3$ to $+9$

Comparing these values with the data given in Table 3.1 for the vinyl skeleton it obvious that for the cyclopropyl fragment, $^3J_{cis}$ is now larger than $^3J_{trans}$ and the value of the geminal coupling constant can occur over a much wider range.

The ^1H spectrum of *chrysanthemic acid (3-phenoxy)-benzyl ester* contains both the cis and the trans isomers.

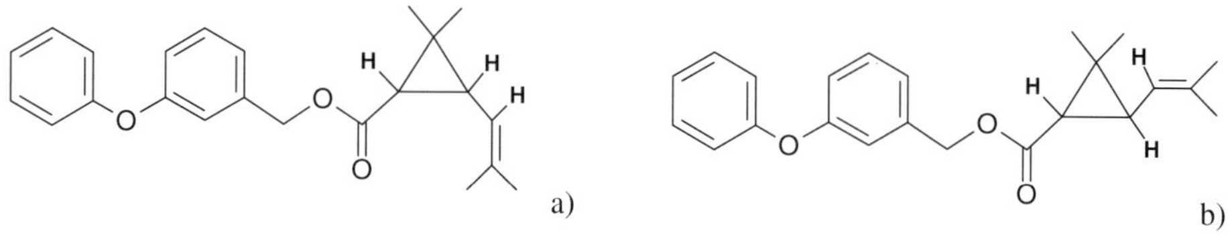

Fig. 3.27: a) Structure of *cis-chrysanthemic acid (3-phenoxy)-benzyl ester*.
b) Structure of *trans-chrysanthemic acid (3-phenoxy)-benzyl ester*.

Check it in WIN-NMR:

Inspect the structures given in Fig. 3.27 and think about the multiplet patterns expected for the three protons printed explicitly in the structure. Take into account the four methyl groups and the possibly that they might couple with the cyclopropyl and vinylic protons.

Open... the experimental ^{1}H spectrum, ...\MIX\002001.1R, in WIN-NMR and try to identify the signals corresponding to the cyclopropyl protons. Keep in mind, that the ratio of the two compounds might not necessarily be 1:1. Open the **Analysis | Integration** mode and integrate the vinylic and cyclopropyl protons and the methyl groups. Theoretically there should be seven signals for each isomer including the methyl groups, but some of the signals overlap and some may not exhibit coupling. **Load...** the predefined iteration **Regions...** to check which signals shall be analyzed. Measure a suitable **Linewidth** and then switch to the **Analysis | Multiplets** mode. Analyze the spectrum and connect the corresponding coupling constants.

Note: In case you wish to optimize the parameters you need to define all the multiplets in the spectral regions. The methyl groups of interest all appear as overlapping doublets. To assign the couplings easily Define Identifier with labels of small and large integrals or print out the multiplets.

The vinylic protons will appear as a doublets of quartets (dqq) due to coupling with one of the cyclopropyl protons ($^3J_{HH}$) and the two vinylic methyl groups (both $^4J_{HHcis}$ are $^4J_{HHtrans}$ are expected to have approximately the same magnitude with a *negative* sign). The signals of the vinylic protons appear at about 5 ppm. The cyclopropyl proton in the allylic position appears at approximately 2 ppm as a doublet of doublets due to coupling with the other cyclopropyl proton and the vinylic proton. Using these signals the cis and trans coupling constant in the cyclopropyl skeleton can be identified and the ratio of the individual components of the mixture determined (refer to Fig. 3.28). The other cyclopropyl proton appears as a simple doublet but only the signal of the trans isomer at 1.44 ppm is clearly visible, in the cis isomer the signal is partly obscured by the vinyl methyl groups.

Check it in WIN-NMR and WIN-DAISY:

Export... the data to WIN-DAISY *with* fragmentation of the spin system (**Yes**). Accept to *include* all non-assigned coupling constants (**Yes**) if there are any. Two 9-spin AKMX$_3$Y$_3$ type spin system will be generated in 5 groups each. In the **Main Parameters** dialog box for the first or the second fragment enter the relative integration value in the **Statistical weight** (the other normalized to 1.0). Open the **Lineshape Parameters** dialog box and deselect the **Global Linewidths** from iteration. Open the **Scalar Coupling Constants** dialog and enter negative sign to the four ^{4}J. Run a spectrum **Simulation** to check if the isomeric ratio and the spin systems have been correctly defined. In case all signals are analyzed you can run an **Iteration** and inspect the result in the **Dual Display** mode of WIN-NMR.

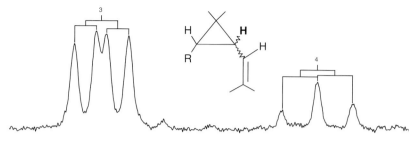

Fig. 3.28: Allylic proton signals showing dd at about 2 ppm.

The complete analysis of the aliphatic spectrum region including the methylene and methyl groups of the compound is stored in the input file CRY.MGS. In case not all methyl doublets are defined in the Multiplet Analysis, WIN-DAISY adds a three-spin group which is disabled from simulation and iteration. Then the coupling constant is merely defined in one signal which is especially not sufficient for the low-abundant isomer to obtain a proper iteration result. Still after iteration the low-abundant isomer low-field signal parameters shows larger standard deviations, because of the low intensity and high signal-to-noise ratio.

Check it in WIN-DAISY:

In case an iteration was performed, open the **Lineshape Parameters** and select the option **Nuclei specific linewidths** and check the **Iterate All** button. Run another **Iteration**. Inspect afterwards the **Iteration Protocol**. Still the standard deviations for the low-concentrated fragment are larger, but the signals give a better coincidence in the **Dual Display**.

3.2.2 Enantiomers and Diastereomers

NMR spectroscopy can also be used to study the configuration of molecules containing an asymmetric carbon atom. In the cases where one asymmetric carbon atom is present the absolute configuration of the molecule is either *(R)* or *(S)*. The two conformations are mirror images of each other and are called enantiomers. Enantiomers have identical physical properties, except for the direction of rotation of polarized light, and cannot be distinguished by normal NMR spectroscopy. However if the molecule is dissolved in a chiral solvent, or a solvent containing a chiral shift reagent or chiral compound, e.g. PIRKLE's reagent, the compound forms diastereomers and can be distinguished by NMR spectroscopy. In contrast to enantiomers, diastereomers can be distinguished by normal NMR spectroscopy. Consider the example of *1,2-dichloropropane*. The enantiomeric forms of the molecule are given in Fig. 3.29. Under standard conditions, both molecules will give an identical NMR spectrum, so it is not possible to determine whether the compound is enantiomeric pure or if it is a racemate.

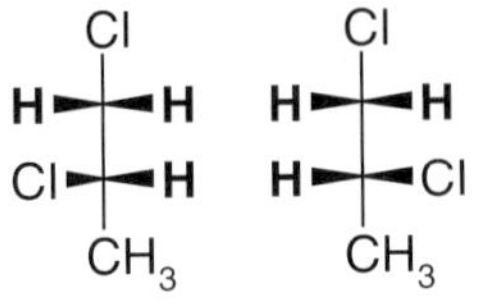

3 Structure and Spin System Parameters

```
       Cl              Cl
       |               |
   H ──┼── H       H ──┼── H
   Cl──┼── H       H ──┼── Cl
       |               |
      CH₃             CH₃
```

Fig. 3.29: (S) and (R) *1,2-dichloropropane*.

For a molecule to form enantiomers, it must contain an asymmetric carbon atom. In *1,2-dichloropropane* a methylene group is directly attached to an asymmetric carbon atom and the geminal protons are *diastereotopic*. This fact can be worked out using the substitution method. The substitution of either one or the other geminal proton leads to diastereomers as illustrated for the *(S)* enantiomer in Fig. 3.30. For each enantiomer a diastereomeric pair is obtained by substitution. The environment of the geminal protons is diastereotopic. Using Fig. 3.30 as a guide, draw the corresponding substitution scheme for the *(R)* enantiomer and compare the result with those obtained for the *(S)* enantiomer.

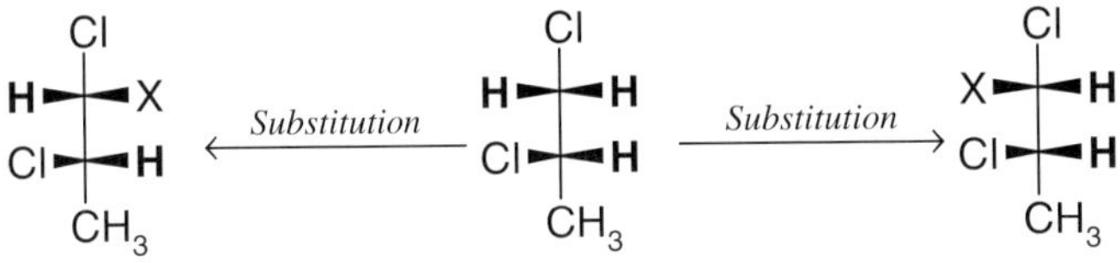

```
       Cl                       Cl                        Cl
       |                        |                         |
   H ──┼── X    Substitution    H──┼──H     Substitution   X──┼──H
   Cl──┼── H   ←──────────      Cl──┼──H    ──────────→   Cl──┼──H
       |                        |                         |
      CH₃                      CH₃                       CH₃
```

Fig. 3.30: Fisher projection of (S) *1,2-dichloropropane* with substitution of protons.

The diastereotopic methylene protons are distinguishable by normal NMR spectroscopy, because they have different chemical shifts. A common misconception is that this difference in chemical shifts is caused by hindered rotation around single bonds. This is not the case, even when the rotation is free the geminal protons still have different chemical shifts because they are diastereotopic!

Draw the NEWMAN projections for the *(R)* and the *(S)* enantiomer of *1,2-dichloropropane*. If necessary look at Fig. 3.30. Is there any orientation where both the geminal protons are in the same environment?

Before discussing related spin systems and their associated NMR parameters, it is useful to consider first of all the conformation of such compounds.

3.3 Conformation and NMR

Conformation isomers may be converted into each other by simple rotation about a single bond, no bond breaking is necessary. The energy needed to rotate around single bonds is normally quite low and in most cases conversion of isomers readily occurs at room temperature. The NMR time scale is long compared with the rate of inter-conversion and it is not possible to detect the individual conformational rotamers, only a time-averaged picture over all of the conformations. This effect has already been briefly mentioned for CH_3 and CH_2 groups in the discussion on composite particles (section 2.4). In such systems the individual conformations should be considered using the NEWMAN projections in order to determine the chemical environment of particular protons. If the protons on the same carbon have the same, time-average chemical environment, they are magnetically equivalent and can be treated as a composite particle. For protons in a composite particle, it is not possible to detect the geminal coupling constant by high-resolution NMR spectroscopy.

In section 2.4.5.2, the freely rotating methyl group has been discussed. The energy barrier to free rotation of the CH_3 group against the core of the residual molecule is so low, that thermal energy is sufficient to overcome it. The value of the coupling constant is therefore the average value over all of the conformers. The protons of a methyl group are *homotopic*, their substitution leads to identical molecules. Homotopic nuclei cannot be distinguished by NMR.

In the present section the phenomenon of *diastereotopy* has been introduced. Based on the substitution criterion it was illustrated that geminal groups bonded to an asymmetrical carbon atom are diastereotopic. Diastereotopic protons have different chemical shifts and do not fulfill the conditions for magnetic equivalence. This can be illustrated using the NEWMAN projections shown in Fig. 3.31 for the *(S)* enantiomer of *1,2-dichloropropane*. To distinguish between the two geminal protons in Fig. 3.31 one is printed as an italic letter. At this point we do not have to consider the different weights of the individual conformations, because the two protons (**H** and ***H***) will never have identical chemical environments either in any particular conformation or by time-averaging. Thus they are not isochronous and not be magnetically equivalent. It might occur only accidentally that their chemical shifts become isochronous.

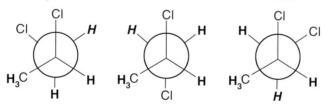

Fig. 3.31: Three possible NEWMAN projections of (S) *1,2-dichloropropane*

176   3 Structure and Spin System Parameters

The three NEWMAN projections of *(R) 1,2-dichloropropane* are mirror images of the the *(S)* form, but the chemical environment of the CH$_2$ protons is the same in both enantiomers. Consequently the two enantiomers give the same NMR spectra and cannot be distinguished by normal NMR techniques.

The *magnitude* of the vicinal coupling constant of the individual methylene protons to the methine proton depends on the weight of each conformer in the dynamic rotation process. The resulting coupling constant is a weighted mean of the coupling constants of the individual conformers.

For free rotation, a typical value for a vicinal coupling constant is about 5–7 Hz. The magnitude of a vicinal coupling constant dependents on the dihedral angle of the protons involved. The relationship between coupling constant and dihedral angle is given by the equation theoretically introduced by KARPLUS [3.9]. Various modifications have been made also by other authors to describe the dependency on the dihedral angle more precisely [3.10]. One example is given below with coefficients A, B and C specified for the current structure fragment.

$$^3J_{HH} = A + B \cdot \cos \varphi + C \cdot \cos 2\varphi$$

The general shape of the KARPLUS function ways fulfills that $J(\varphi=0°) < J(\varphi=180°)$ and J_{gauche} ($\varphi = \pm 60°$) $< J_{trans}(\varphi = 180°)$. In many investigations it was shown early that e.g. the sum of electronegativities of the substituents correlate with the averaged coupling constant values in ethanes [3.11]. Attempts were made according to compound-specific relations [3.12]. The combination of dihedral and substituent effects via electronegativities in the formulae [3.13], [3.14] and its expansions to coupling constants between other isotopes were introduced [3.15].

Table 3.3:   H-H coupling constant range of acyclic saturated aliphatic compounds.

Coupling Constant	Coupling range [Hz]
$^2J_{gem}$	– 20 to – 9
$^3J_{vic}$	+ 0 to + 12

Check it in WIN-NMR and WIN-DAISY:

Open... the spectrum ...\AMNX3\001001.1R. **Integrate** the peaks and then switch to the **Analysis | Multiplets** mode and analyze the spectrum. Only the three-bond couplings are visible, the four-bond couplings are not resolved.

1,2-dichloropropane

By analyzing the methyl and methylene signals the four coupling constants can be determined directly from the spectrum. These coupling constants can also be found in the methine signal. To a first approximation, this signal is a doublet of doublets of quartets (ddq). Start by defining either a quartet or a doublet splitting pattern but remember that for a higher multiplicity you have to start with an inner line of the selected pattern. (If the mouse cursor is placed on top of the signal at 4.20 ppm and define **Free Grid 4 Distance Lines** selected, some of the quartet lines will fall outside the signal multiplet.) Fig.

3.32 shows the quartet pattern with its two-fold duplication. **Export...** the spin system to **WIN-DAISY**.

Open the Scalar Coupling Constants and apply a negative sign to the geminal coupling constant (J_{23}). Run an **Iteration** and inspect the result in **Dual Display**.

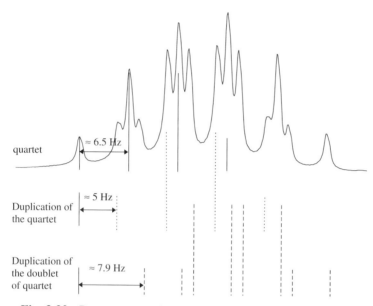

Fig. 3.32: Pattern composition of a doublet of doublet of quartets.

Diastereotopy can occur even when the chiral center is not the nearest neighbor. If diastereotopic protons are observed in a spectrum, the molecule must have an asymmetry center. However the converse it not true, the presence of magnetically equivalence protons does not mean that the molecule has no asymmetry center, it may be that the shifts are simply accidentally coincident even though their chemical environments are different.

Check it in WIN-NMR:

Open... the spectrum ...\AMNXY\001001.1R. Analyze the methylene signals in the **Multiplets** mode as a doublets of doublets. Take into account the values of the aliphatic coupling constants (refer to Table 3.3) and manually connect the geminal coupling constants. The remaining coupling constants, all doublet splittings, are to the methine proton. Analyze the methine signal and connect all the remaining couplings. **Export...** the spin system data to **WIN-DAISY**.

1,2-dibromo-3-chloro-propane

3 Structure and Spin System Parameters

Check it in WIN-DAISY:

Run a spectrum **Simulation** first and compare the spectra in WIN-NMR **Dual Display**. It is clear, that the diastereotopic proton signals are shifted. Apply a negative sign to the geminal **Scalar Coupling Constants**. Optimize the parameters by **Iteration** and examine the result in WIN-NMR **Dual Display**. The best result refers to an R-Factor of ca. 1.2% in the **Iteration Protocol**.

A comparison of experimental and simulated spectrum in WIN-NMR shows that the calculated linewidth of the methylene protons are larger than the experimental value, even though there is good agreement for the methine proton. This effect can be caused by long-range coupling over four bonds which, although not zero, is too small to be resolved and just contributes to the natural linewidth.

Check it in WIN-DAISY:

Open the **Scalar Coupling Constants** dialog box and press the **Iterate All** button. Modify the **global linewidth** parameter value in the **Lineshape Parameters** to <0.5> Hz. Perform an **Iteration** and display the spectrum in the **Dual Display** mode of WIN-NMR. Inspect the values of the long-range coupling constants in the **Scalar Coupling Constants** dialog box. All the long range couplings are smaller than the experiment linewidth and cannot be determined with any degree of accuracy. In practice the effect of this long range coupling is to broaden the signals of the proton they couple with, but the presence of long range couplings can be proved. The R-Factor is below 0.17% as reported in the **Iteration Protocol** file.

The following example will deal with another spin system of type RCH_2-R'CH-CH_2X but in this compound R and R' are part of a norbornene type structure. One of the two methylene groups of the molecule is acyclic and the other cyclic. The norbornen structure fragment is rigid and similar to the cyclopropyl example discussed earlier. (refer to Table 3.2). In a similar manner to the cyclopropyl fragment there exists a cis and trans arrangement of protons in this structure and $J_{cis} > J_{trans}$

Check it in WIN-NMR and WIN-DAISY:

Open... the spectrum ...\AKMRX\001001.1R in WIN-NMR. The exocyclic methylene group is directly attached to the chiral centre. The other methylene group is fixed in the norbornen structure and with the methine proton make up a cyclopropyl type structure.

Measure a **Linewidth** and analyze the spin system in the **Multiplet** mode. **Export...** the data to **WIN-DAISY**, apply a negative sign to both the geminal coupling and perform an **Iteration**. The R-Factor in the **Iteration Protocol** will be below 0.7%.

5-bromomethyl-1,2,3,4,7,7-hexachloro bicyclo [2.2.1] hept-2-ene

3.3 Conformation and NMR 179

Check it in WIN-DAISY:

Open... again the experimental spectrum ...\AKMRX\001001.1R in WIN-NMR and measure again a suitable **Linewidth**. **Delete all** the previous defined **Multiplets**. Determine the chemical shift of each signal by using the multiplet analysis to define a doublet based on the two outer lines of each signal pattern. **Export...** the data do WIN-DAISY. Do *not* fragment the spin system and do *not* include all the non-assigned coupling constants. In this way only the estimated chemical shifts are used to set up the basic structure of the spin system. Display the program **Options** and enter the iteration limits for **Scalar Coupling Constants** of ±<15> Hz. Open the **Control Parameters** and select smoothing parameters (e.g. **medium Broadening** and **medium No. of Cycles**) and increase the **Convergence Criterion** to about <0.05>.

Press the **Scalar Coupl.** button and click in the dialog box the **Select all** button to select all couplings for iteration and then deselect the four-bond couplings J_{14}, J_{15}, J_{34} and J_{35}. Perform an **Iteration**. Then you may apply a negative sign to the geminal coupling constants and select none Broadening in the Control Parameters for the final fit (**IT**).

It is not always possible to achieve convergence using very poor starting parameters. In practice, the best way to proceed is to perform an initial iteration and then to inspect the results using the dual display mode of WIN-NMR. Based on this inspection either change some of the parameter values in WIN-DAISY and / or modify the iteration control values before performing another parameter optimization. By repeating this process, a satisfactory result may be obtained from a set of poor starting parameters. In such cases it is important that the iteration limits are large enough to encompass the expected result, if the real value falls outside these limits the program is not able to achieve the correct result. If a parameter does reach its iteration limits a warning is printed in the protocol file, if this occurs the iteration limits should be increased and another iteration performed.

Check it in WIN-DAISY:

Open... again the experimental spectrum ...\AKMRX\001001.1R in WIN-NMR and **Export...** again the last **Multiplets Analysis** in the same way to **WIN-DAISY**. (Do *not* fragment the spin system and include *not* the non-assigned coupling constants.) Display the Program **Options** and enter again new default limits for **Scalar Coupling Constants** of ±<12> Hz. Open the **Scalar Coupling Constants** dialog box and enter rough values of coupling constants, e.g. vicinal <2> Hz, geminal <–5> Hz and check the **Iterate** option for these couplings. Open the **Control Parameters** and select the **Advanced Iteration Mode** with smoothing parameters (e.g. **medium Broadening / low No. of Cycles**), check the **Starting Frequencies poor** option and increase the **Convergence Criterion** to <0.05>.

Execute an **Iteration**.

These are different ways how to analyze the spectrum: either to use the Multiplets Analysis mode to set up the complete spin system or to determine merely chemical shifts and use the iteration algorithm of WIN-DAISY for poor trial parameters with smoothing.

3.3.1 Taking Advantage of Field and Solvent Effects

Changing the magnetic field strength a sample is measured at or the solvent it is dissolved in can have a profound effect of the appearance of the NMR spectrum. Going to higher magnetic field can simplify a spectrum by converting it from a second order to a first order spin system. Changing solvent can alter chemical shifts, make couplings appear / disappear and increase / decrease the linewidths of signals. The use of field and solvent effects in spectral analysis will be discussed using the model compound *2-bromo-2-nitropropane-1,3-diol*, but first of all it is necessary to define the model compound spin system.

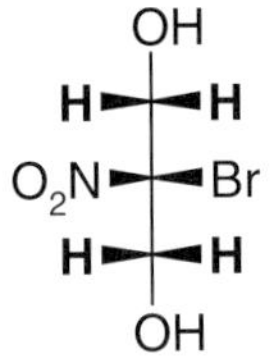

$$\begin{array}{c} \text{OH} \\ \text{H} \blacktriangleright\!\!\blacktriangleleft \text{H} \\ \text{O}_2\text{N} \blacktriangleright\!\!\blacktriangleleft \text{Br} \\ \text{H} \blacktriangleright\!\!\blacktriangleleft \text{H} \\ \text{OH} \end{array}$$

Fig. 3.33: FISHER projection of *2-bromo,2-nitropropane-1,3-diol*.

The structure features a mirror plane through the central carbon atom (which is prochiral), with the nitro- and bromine-substituent lying within the plane. The protons of each CH_2 group are diastereotopic, because they are attached to a prochiral carbon atom. (This last point may be proved by constructing the substitution products.) Due to the mirror plane in the center of the molecule the protons on each side of the prochiral carbon atom are enantiomeric. This may also be proved using the substitution criterion, substitution of the enantiotopic protons lead to enantiomeric compounds. As shown previously, enantiotopic protons cannot be distinguished by NMR because they have the same chemical shift - the latter part of this chapter will look at examples containing enantiotopic protons. In the present case, the problem of enantiotopic protons can be ignored because there is *no* long range coupling interaction between the CH_2 groups.

Taking into account all the relevant information it is sufficient to determine the spin system of only half of the compound, because the second half is identical. There will be three different proton signals in the NMR spectrum, the two protons from the diastereotopic CH_2 group and the OH proton - it is not possible to estimate the difference in chemical shift between the diastereotopic protons. The resulting spin system will be either AXY or ABX, depending on whether the resonance frequency of the OH is higher or lower than that of the CH_2 protons. Depending on the solvent the coupling to the OH proton may or may not be visible. In cases where the solvent or protic residues within the solvent exchanges with the OH proton, the coupling vanishes, because only permanently bonded protons can display coupling constants. Exchange processes will be discussed in detail in chapter 5.3, but it should be mentioned that in many spectra, due to exchange with the residual water in the solvent, a line-broadened OH signal can often be detected and easily identified. For the present spin system, if no coupling occurs between the OH and CH_2 protons, a line-broadened OH peak and an AB spin system will be observed in the spectrum. Finally, in the case where the shifts of both CH_2 protons are accidentally identical they would give rise to only a singlet.

Check it in WIN-NMR and WIN-DAISY:

Open... the spectrum, ...\AB\003001.1R, of the propanediol derivative in WIN-NMR. The spectrum has been measured in wet DMSO. As expected for the structure, three different signals are observed. Using the **Analysis | Integration** mode integrate the spectrum and, taking into account the linewidth, assign the signals.

2-bromo-2-nitropropane-1,3-diol

Measure a **Linewidth** and analyze the spin system excluding the OH signal. **Export...** the data to **WIN-DAISY**, apply a negative sign to the geminal coupling constant and optimize the parameters. Open the **Iteration Protocol** file and note the value of *R-Factor (%)*. Keep the WIN-DAISY document open.

Check it in WIN-DAISY:

Using the optimized parameters in the WIN-DAISY **Frequencies** and **Scalar Coupling Constants** dialog boxes, **Calculate** the perturbation parameter λ_{AB}. **Print...** the spectrum and estimate graphically the angle 2θ (refer to section 2.3.2.1) and calculate the relative signal intensities based on the angle 2θ.

Looking at the AB pattern in the spectrum, and ignoring any structural consideration, the four peaks could be interpreted as a quartet. If the structure of the measured compound was unknown, this possibility would also have to be considered.

Check it in WIN-NMR:

Delete the AB multiplet analysis of the present experimental spectrum ...\AB\003001.1R and analyze the four peaks as a quartet. **Export...** the data to WIN-DAISY and include the non-assigned coupling constant (**Yes**). Open the **Frequencies** dialog box, WIN-DAISY has assumed a first order spin system consisting of three equivalent spin ½ nuclei coupled to one spin ½ nucleus. Execute a **Iteration** and compare the result in the **Dual Display** mode of WIN-NMR. Compared to the AB analysis there appears to be very little difference in the spectra but examining the **Iteration Protocol** file shows that the value of *R-Factor (%)* is over five times larger. Keep this WIN-DAISY document open as well.

The special case of an AB spin system looking like a quartet pattern is defined by the parameters:

$$\lambda_{AB} = \frac{1}{\sqrt{3}}$$

$$2\theta = 30° = \frac{\pi}{6}$$

Intensities of outer lines: $1 - \sin 30° = 0.5$

Intensities of inner lines: $1 + \sin 30° = 1.5$

$$\Delta_{AB} = \sqrt{3} \cdot J_{AB}$$
$$D = 2 \cdot J_{AB}$$

Comparing the results of the two methods of analysis, particularly the R-Factor (%) values, would indicate that the spin system is correctly described as an AB and not a quartet pattern. There are two ways to confirm which is the correct spin system; change the external magnetic field strength or change the solvent used to dissolve the diol. It is one of the fundamental principles of NMR spectroscopy that chemical shifts, when measured in Hertz, are *field dependent* while coupling constants are *field independent*. Measurement at a higher magnetic field strength would not change the splitting in the quartet pattern because the splitting are caused by the coupling between spins and are therefore independent of field. But the appearance of an AB pattern is inversely proportional to the value, in Hertz, of $|v_A - v_B|$ and changing the magnetic field strength would change this value and hence the appearance of the spectrum.

Check it in WIN-DAISY and WIN-NMR:

Open the **Control Parameters** of the AB input document and select the **Mode Single Simulation**. Switch with Previous into the **Spectrum Parameters** dialog box and change the **Frequencies adapted to Spectrometer Frequency** selection to from "300" to **600** MHz. Answer the question with **Yes**. **Change** the spectrum name, e.g. to …\003998.1R and run a spectrum **Simulation** and display experimental 300 MHz spectrum and calculated 600 MHz spectrum in WIN-NMR **Dual Display**.

Use the **4 spin system in 2 groups** WIN-DAISY input document and modify here as well the **Spectrometer** in the same way. **Change** the calculated spectrum name, e.g. to … \003997.1R and run a **Simulation**. Toggle to **WIN-NMR**.

Switch into the **Multiple Display** mode and select the experimental …\AB\003001.1R and the two calculated spectra …\AB\003998.1R and …\AB\003997.1R, zoom in the multiplet spectral region and select the **Equal X** option. Use <Ctrl + X> to change the x-axis units into Hz to make clear the different frequencies of the top and the other two traces. Select by mouse click the lowest trace and expand the x-axis by the factor of two using the <> panel button.

In double expansion the signal pattern of the lowest trace (the quartet pattern) has identical characteristics as the experimental spectrum. But in the middle trace the signal pattern itself has changed its characteristics: The spin system characteristics are changed.

It has already been discussed that changing solvent can effect the values of chemical shifts and coupling constants especially in aliphatic spin systems. Additionally, changing solvent can also alter the nature of the spin system. In the present example the OH proton is not coupled to the AB spin system due to exchange. Other solvents or solvent mixtures could prevent this exchange making the coupling visible, so that the spin system changes from AB to AXY or even ABX and so prove that the spin system was an AB and not a quartet type.

Check it in WIN-NMR and WIN-DAISY:

Open... the spectrum ...\ABX\004001.1R into WIN-NMR. This is the same compound and field strength (300 MHz) as used in the original spectrum ...\AB\003001.1R but the composition of the solvent is different. **Analyze** the spin system and comment of the chemical shifts, coupling patterns and coupling constants and perform an **Iteration** with **WIN-DAISY**.

2-bromo-2-nitropropane-1,3-diol

Open... the spectrum ...\ABX\005001.1R in WIN-NMR. This is the same sample and field strength as used in the first part of this "Check it" but measured using another solvent mixture. **Analyze** the spectrum. Compare the WIN-DAISY results of both parts of this "Check it". What influence does the solvent have on the chemical shifts and coupling constants?

The next example also illustrates the strong influence of solvent effects on aliphatic spin systems.

Check it in WIN-NMR and WIN-DAISY:

Open... the proton spectrum of *2,3-dibromo-propionic acid* in C_6D_6...\AMX\008001.1R. This compound is used as a standard for AMX spectra.

Measure a **Linewidth** and **analyze** the spin system. **Export...** the data to **WIN-DAISY** and optimize the parameters. Try to assign the signals in the spectrum to the protons in the structure.

2,3-dibromo-propionic acid

The protons in *2,3-dibromo-propionic acid* all have different chemical shifts, because the methylene protons are diastereotopic. This spin system has one geminal and two vicinal coupling constants. A more reliable assignment would be possible, if the geminal coupling constant could be found but this is not obvious because two of the coupling constants are quite similar in magnitude. One possible method for assigning the geminal coupling constant is to determine the relative signs of the coupling constants using a spin tickling experiment as discussed in the previous section.

Check it in WIN-DR and WIN-NMR:

Export the WIN-DAISY optimized data to WIN-DR. Open the **Parameters | Scalar Coupling Constants...** dialog box and apply a negative sign to J_{12}, <Home + –>. Open the **Parameters | Perturbation Parameters...** dialog box and select the **Double Resonance** option. Click the **Tickling** experiment radio button and then the **Calc. Transitions: Unperturbed** button. Double click in the list box on the highest frequency transition (indices 6 - 4). Set **Field in B2** to the experimental linewidth e.g. <1> Hz. To ignore any frequency "spikes" set **Min. Absol. Linewidth [Hz]** to <0.1> Hz. **Simulate** the spectrum and examine the results in WIN-NMR **Dual Display**. Pressing this button will display the reference (unperturbed) calculation with the double resonance simulation. Determine the progressive and regressive transitions in the M and X region of the spectrum.

Toggle back to **WIN-DR**. Open the **Parameters | Scalar Coupling Constants...** dialog box again, change the sign of J_{12}, <Home + Del> and apply a negative sign to J_{23}. The values in the **Perturbation Parameters** dialog box need not to be changed, because the lowest field transition shall be irradiated still. Execute a **Simulation**. Inspect the result in WIN-NMR **Dual Display** to determine the progressive and regressive transitions.

Toggle back to **WIN-DR** and display only the perturbed spectrum (red peak button). In WIN-NMR call the **Dual Display** and load the experimental spin tickling spectrum ...\AMX\009001.1R as the **second file**. Use <Ctrl + Y> and **ALL** to overlay the signals properly. Decide whether J_{12} or J_{23} is the geminal coupling constant.

Another possible method for determining the relative signs of the coupling constants is to change the spin system to second order, currently the system is approximately first order. This might be achieved by changing the solvent. However, it has to be considered that not only will the resonance frequencies be shifted, they might also be interchanged compared with the original benzene solution spectrum.

Check it in WIN-NMR and WIN-DAISY:

Open... the experimental spectrum ...\ABX\006001.1R in WIN-NMR. **Analyze** the spin system (in practice an AXY) taking into account that the second order part is *deceptively simple*. Such patterns look like a singlet and an (ab) subsystem and arise because the inner lines of the second (ab) subsystem collapse to give a singlet while the outer lines have disappeared. It is still possible to perform a first order analysis on this part of the spectrum starting with the three intense signals. Define a doublet based on the most intense line and its right hand side neighbor, move the **Coupled Grid** so that the right hand grid line coincides with the weak line to the left of the main singlet. Again, define a doublet based on the most intense line and the remaining peak, move the **Coupled Grid** so that the right hand grid line coincides with the weak outer line on the right-hand side. The two sets of doublets of doublets are shown in Fig. 3.34. The connection of the coupling constants is then straight forward. **Export...** the data to **WIN-DAISY** and open the **Scalar Coupling Constants** dialog box to change the sign of J_{23} and remove the **iterate** option for this coupling. Optimize the NMR parameters via **Iteration** with default values and look at the result in WIN-NMR **Dual Display**. Then open the Lineshape Parameters and fix the **global linewidth** to <1> Hz and remove the **iterate** option. Open the **Scalar Coupling Constants** and click on the **Iterate all** button. Run another **Iteration**.

Hint: Another way of optimizing the parameters is as follows: Open the **Control Parameters** dialog box and select **Advance Iteration Mode** with no smoothing parameters and the **Starting Frequencies poor** option. Perform an **Iteration** and examine the result in WIN-NMR.

3.3 Conformation and NMR

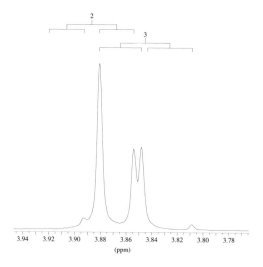

Fig. 3.34: First order analysis of deceptively simple AB part of ABX type spectra.

Check it in WIN-NMR and WIN-DAISY:

In Section 2.3.3. it has been shown that in an ABX spin systems the relative signs of J_{AX} and J_{BX} can be determined from the spectrum. For the current example, guess which might be the methylene protons and work out the relative sign combination. Open the Options dialog and check **Allow Sign change**. To confirm your choice open the **Scalar Coupling Constants** dialog box and change the sign(s) of the appropriate coupling constant(s). Perform a spectrum simulation. Compare the experiment and calculated spectrum in the **Dual Display** mode of WIN-NMR. If the fit is not particularly good, try a different sign combination and repeat the simulation until a satisfactory result is obtained.

It is most likely that the geminal protons exhibit a second order relation. For an ABX system, it is only possible to determine the magnitude of J_{AB} from the spectrum. However the relative signs of J_{AX} and J_{BX} *can* be determined. By changing the signs of the coupling constants in WIN-DAISY until there is good fit between the experimental and calculated spectrum it is possible to determine which of the three couplings is the geminal coupling J_{AB}, the sign of J_{AB} having no effect of the appearance of the X signal.

Check it in WIN-NMR and WIN-DAISY:

Open... the spectrum, ...\ABX\007001.1R, of *1-phenyl-1,2-ethanediol* in CDCl$_3$ in WIN-NMR. **Analyze** the signals between 3 and 6 ppm ignoring the broad hydroxyl signals and **Export...** the data to **WIN-DAISY**. Optimize the parameters via **Iteration**. Note the large R-Factor (%) in the **iteration protocol** file due to the neglected OH signals.

1-phenyl-1,2-ethanediol

Open... the spectrum, ...\ABX\008001.1R, of the same compound dissolved in DMSO in WIN-NMR. Again, analyze the signals in the region 3 to 6 ppm. Compare both spectra and the determined NMR parameters. What are the differences?

Check it in WIN-NMR and WIN-DAISY:

Open... the spectrum ...\AKMRX\002001.1R of the *allyl-2,3-expoxypropyl ether* in WIN-NMR. The spin system contains 10 protons. Decide first of all if the methylene protons are magnetically equivalent? (Is the molecule chiral?) Decide what signal multiplicities should be present in the spectrum?

allyl-2,3-epoxypropylether

Integrate the spectrum to identify the number of protons per signal. Determine a **Linewidth** and then switch to the **Analysis | Multiplets** mode and analyze the spectrum.

Export... the data to **WIN-DAISY** with fragmentation into two five-spin systems. Apply a negative sign to both the geminal **Scalar Coupling Constants** in both the fragments. Run a spectrum **Simulation** and compare the calculated and experimental spectrum in the **Dual Display** mode of WIN-NMR focusing in particular on the signal at ca. 4 ppm.

Open the **Control Parameters** dialog box and select **Advanced Iteration Mode** with no smoothing parameters (**none Broadening**) and the **Starting Frequencies poor** option. Execute an **Iteration**. Think about the signs of coupling constants and modify them in the **Scalar Coupling Constants**. Enter small values (sign!) for the non-determined long range couplings, **Iterate all** and run another **Iteration** till the *R-Factor* is about *0.21%*.

Because the allyl expoxypropyl ether is chiral, no magnetic equivalence is present and each of the ten protons exhibits a different chemical shift. There is no signal overlap in the spectrum except for the two geminal methylene protons of the allyl fragment at about 4 ppm. In this situation, the standard iteration mode is not able to obtain a satisfactory result. Even without any smoothing parameters, the advanced iteration mode is able to achieve the correct result provided that the additional option of poor starting frequencies, which is more sensitive to frequency shifts, is selected.

3.3.1.1 Linewidths

Most of the spin systems discussed up to this point were solved using one common linewidth for the whole spin system. There have been quite some examples where the linewidth differences arose from long range coupling constants which did not lead to a visible splitting, but to a line broadening in some of the signals. Generally spin systems should be treated with only one linewidth parameter, at least in the first steps. In case no improvement can be made nuclei specific linewidths can be introduced.

It has been mentioned already that for example OH protons often do not show any visible coupling across the oxygen, but feature a broad linewidth parameter. The absence of a splitting in some solvents is a result from a dynamic exchange process between water impurities and OH-protons (refer to section 5.3). To include these signals as well into the spin system data either nuclei specific linewidths or different fragments can be used.

Check it in WIN-NMR and WIN-DAISY:

Open.. the spectrum ...\AKMX3\001001.1R in WIN-NMR. The spin system of the C-H protons is quite clear. For the NH$_2$ and OH protons only one signal is found in the spectrum.

2-amino-1-propanol

Measure the **Linewidth** of a well resolved sharp line. Because this signal is very broad and overlaps with the **Multiplets** of the other spin system define it as a singlet in the multiplet analysis.

Export... the data to **WIN-DAISY** with fragmentation of the spin system. WIN-DAISY will appear with a document containing two fragments: the first containing the AKMX$_3$ spin system and the second consisting of only the singlet information. Open the **Frequencies** dialog box of the **1-spin system** fragment and alter the **spins per group** from "1" to <3> representing the three NH$_2$/OH protons. Open the **Lineshape Parameters** dialog box and change the **Global Linewidth** to the large value, e.g. to <50> Hz (the **upper limit** will be adapted accordingly). Run an **Iteration** and display the result in the **Dual Display** mode of WIN-NMR.

Check it in WIN-NMR and WIN-DAISY:

Open... the proton spectrum ...\ABX\007001.1R, of *1-phenyl-1,2-ethandiol* used in a previous "Check it" in WIN-NMR. Enter the **Linewidth** Analysis mode and measure a linewidth of the left-hand OH proton and note the value. Then measure the linewidth of a sharp signal and switch to the **Multiplets** mode. If not done already, analyze the signals between 3 and 6 ppm but this time include the two broad hydroxy signals in the analysis as singlets. **Export...** the data to **WIN-DAISY** with fragmentation of the spin system. Proceed as in the previous "Check it" using different **Global Linewidths** for the hydroxy fragments (about <17> Hz). Optimize the parameters via **Iteration** and compare the *R-Factor (%)* values obtained with and without the two hydroxy signals.

Both these examples show that the different linewidth(s) of non-coupled spectrum components can be treated as a global linewidth in different fragments. Including broad signals in the analysis this way achieves a better R-Factor (%) value as illustrated in the second "Check it" where the value decreases by over two orders of magnitude when the hydroxyl signals are included. Some of the signals may still show a different linewidth indicating that there might be a small coupling constant present which does not lead to a splitting in any signal.

The next example will demonstrate a different situation where coupling constants are hidden under the lineshape. The compound has already been discussed in the section about second order spin systems as an ABX type spin system surrounded by an AX type system. The following example use the same compound but dissolved in a different solvent.

Check it in WIN-NMR and WIN-DAISY:

Open... the spectrum ...\AMX\010001.1R, of the compound opposite with WIN-NMR. First **Integrate** the signals, then measure a **Linewidth** and finally enter the **Multiplets** mode and analyze the signals.

4-(2,3-dichlorophenyl)-1H-pyrrole carbonitrile

The benzene fragment no longer gives a ABX type system, but an AMX pattern with one of its signals overlapped with one of the pyrrole signals (Fig. 3.35).

The splitting pattern of all the signals can be interpreted only when assuming a coupling to the NH proton in the pyrrole fragment, but there is no splitting visible in the NH proton signal at low field.

There are now two ways to analyze the system: One way is to define a singlet for the NH signal and leave two constants unassigned, the other is to define an artificial doublet of doublet using the **Free Grid** and **Coupled Grid** lines to form the correct coupling tree (because the peak is so broad, the peak picking algorithm cannot find any individual peaks caused by the splitting).

The simplest way is the first method, proceed as follows. **Define** a **Singlet** for the NH signal. **Export...** the data to **WIN-DAISY** *with* fragmentation of the spin system and include the non-assigned couplings constants. WIN-DAISY will generate three fragments; the first fragment contains the one spin NH singlet, the second the three spin system of the benzene and the third a three spin system made up of the two pyrrole protons and an additional 'X' nucleus providing the non-assigned couplings. Open the **Frequencies** dialog box of the first fragment, place the cursor in the edit field of the resonance frequency and double click the left mouse button. Copy the highlighted value into the clip board using <Ctrl + C>. Close the dialog box. Select the third fragment containing the pyrrole data and open the **Frequencies** dialog box. Place the cursor in the third frequency field belong to the X spin and type <Ctrl + V> to insert the NH chemical shift value. *Remove* the **Disable** flag and check the **iterate** option. Open the **Scalar Coupling Constants** dialog box for the third fragment. The two non-assigned coupling constants have been assigned to the NH proton.

Now select the first fragment in the list box of the document main window and delete the NH fragment using either the **Edit | Delete fragment** option or the **Delete fragment** (recycle bin) button. Now the benzene fragment is the first fragment and the extended pyrrole fragment is the second fragment. Open the **Lineshape** Parameters of the second fragment and check the option **Nuclei specific linewidths** and press the **Iterate All** button, change the linewidth value of the third spin to <30> Hz. Execute an **Iteration**.

3.3 Conformation and NMR

[Figure: NMR spectrum showing multiplet peaks between 7.360 and 7.410 ppm, with bracketed analysis labels 3 and 4]

Fig. 3.35: First order multiplet analysis of overlapped signals.

3.3.2 Symmetry of Rigid and Non-Rigid Compounds

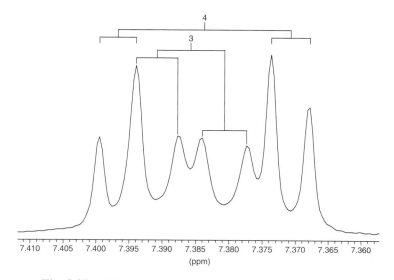

α-endosulfan

Check it in WIN-NMR:

Open... the proton spectrum of α-endosulfan ...\SYMMETRY\009001.1R. The molecule contains six protons but exhibits twofold symmetry and therefore only three chemical shifts are found in the spectrum.

Measure a **Linewidth** and switch into the **Multiplets** mode. Zoom in and inspect the signals.

Ignoring the weak outer lines, the signals at 3.94 ppm and 4.76 ppm look like two doublets of doublets, reminiscent of an AMX spin system.

In fact the three shifts A, M and X have basically a first order relationship, but with the added complication that both the X spins (the methine protons) are coupled with each other. Because of the isochronicity of the methine protons there is a very strong second order effect. Looking at the signal at 3.43 ppm the doublet of doublet pattern is not obvious.

Start the analysis with the methylene protons at low field, define both signals as doublets of doublets. Zoom in on the methine signal at 3.4 ppm. The pattern is symmetrical and shows about 14 resolved lines. Define a doublet using the first weakest outer line on the left hand side of the signal and the most intense line on the left hand side of the signal. Move the **Coupled Grid** doublet until the grid-lines matches the remaining outer line and the remaining inner line of the signal that is closest to the center. Finally define a third coupling constant by duplicating the pattern in the right hand side of the signal (Fig. 3.36).

Open the **Report...** box. Manually connect the geminal coupling constant between shift 1 and 2 and then the 3.4 Hz coupling between shift 2 and 3. The remaining coupling constant in signal 1 of about 12 Hz has to be connected with one of the remaining couplings in signal 3 (either 8.5 Hz or 15.5 Hz). Connect it to the large coupling constant and leave the 8.5 Hz coupling unassigned. **Export...** the data to **WIN-DAISY** and include the non-assigned coupling in the analysis.

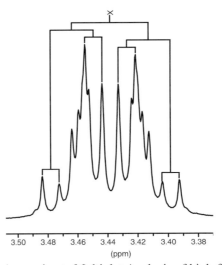

Fig. 3.36: Approximate Multiplet Analysis of high-field signal X.

Check it in WIN-DAISY and WIN-NMR:

The WIN-DAISY document contains a four spin system. Open the **Frequencies** dialog box. The non-assigned coupling is caused by a 'X' nucleus with a shift outside the spectral region. Using these initial parameters run a spectrum **Simulation** and display both spectra in the **Dual Display** mode of WIN-NMR. As expected for the low-field signals only the intense doublets of doublets are reproduced (the signal at 4.76 ppm has been shifted slightly because WIN-DAISY takes the arithmetic mean of the assigned coupling constants which were not quite right in this case). The high-field signal does not fit very well because the second order coupling is calculated as an 'X' coupling (compare Fig. 3.37e).

Toggle back to **WIN-DAISY**. Open the **Main Parameters** dialog box and change the **Number of spins or groups with magnetical equivalent spins** to the correct number <6>. Click the **Next** button and open **the Symmetry Group** dialog box, select a twofold symmetry **Cs**. Click the **OK** or **Next** button to display the **Symmetry Description**. Enter the symmetry description, <6> <5> <4> <3> <2> and <1>. The **Frequencies** dialog box now only shows three independent resonance frequencies. The **Scalar Coupling Constants** dialog box also only shows the independent coupling constant values, all others are hidden. Apply a negative sign to the geminal coupling constant (J_{12}) and execute a **Simulation**. Compare experimental and calculated spectrum in the **Dual Display** mode of WIN-NMR. There is now a much better fit between calculated and experimental data (compare Fig. 3.37d).

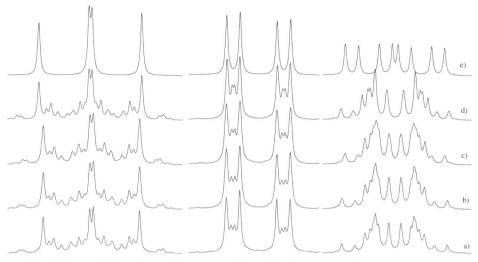

Fig. 3.37: a) Experimental spectrum (X-axis in the same expansion).
b) final iteration result including long-range couplings.
c) first iteration result ignoring long-range couplings.
d) initial simulation of the symmetric six spin system.
e) Initial simulation of the four spin system.

Execute an **Iteration** and display the result in the **Dual Display** mode of WIN-NMR. After the parameter optimization the pattern matches (compare Fig. 3.37c), but some lines are still shifted particularly the very weak outer lines at 4.76 ppm. The *R-Factor* in the **Iteration Protocol** will be about 1.64% and cannot be lowered any more, neither by changing the parameter values in the **Frequencies** or **Scalar Coupling Constants**, nor by selecting smoothing parameters or modification of parameter limits in the **Control Parameters**.

This effect is due to some long range coupling not yet taken into account by the iteration. Display the program **Options** and check the option **Allow sign changes**. Open the **Scalar Coupling Constants** dialog box and select the coupling constants J_{14} and J_{24} for **iterat**ion. (You may also **Select all** but the others will stay almost about zero). Run another **Iteration** and inspect the result in the **Dual Display** mode of WIN-NMR and the **Iteration Protocol** (*R-Factor* about *0.65%*). The fit should be perfect. Open the **Scalar Coupling Constants** dialog box containing the iteration result, the long-range couplings *must* be negative to fit the experimental spectrum. **Export...** the iteration result to the **exp**erimental WIN-NMR spectrum. Toggle back to **WIN-DAISY**.

Open the **Scalar Coupling Constants** dialog box and change the sign(s) of the long range coupling(s). Run a **Simulation** and compare the spectra in WIN-NMR **Dual Display**. The signal pattern differ distinctively. Compare the magnitude of the long range **Scalar Coupling Constants** with the linewidth in the **Lineshape Parameters**.

192   3 Structure and Spin System Parameters

Although the magnitude of the long range coupling constants is smaller than the linewidth parameter, the result on the spectrum is not a line broadening, but a shifting of the transition frequencies. Additionally the spectrum is sensitive to the *sign* of the long range coupling constant. All these fact result from the symmetry of the spin system and the strong second order effects involved!

It is instructive to compare the optimized coupling constant values with the range of values found in six-membered rings. In a non-rigid structure, as in a six-membered ring, the vicinal coupling constants are averaged over all possible dihedral angles. Using the KARPLUS equation a vicinal coupling constant value of about 5 - 7 Hz is expected in non-rigid structures, but if the compound is rigid the values should be quite different. Even though the carbon skeleton in the current example is not a six-membered ring, the vicinal coupling constants of 12 Hz and 4 Hz suggest a more rigid structure.

Check it in WIN-NMR and WIN-DAISY:

Open... the spectrum ...\SYMMETRY \010001.1R into WIN-NMR. The compound structure is similar to that discussed in the previous "Check it".

Measure a **Linewidth** and switch into the **Analysis | Multiplets** mode. Zoom in and inspect the signals. The shifts are moved a bit closer together so that the two low-field signals show a definite roof effect. The high-field signal pattern is not so well-resolved as in the previous example.

1,2,3,4,7,7-hexachloro-5,6-bis (chloromethyl)bicyclo[2.2.1]-hept-2-ene

Analyze the spectrum in the same way as in the previous "Check it" and **Export...** the spin system data to **WIN-DAISY**. Modify the spin system as in the previous "Check it" and run an **Iteration** with the default values. Then include the long range couplings and perform a final optimization. For this last step it will be necessary to extend the iteration **Lower limit** of J_{34} in the **Control Parameters Scalar Coupl.** dialog box, because the current value is very close to the lower limit.

In the present example the vicinal coupling constants are about 5.6 Hz and 7.6 Hz respectively, close to the an average value indicating that the structure is more flexible resulting into averaged coupling constants according to the KARPLUS equation, compared with that in the previous example (12.2 Hz and 3.8 Hz) where quite different values are obtained relating to a more rigid structure.

Check it in WIN-NMR and WIN-DAISY:

Open... the spectrum of morpholine, ...\MORPHOL\001001.1R with WIN-NMR.

Define the spin system, predict the number of signals expected in the spectrum and compare the predicted result with the experimental data.

morpholine

3.3 Conformation and NMR

Only two chemical shifts are present in the spectrum, why? First of all the molecule exhibits a mirror plane (through the nitrogen and oxygen) dividing the compound into two identical halves. It is often forgotten that most molecules are a three dimensional structure and that a six-membered saturated carbon skeleton is not planar. This last point is obvious if the structure is rewritten as a NEWMAN projection (Fig. 3.38).

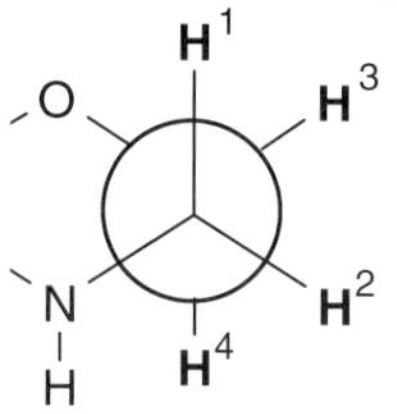

Fig. 3.38: NEWMAN projection of half of the *morpholine* molecule.

From the projection four non-isochronous spins are expected (provided that they are not accidentally isochronous). In fact the molecule undergoes rapid exchange between two chair conformations at room temperature (Fig. 3.39), refer also to chapter 5.3 [3.16].

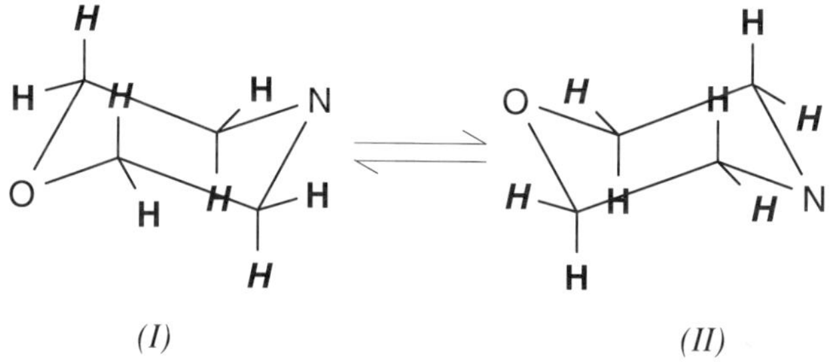

(I)   *(II)*

Fig. 3.39: Ring inversion of morpholine; italic labeled axial protons in (I) become equatorial in (II).

Due to this equilibrium two instead of four different chemical shifts are observed, the ring-inversion converts proton 1 into 2 and proton 3 into 4. In this example the symmetry of the spin system is caused by molecular dynamics and not molecular geometry.

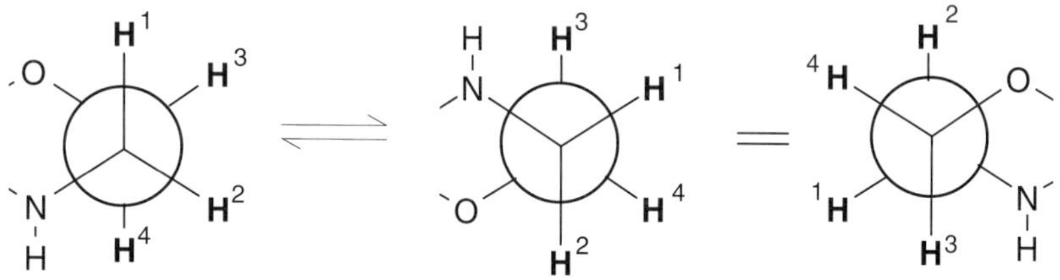

Fig. 3.40: NEWMAN projection of half the *morpholine* molecule.

Check it in WIN-NMR and WIN-DAISY:

Measure a **Linewidth** and switch into the **Analysis | Multiplets** mode. Define two singlets to mark the two chemical shifts and **Export...** the data to WIN-DAISY without fragmentation. Extend the spin system to <4> spins in the **Main Parameters** dialog box and select a twofold symmetry, e.g. **Cs** in the **Symmetry Group** window. Enter the **Symmetry Description** row based on the two different chemical shifts 1 and 2 as follows: <4> <3> <2> <1>. Open the **Scalar Coupling Constants** dialog box. Enter two different negative geminal coupling constants for J_{23} and J_{14} and give small values for the vicinal coupling constants. Finally **Select all** for iteration. Open the **Control Parameters** and select some smoothing parameters (e.g. **low Broadening** and **low No. of Cycles**). Click on the **Scalar Coupl.** button and enter appropriate iteration limits for the vicinal coupling constants J_{12} and J_{13} (e.g. from <0> to <10> Hz). Execute one or more **Iteration**s and inspect the **Iteration Protocol**(s). Compare also the results in WIN-NMR **Dual Display**.

It is conspicuous, that the standard deviations of the two geminal coupling constants are much larger than for all the other parameters and that their values tend to the upper parameter limits, even if their values are changed to lower (more negative) values and another iteration is run. Additionally the lineshape shows different linewidths for the N-lines (the large doublets) and for the other appearing signals respectively.

Check it in WIN-DAISY and WIN-NMR:

In the **Dual Display** of WIN-NMR zoom in the base line region of one of the signals. There are weak outer lines present and they might not coincide between the experimental and calculated lineshape as they are supposed to (as shown in Fig. 3.41).

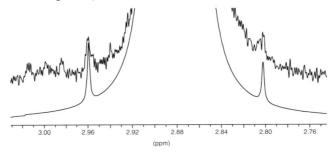

Fig. 3.41: Zoomed baseline of high-field *morpholine* signal with experimental and calculated lineshape.

The N-lines separation is defined by the value $| J_{AX} + J_{A'X} |$ which is the sum of the two vicinal coupling constants and well defined in the spectrum. Therefore the values of the coupling constants J_{12} and J_{13} do not vary very much in different iterations.

Check it in WIN-DAISY:

Open the **Main Parameters** and select the **Output Option Output of subspectra**. Run a **Simulation** and afterwards **Generate Subspectra**.

Toggle to **WIN-NMR** and load the first subspectrum ...\001998.1R and in the **Dual Display** mode ...\001998.1R as the **second file**.

The weak outer lines and the central line of each signal are composed of the symmetrical (ab) subspectrum, where the inner lines appear as to be about degenerate to a singlet. Thus the distance between the weak outer lines is about $S_{outer} = 2 \cdot J_{(ab)+} = 2 \cdot |J_{AA'} + J_{XX'}|$. The distance S_{inner} cannot be determined from the spectrum, so the parameter L cannot be calculated. The antisymmetrical (ab) subspectrum forms the second subspectrum and refers to the remaining four inner lines, defining the auxiliary parameter M. The problem why WIN-DAISY cannot optimize properly the weak outer lines is based on the very low intensity contribution. In addition the N-lines show a distinct smaller linewidth [3.17] that makes the problem even more complicated. Thus there is no way to obtain a reliable result by optimizing all spin system parameters via total-lineshape fitting. In various papers the different linewidths for N-lines in symmetrical spin systems have been reported [3.18],[3.19] but no theoretical reason is given yet.

Check it in WIN-NMR:

Determine in the **Multiplets** Mode the S_{outer} of (ab)+ subspectrum and the parameter M. Toggle to **WIN-DAISY** and run the Calculator. Determine the geminal coupling constants and enter these values into the corresponding edit fields of **Scalar Coupling Constants** and remove the **iterate** flag for both parameters. Run an **Iteration** and inspect the result in WIN-NMR **Dual Display**.

3.3.3 Advanced Examples

Check it in WIN-NMR:

Open... the 360 MHz proton spectrum ...\ABMX\002001.1R of *strychnine* in WIN-NMR. The four spin aliphatic system of interest is printed in bold in the structure opposite.

The corresponding signals are found at 3.23 ppm, 2.88 ppm and 1.89 ppm, the latter representing two strongly coupled protons [3.20].

L-strychnine

Identify the signals and examine the multiplicities. Perform a first order analysis in the **Multiplets** mode ignoring any weak outer lines. There is one problem, the signals at 1.89 ppm are strongly coupled with each other and a first order analysis is not easy at all. Ignoring any weak outer lines will miss out the geminal coupling constant (refer to Fig. 3.42).

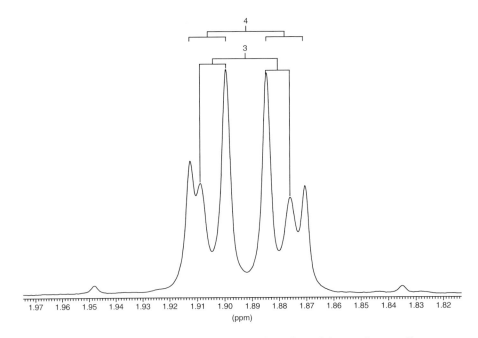

Fig. 3.42: Analysis of the signal at 1.89 ppm ignoring of the weak outer lines.

The presence of weak outer lines around each of the signals of interest proves that there is a strong second order relation in the spin system.

Check it in WIN-NMR and WIN-DAISY:

Connect the coupling constants and **Export...** the data to **WIN-DAISY**. Open the **Scalar Coupling Constants** dialog box and apply a negative sign to the geminal coupling constant J_{12}. Enter a value for the geminal coupling J_{34} of about <–14> Hz., but do *not* select J_{34} for iteration. Using the default values, run an **Iteration** and compare the calculation and experiment in the **Dual Display** mode of WIN-NMR. In case the main signal pattern does not fit, toggle back to **WIN-DAISY** and check in the **Control Parameters Scalar Coupl.** dialog box the **limits** for the coupling constants. It might be necessary to enlarge some limits when the actual parameter value is very close. Then run another **Iteration** until the *R-Factor* in the **Iteration Protocol** file reaches about *0.95%*.

Then select in the **Scalar Coupling Constants** J_{34} for **iter**ation and perform another parameter optimization via **Iteration** (*R-Factor ca. 0.84%*). Display the result in WIN-NMR **Dual Display**. **Zoom** in the signals and inspect the result.

In this example the weak outer lines represent enough intensity to be optimized in this last phase of iteration when the other couplings are already fitted. There is still a small difference in the linewidths of the different signals.

3.3 Conformation and NMR

Check it in WIN-NMR and WIN-DAISY:

Toggle back to **WIN-DAISY** and open the **Lineshape Parameters**. Select the option **Nuclei specific linewidths** and click the button **Iterate all**. Run another **Iteration** and display the result in WIN-NMR **Dual Display**. The best result refers to the *R-Factor* of ca. *0.42%* in the **Iteration Protocol**.

To identify the transitions form the individual high-field nuclei, **Display** the **Calculated Spectrum**. Then toggle back to **WIN-DAISY** and open the **Frequencies** dialog box and **disable** either spin **3** or **4**. **Change** the calculated spectrum name, e.g. to ...\002998.1R and run a **Simulation**. Then toggle to **WIN-NMR**. In WIN-NMR call the **Dual Display** and select the new calculation ...\002998.1R as the **second file**.

The chemical shifts of the geminal high-field protons are very similar. Would it be possible to turn the spin system to first order by increasing the spectrometer frequency?

Check it in WIN-DAISY:

Remove the **disable** flag from the shift in the **Frequencies** dialog. **Calculate** the perturbation parameter. **Change** the calculated spectrum name, e.g. to ...\002996.1R. Open the **Control Parameters** and switch the **Mode** to **Single Simulation** and enter with **Previous** the **Spectrum Parameters** dialog. Select **800 MHz** for **Frequencies adapted to Spectrometer Frequency**, quit the dialog with **Ok** and answer the question with **Yes**. Run a **Simulation** and display both spectra (experimental 360 MHz and calculated 800 MHz) in the **Dual Display**. **Calculate** the perturbation parameter for 800 MHz [3.21].

You may overlay the high-field methylene group signals in the frequency scale to decide whether the spin system order changed significantly or not.

Check it in WIN-NMR:

Open... the spectrum ...\ABMX\003001.1R in WIN-NMR, the proton spectrum of *ascorbic acid* (vitamin C). The structure is opposite without the interesting protons of the carbon skeleton marked (due to clarity reasons).

ascorbic acid

Measure a **Linewidth** and analyze the spectrum in the **Multiplets** mode, **Export...** the data to **WIN-DAISY** and optimize the parameters via **Iteration**. What can be concluded about the geminal coupling constant? Can it properly determined from this 80 MHz spectrum?

3 Structure and Spin System Parameters

Check it in WIN-NMR:

Open... the spectrum ...\EXTENDED\ 001001.1R in WIN-NMR. Using the given structure identify the spin system and decide whether it can be fragmented. Is there an asymmetric carbon atom present?

Measure a **Linewidth** and analyze the spectrum in the **Multiplets** mode and comment on the appearance of the allylic methylene protons.

1-(β-allyloxy-2,4-di-chlorophenylethyl)-imidazole

The spin system consists of 14 magnetic non-equivalent spins. The protons of both methylene groups in the molecule are diastereotopic. The allylic methylene group is not directly attached to the chiral centre but the effect of the chiral center propagates through the molecule making the protons non-equivalent. The signal at 5.75 ppm (Fig. 3.43) consists of sixteen resolved lines and appears as a dddd caused by four coupling constants of different magnitude. If the two protons of the allylic methylene group were magnetically equivalent, the pattern would be a ddt. The pattern shown in Fig. 3.43 is easily analyzed starting with the outermost signal and the smallest coupling constant and taking repeated spacings.

The consequence for the whole spin system is that all the patterns appearing in the spectrum are based on doublet splittings. A problem may occur when analyzing the two signals in the region of about 5.2 ppm. One signal corresponds to the proton cis to the methylene group and the other to the proton trans. The cis and trans vinylic couplings found in the multiplet at 5.75 ppm (Fig. 3.43) can be used to analyze these signals and generate the relevant coupling tree as shown in Fig. 3.44. The best way to build up the multiplet tree is to start on the left hand part of the signal.

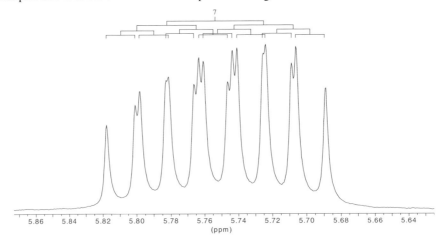

Fig. 3.43: Multiplet analysis of the signal at 5.75 ppm.

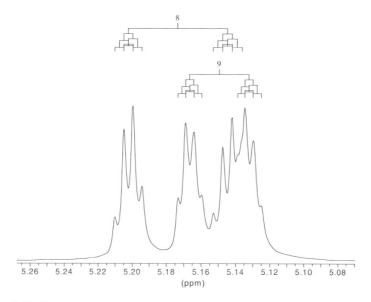

Fig. 3.44: Overlapped signal pattern of cis and trans protons of the double bond.

Check it in WIN-NMR and WIN-DAISY:

Define Identifier for the individual signals for ease the forming of the spin systems, e.g. label the first signal <A-1> ('A' for lowest field shift, '1" for first spin system), and the second <A-2> ('A' again for lowest field shift of the spin system no. '2') etc. Connect all the coupling constants to define the complete spin system. **Export...** the data to **WIN-DAISY** *with* fragmentation of the spin system.

This spin system can only be calculated in a reasonable time provided that the system is subdivided into four independent subsystems. This is only possible when there is no coupling between the different subsystems as in the present case. There are two aromatic subsystems (the dichlorophenyl and imidazol spin systems) each containing three nuclei, one aliphatic three spin system and the finally the five spin allylic system.

Check it in WIN-DAISY:

Open the **Scalar Coupling Constants** dialog box and scroll through the fragments using the **Previous** and **Next** buttons in the head line. Apply proper signs and select non-determined constants for iteration. An iteration will take considerably more time than earlier calculations, because there are 37 parameters to be adjusted. Run a parameter **Iteration** and examine the result in the **Dual Display** mode of WIN-NMR. The best result is obtained when the R-Factor in the **Iteration Protocol** is below *0.66%*. Is it possible to determine the *sign* of the geminal coupling constant in the vinyl fragment?

Check it in WIN-NMR and WIN-DAISY:

Open the proton spectrum ...\FURORE\001001.1R with WIN-NMR. Measure a suitable **Linewidth** and analyze the complete spectrum in the **Multiplets** mode. (The aliphatic part of the spectrum has been analyzed already in section 2.4).

Export... the data to **WIN-DAISY** and complete the spin system. Modify the spin system data in a way that they can be fitted with identical smoothing parameters. Run one or more **Iterations**. (*R-Factor* ca. 0.32%).

2-[4-(6-chloro-2-benzoxazolyloxy)-phenoxy] propionic acid ethyl ester

To optimize the NMR parameters one after another it would be necessary to get all the signals in separate spectral regions what is not easy in this case. The best way is to extend the symmetrical aromatic spin system as done many times before and give initial coupling constant values for para-substituted benzene to fit all four spin systems without smoothing parameters.

Check it in WIN-NMR and WIN-DAISY:

Open... the proton ...\MULTI_CP \001001.1R in WIN-NMR. Measure a suitable **Linewidth** and analyze the complete spectrum in the **Multiplets** mode.

Export... the data to **WIN-DAISY** and complete the spin system. Optimize the parameters.

carbonic acid 2-sec.-butyl-4,6-dinitrophenyl-isopropyl ester

The basic question of the present example is whether the methylene group of the sec-butyl ester consist of two magnetically equivalent nuclei or of two diastereotopic protons. Inspect the signal at ≈ 1.68 ppm. What kind of multiplicities should be present in both cases? Are there any weak outer lines visible?

3.3 Conformation and NMR 201

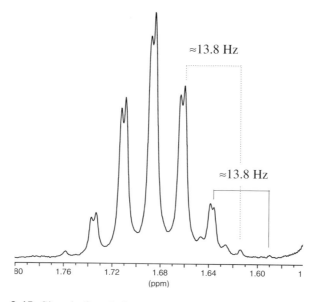

Fig. 3.45: Signal of methylene group with geminal coupling constant.

Compare as well Fig. 3.45 for the location of the weak outer lines which identify clearly the lines proving that the protons are diatereotopic. Define the Multiplets without taking into account the weak outer lines, then the resonance frequencies are sufficient to be obtain proper results without smoothing. For optimization the geminal coupling should be entered and not optimized since the other constants are adjusted. The best result is determined when the R-factor falls below 0.12%.

Check it in WIN-NMR and WIN-DAISY:

Open... the proton spectrum ...\ETHYLB\001001.1R. Measure a suitable **Linewidth**.

What kind of spin systems are present (notation)?

ethyl benzene

Analyze in the **Multiplets** mode the aliphatic part ignoring the fine structure. In the aromatic part merely **Define Singlets**.

The aliphatic spin system is of A_2X_3 type, the aromatic spin system is $[AB]_2C$.

Check it in WIN-NMR and WIN-DAISY:

Export... the data to **WIN-DAISY** and complete the aromatic spin system as follows: When the first fragment is selected execute the Merge fragments command with default destination (fragment 2), afterwards accept the message with **Yes**. Call again the Merge fragments command with default destination (fragment 3) and accept the message with **Yes**. Open the **Frequencies** dialog to check if the shifts are ordered in descending order. Then extend the **Number of spins...** in the **Main Parameters** and switch to the **Symmetry Group** dialog (select **C2** or **Cs**) and enter the **Symmetry Description** dialog and enter the permutations <5>, <4>, <3>, <2> and <1>. Now display the **Frequencies** dialog to check that only three shifts are available end switch with **Next** into the **Scalar Coupling Constants** window. Therein six independent coupling constants shall be available. Enter approximate values and run a **Simulation** to check in the WIN-NMR **Dual Display** if the shift assignment is correct, otherwise exchange one ortho and one meta coupling constant in the **Scalar Coupling Constants** dialog box. Therein select the **Iterate all** option. Open the **Control Parameters** and select some smoothing parameters. The spectrum cannot be properly fitted in this way, the *R-Factor* reported in the **Iteration Protocol** stays bigger than *6%*.

The remaining differences occur due to a long range coupling between the methylene group and all the aromatic protons which gives additional triplet splitting. To include the long range couplings the two fragments need to be merged. In addition the corresponding $^4J_{HH}$ and $^6J_{HH}$ are supposed to be negative, whilst the $^5J_{HH}$ is reported to be of positive sign.

Check it in WIN-DAISY:

To merge symmetrical with non-symmetrical spin systems the symmetry has to be removed in the **Symmetry Group** dialog box. Then select again the **Merge fragments** command (answer message with **Yes**). Restore the **Symmetry Group** and **Description**. Enter according the sign information small values for the three available long range couplings of the CH_2 group (keep in mind which shift belongs to the ortho and which to the meta protons!) Fix the **global linewdith** in the **Lineshape Parameters** to <0.3> Hz - remove the **iterate** flag - and run an **Iteration** without smoothing. The final *R-Factor* shall be about *12.53%*.

3.3.3.1 Glucose

The examples in this section have been taken from references [3.22] and [3.23], they have been chosen to illustrate two points. Firstly that the analysis of related compounds can help in the elucidation of an unknown structure and secondly how to combine individual WIN-DAISY documents. The structures of α- and β-D-glucose pentaacetate are shown in Fig. 3.46. Although the glucose ring has a carbon-oxygen skeleton it is still instructive to compare the experimentally observed coupling constants with those expected in cyclohexane derivatives; Fig. 3.47a shows the notation used to label the protons in cyclohexanes, Fig. 3.47b shows the NEWMAN project and Table 2.5 gives the typical range of coupling constants.

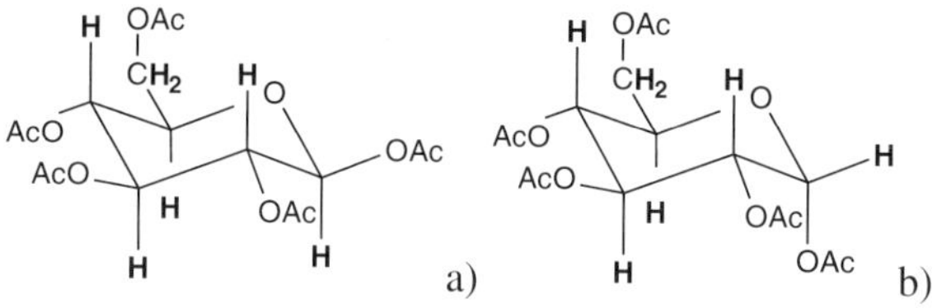

Fig. 3.46: a) β-D-glucose pentaacetate and b) α-D-glucose pentaacetate.

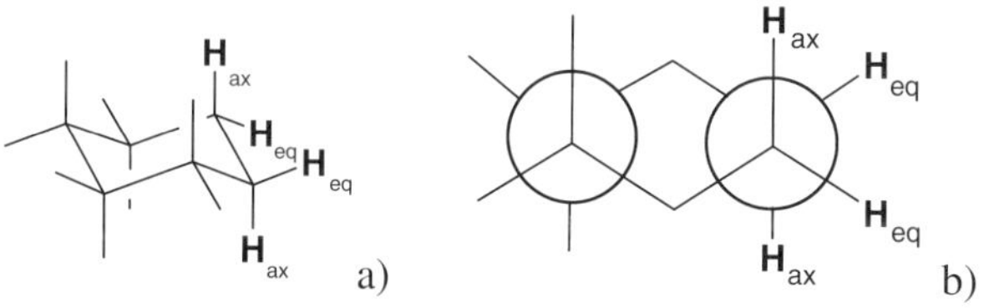

Fig. 3.47: a) Notation of axial (H_{ax}) and equatorial (H_{eq}) protons in cyclohexane fragment, b) NEWMAN projection.

Table 3.4: Range of H-H coupling constants in cyclohexane derivatives.

Coupling Constant	Coupling range [Hz]
$^2J_{gem}$	−14 to −11
$^3J_{eq,eq}$	+ 2 to + 5
$^3J_{eq,ax}$	+ 2 to + 5
$^3J_{ax,ax}$	+10 to +13

When the compound is non-rigid the vicinal H-H coupling constants are averaged over all possible dihedral angles and, according to the KARPLUS relationship, a value of 5 – 7 Hz is expected. If the structure is rigid, the vicinal H-H coupling constants are quite different as shown in Table 3.4.

Comparing the spectrum of α- and β-D-glucose pentaacetate (Fig. 3.48) the main differences are clear; the anomeric proton in the α-form is deshielded by approximately 0.6 ppm with respect to the β-form and the coupling displayed by the anomeric proton is different in the two forms. In the α-form this coupling has an equatorial-axial (ea) relationship and is smaller than in the β-form which has an axial-axial (aa) relationship (Table 2.5). Consequently it is possible to distinguish between α- and β-forms using the value of the vicinal coupling constant of the anomeric proton. The coupling constants in these six-membered rings are very different from those of aromatic systems discussed in Section 1.2.3. Finally, since both spectra are recorded in $CDCl_3$ there is no possibility of an equilibrium between the two forms as can occur in protic solvents.

204   3 Structure and Spin System Parameters

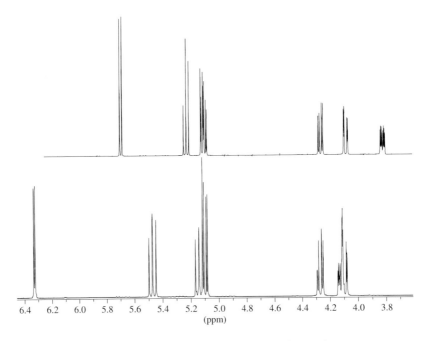

Fig. 3.48: Upper trace: β-D-glucose pentaacetate experimental spectrum.
Lower trace: α-D-glucose pentaacetate experimental spectrum.

Check it in WIN-NMR:

Open... the spectrum of β-D-glucose pentaacetate ...\GLUCOSE\001001.1R into WIN-NMR. Measure a **Linewidth** and then switch to the **Analysis | Multiplets** mode and perform a first order analysis of the signals. The only overlapping signals appear at 5.1 ppm, the signals are not strongly coupled and may be analyzed as two overlapping doublet of doublets (refer to Fig. 3.49). The connection of the coupling constants is not difficult because all the couplings, even the axial-axial, differ significantly.

The assignment of each multiplet to the structure shown in Fig. 3.46a is easily completed starting with the anomeric proton. Annotate the multiplets with the IUPAC nomenclature. To annotate a multiplet, mark the desired multiplet with the mouse in the **rectangular cursor** mode. Click the right mouse button within the spectrum window and select the option **Define Identifier** in the context menu. In the **Identifier** edit window replace the question mark '?' by the IUPAC notation of the corresponding proton in the structure, e.g. place the mouse cursor in the edit field, double click the left mouse button and then enter <C1-H> for the anomeric proton. The annotation text should not exceed eight characters. Mark the signal at highest field with the **maximum cursor**. **Export...** the data to **WIN-DAISY**.

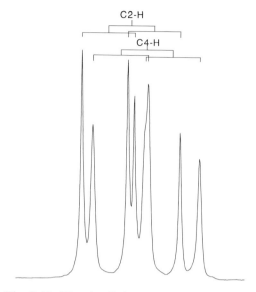

Fig. 3.49: Signal splitting of the signals at 5.1 ppm.

Check it in WIN-DAISY:

Open the **Frequencies** dialog box and click the **>>>** button in the footer to enlarge the dialog box. The annotation entered for the multiplet analysis in WIN-NMR is also displayed in WIN-DAISY. In the Scalar Coupling Constants dialog box apply a *negative* sign to the geminal coupling constant. Run a spectrum **Iteration** and display the results using the **Dual Display** mode of WIN-NMR. The *R-Factor* in the **Iteration Protocol** is about 2.64% due to different linewidths in the spectrum.

Toggle back to **WIN-DAISY** and **Export** to **WIN-NMR exp.** the WIN-DAISY result for use in future "Check its". Keep the WIN-DAISY document open.

Before continuing with the analysis the assignments in β-D-glucose pentaacetate can be confirmed using a simple homo-nuclear decoupling experiment. Irradiation of the anomeric proton will confirm which of the overlapping signals at ca. 5.1 ppm is C2-H, while irradiation of C5-H will confirm the position of C4-H and the C6 methylene group.

Check it in WIN-NMR and WIN-DAISY:

Open... the experimental off-resonance reference spectrum ...\GLUCOSE \003001.1R into WIN-NMR and load as the **second file** the proton spectrum ...\GLUCOSE\001001.1R into the **Dual Display**. Use <Ctrl + Y> to align the Y-scales of both spectra. Use the **Move Trace** option and overlay the signals. Switch with <Ctrl + X> the x-axis units into Hz to note the frequency shift in the header status line. Toggle to **WIN-DAISY** and open the **Spectrum Parameters** dialog box. Enter the frequency shift, e.g. <–16> Hz into the **Offset** edit field and press the **Apply!** button.

Check it in WIN-DR and WIN-NMR:

Execute the command **Export WIN-DR**. Prepare the input document for decoupling of the anomeric proton. Open the **Perturbation Parameters** dialog box is opened. Select **Experiment: Decoupling**, set the **Field in B2** to ca. <17> Hz and check **Effected Spin(s)** <1>. Enter a **min. absol. linewidth** of <0.5> Hz. Run a **Simulation** and **Display** the **Calculated Spectrum** (red peak button) in WIN-NMR. In WIN-NMR switch to the **Display | Dual Display** mode and load the experimental spectrum ...\GLUCOSE\004001.1R as the **second file**. Click the **Move Trace** button and align the X- and Y-axes. Zoom in the two overlapping signals at ca. 5.07 ppm.

The decoupling of the anomeric proton only effects C2-H which is coupled to C1-H. The irradiation leaves signal C4-H unaffected.

Check it in WIN-DR:

Experiment with different **Field in B2** values and inspect the results in the WIN-NMR **Dual Display** to get a feeling for the influence of this parameter.

To compare the double resonance simulation with the unperturbed simulation use the **Dual Display** (blue and red peak button) of WIN-DR. **Zoom** in the region at 5.1 ppm and examine the effect of the decoupler field. It is possible to distinguish between the signals belonging to C2-H (affected) and C4-H (unaffected).

Check it in WIN-DR and WIN-NMR:

Open the **Perturbation Parameters** dialog box change the **Effected Spin(s)** from "1" to <5>. Quit the dialog with **Ok**. Run a **Simulation** and **Display** the **Calculated Spectrum** (red peak button) in WIN-NMR. Switch to the **Display | Dual Display** mode and load the experimental spectrum ...\GLUCOSE\005001.1R as the second trace. Zoom in on the two protons of the C6 methylene group at 4.15 ppm. Examine the signal at 5.07 ppm where C2-H and C4-H are situated. This time the signal arising from C4-H is disturbed. Due to the coupling to the methylene C6-H coupling the corresponding signals are disturbed as well. To make sure that this signal is C4-H compare the perturbed and unperturbed spectrum in the **Dual Display** (blue and red peak button) of WIN-NMR.

Homo-nuclear decoupling experiments can be used to establish the coupling network within a molecule and consequently its structure. The simulation of these two double resonance spectra clearly shows that C2-H is the left multiplet of the overlapped signal at 5.1 ppm (disturbed by irradiation of C1-H) and C4-H is the right hand side signal (disturbed by irradiation C5-H). The last calculation shows some differences in the simulated pattern which is a sign, that there might be some long range couplings present.

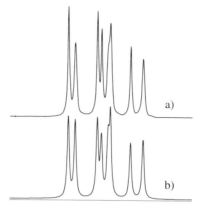

Fig. 3.50: a) Experimental spectrum signal at 5.1 ppm.
b) Calculated spectrum signal at 5.1 ppm (only up to 3-bond coupl.).

A comparison of the experimental and calculated spectra for the overlapping signals at 5.1 ppm (Fig. 3.50) shows a stronger roof effect in the experimental spectrum. This difference in the intensity of the peaks results from long range couplings over four bonds which are so small they cannot be resolved and they contribute to the overall linewidth.

Check it in WIN-NMR:

Expand the signal at 5.1 ppm, change to the **Analysis | Multiplets** mode and plot the spectrum ensuring that the **Multiplets** option is selected in the **Page Layout...** Switch to the **Analysis | Linewidth** mode and measure the linewidths for the individual lines and write them on the spectrum.

Check it in WIN-NMR and WIN-DAISY:

Open... the experimental spectrum ...\GLUCOSE\003001.1R and enter the **Interval** mode and **Export...** any region to **WIN-DAISY**. In WIN-DAISY press the **Connect** button and run a **Simulation**. Enter a small value for the long-range coupling for nucleus 4 (C4-H) to the methylene protons (C6-Ha, C6-Hb), e.g. <−0.3> Hz and/or <0.3> Hz. Run a spectrum **Simulation** and examine both spectra in the **Dual Display** mode of WIN-NMR. Try out different sign combinations. **Close** the input document.

The long-range coupling constants do not lead to any visible splitting of signals so consequently it is not possible to get reliable values for neither the magnitude nor the sign of these couplings from the spectrum. However, the effect of long-range coupling should be kept in mind in future "Check its".

Check it in WIN-DAISY and WIN-NMR:

Toggle to **WIN-NMR** and switch to the **Analysis | Multiplets** mode. Click the **WinDaisy** button and **Export...** the WIN-DAISY result. Toggle to **WIN-NMR** and load the experimental spectrum of the corresponding α-isomer ...\GLUCOSE\002001.1R. Switch to the **Analysis | Interval** mode and press the **Regions**... button. **Load...** the predefined iteration regions. Mark one peak with the **maximum cursor** and then **Export...** the interval data to WIN-DAISY, do not export any multiplet data even if it exists. Accept the message box WIN-DAISY displays and press the **Connect** button.

WIN-DAISY will detect that the new experimental data was recorded at a different field strength (400 MHz in contrast to 500 MHz for the previous experiment). Answer **Yes** the question regarding adjusting the resonance frequency in order to adapt the resonance frequencies to the lower field strength (to keep the ppm values). Run a spectrum **Simulation** and display the experimental and calculated spectra using the **Dual Display** mode of WIN-NMR. It will be necessary to switch the Y-axis into relative units, <Ctrl + Y>, to allow a proper comparison for the Y-scale.

The method just described is the correct way to make use of previously calculated data from spectra of similar compounds - even if they were recorded at different field strength. The WIN-NMR Dual Display will currently show the calculated 400 MHz spectrum of β-D-glucose pentaacetate with the experimental 400 MHz spectrum of α-D-glucose pentaacetate. It is possible to perform a complete multiplet analysis on the α-D-glucose spectrum, but it is simpler to adapt the calculated β-D-glucose data. The coupling constants are similar but more importantly, second order effects are stronger in α-D-glucose so that it is possible that the multiplet analysis will be much more difficult.

It is highly unlikely that the order of the protons in the spectrum will be exactly the same in the α- and β-D-glucose. These differences in ordering could be left for the iteration algorithm to sort out by defining iteration regions large enough to cover the whole range of experimental and simulated signals. This would enable the algorithm to move the shifts into the correct positions but would take a lot of calculation time. The quickest and most instructive method is to correct the resonance frequencies manually using the Dual Display mode of WIN-NMR. Before starting any calculation change the value of the coupling constant for the anomeric proton (J_{axeq} in α-D-glucose instead of J_{axax} in β-D-glucose).

Check it WIN-DAISY and WIN-NMR:

Toggle back to **WIN-DAISY** and open the **Scalar Coupling Constants** dialog box. Change the value of J_{12} corresponding to the anomeric coupling constant, e.g. to <3.5> Hz. Run a spectrum **Simulation** and display both spectra in the **Dual Display** mode of WIN-NMR. Change the Y-scale into relative units, <Ctrl + Y> and the X-scale into frequency units, <Ctrl + X>. Plot both spectra and annotate the signals in the simulated spectrum using the same labels as used in the experimental spectrum of β-D-glucose pentaacetate.

Click on the **Move Trace** button. Holding down the <Shift> key drag the calculated spectrum across the experimental spectrum until the anomeric proton matches. It will be necessary to **Zoom** in on this region to ensure a proper match. The WIN-NMR status header line displays the shift of the experimental signal with respect to the calculated one. Make a note of this value, "–256 Hz", next to the anomeric proton on the plotted spectrum. Repeat this process for all the other signals. The relative positions of some signals will change, but the signals coupling patterns will not change dramatically.

Approximate corrections for the resonance frequencies are given in Table 3.5. Note that the order of the entries in the table corresponds to the order of the entries in the WIN-DAISY Frequencies dialog box and is different from that in the WIN-NMR multiplet table. The noticeable differences are the large high-frequency shifts for C1-H, C3-H and C5-H; the other protons having smaller shifts. The relative order of the protons also changes with C2-H exchanging position with C4-H, and C6-H(B) exchanging position with C5-H.

Table 3.5: Approximate resonance frequency corrections for β-isomer nuclei.

proton IUPAC notation	resonance frequencies of β-D-glucose pentaacetate (adapted to 400 MHz)	corrections to apply $\Delta = \alpha - \beta$	resulting α shifts $\alpha = \beta + \Delta$
C1-H	2277 Hz	+ 256 Hz	2533 Hz
C2-H	2043 Hz	− 3 Hz	2040 Hz
C3-H	2090 Hz	+ 100 Hz	2190 Hz
C4-H	2040 Hz	+ 18 Hz	2058 Hz
C5-H	1528 Hz	+ 122 Hz	1650 Hz
C6-H(A)	1705 Hz	+ 3 Hz	1708 Hz
C6-H(B)	1633 Hz	+ 7 Hz	1640 Hz

Check it in WIN-DAISY, Calculator and WIN-NMR:

Adapt the values in the WIN-DAISY **Frequencies** dialog box to match the signal location of the α-isomer. Ensure that the chemical shifts are in **Hz**. Start the Windows **Calculator** by clicking the button in the tool bar. Adjust the position and size of the WIN-DAISY and the Calculator windows so it is easy to switch between them. Open the **Frequencies** dialog box and position the mouse cursor in the edit field for spin 1, double click the left hand mouse button and copy the value into the clip board <Ctrl + C>. Switch to the Calculator window, position the cursor in the edit field, double click the left hand mouse button again and paste the shift from the clip board <Ctrl + V>. Add 256 to the original chemical shift by entering, <+>, <2>, <5>, <6> and <=> (or click the corresponding buttons). Copy the answer into the clip board and paste it into the WIN-DAISY edit field for spin 1. Repeat this for all the other spins. Open the **Scalar Coupling Constants** dialog box and make sure that all long range coupling constants are zero.

Run a spectrum **Simulation** and compare simulated and experiment spectra in the **Dual Display** mode of WIN-NMR. There is a reasonable agreement between the simulated and experiment data, if there is a significant difference in chemical shift return to WIN-DAISY and modify the chemical shift of the appropriate spin in the **Frequencies** dialog accordingly. Toggle back to **WIN-DAISY** and check in the **Control Parameters** dialog box the parameter limits in the **group parameters** dialogs. Optimize the NMR parameters via **Iteration** and compare the experimental and calculated spectra in the **Dual Display** mode of WIN-NMR.

The value of *R-Factor* should drop to ca. *3.5%* in the **Iteration Protocol**.

In the β-isomer, the effect of long range couplings was discussed but discounted as it appeared to have very little effect on the spectrum. Nevertheless, in the current example the long-range coupling should be considered in the calculation of the lineshape.

Open the **Scalar Coupling Constants** dialog box and enter the value <0.3> Hz for the four-bond coupling constants J_{45} and J_{46}. Simulate the spectrum and display the experimental and calculated spectra using the **Dual Display** mode of WIN-NMR (Fig. 3.51).

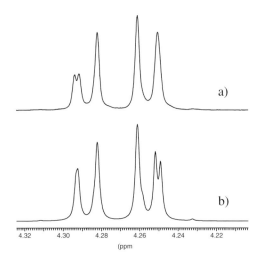

Fig. 3.51: a) Experimental spectrum signal at 4.27 ppm.
b) Calculated spectrum including a positive long-range coupling.

It is obvious from Fig. 3.51 that the effect of the long-range coupling is the wrong way around. In the experimental spectrum (Fig. 3.51a) the left outer line is split and the right outer line is broadened while in the calculated spectrum (Fig. 3.51b) the effect is reversed.

Check it in WIN-DAISY and WIN-NMR:

Toggle back to **WIN-DAISY**. Open the **Options** dialog and select the option **Allow sign change**. Then display the **Scalar Coupling Constants** dialog box and select the long-range coupling constants J_{45} and J_{46} for iteration.

Run a **Standard Iteration** without smoothing parameters (**none Broadening / any No. of cycles** in the **Control Parameters** dialog box). Display the result in the **Dual Display** mode of WIN-NMR. Zoom in on the region at 4.27 ppm, the splitting of the signal should now be reproduced exactly.

Toggle back to **WIN-DAISY**. Open the **Scalar Coupling Constants** dialog box and inspect the optimized values of the long-range coupling constants. They are *negative*. This spectrum also has additional long-range coupling constants but they are not easy to determine because they mainly led to line broadening rather than to line splitting. In addition, the experimental data shows some base line distortions further complicating the calculation. To try and determine these other long range coupling constants perform a step by step iteration without any smoothing parameters starting with all the other couplings effecting spin 1. Select them for iteration and give initial small parameter values. Ensure that the linewidth is not iterated: In the **Lineshape Parameters** *remove* the **iterate** option and fix the **global** linewidth to <0.5> Hz. Initially the value of *R-Factor* will increase as the effects of the long range coupling is no longer masked by the large global linewidth parameter. As more long range couplings are introduced the value of *R-Factor* should drop to ca. 0.73%. The **Scalar Coupling Constant** values being less than 0.2 Hz shall be changed back to <0> Hz (J_{13}, J_{24}, J_{25}, J_{26}, J_{27}). **Export** the result to **WIN-NMR exp.**.

(A final **Iteration** using **Nuclei specific linewidth** and **Iterate all** in the **Lineshape Parameters** would give an *R-Factor* of ca. 0.23%, but in the same way it would abolish the influence of the long-range couplings and decrease their values.)

The examples of α- and β-D-glucopyranose pentaacetate show two interesting features; how it is possible to use previously analyzed data in the analysis of similar compounds and how to decide whether the magnitude and sign of long-range couplings have any influence on the spectrum.

Check it in WIN-NMR:

Open... the spectrum ...\TOCSY1D\001001.1R in WIN-NMR. The spectrum only shows the ring protons of a peracetylated oligosaccharide. Using the information worked out in the previous examples try and decide how many sugar molecules are condensed and whether the saccharide includes α- and / or β-pyranosidic forms. Concentrate on the shifts of the anomeric protons of the sugar fragments. These signals are easily recognized because they show only one large coupling constant. The size of this coupling constant corresponds to either $^3J_{ae}$ in case of an α-sugar or $^3J_{aa}$ for a β-sugar. Use the **Analysis | Multiplets** mode to determine these three coupling constants. You may also take advantage of the **Integration** of the signals to determine the number of protons involved.

Compare the oligosaccharide spectrum with the spectra of the α- and β-glucose, ...\GLUCOSE\001001.1R and ...\GLUCOSE\002001.1R respectively, using the **Multiple Display** mode of WIN-NMR.

An analysis of the spectrum of this trisaccharide consisting of ß-D-glucose elements, using the Multiplets mode of WIN-NMR, would give information about the position of the glycosidic bonding of the three sugars in the structure.

Check it in WIN-NMR:

With the assumption that the sugars are all β-D-glucose elements switch to the **Analysis | Multiplets** mode and try to analysis the spectrum as done for the ß-D-glucose pentaacetate. The analysis is not easy, especially the building of the coupling connections because a lot of the couplings are quite similar and there is a lot of line overlap. It is not possible to complete the analysis and this illustrates how difficult the analysis of spectra can be. **Return** to the normal display without saving the result (**No**).

It would simplify the analysis greatly if it was possible to record the spectrum of each sugar fragment separately in a 1D experiment which could then be analyzed and compared with the plain β-D-glucose spectrum. This may be achieved using the 1D TOCSY experiment (refer to [3.22]).

Check it in WIN-NMR:

Load the three 1D TOCSY spectra ...\TOCSY1D\002001.1R, ...\TOCSY1D\003001.1R and ...\TOCSY1D\004001.1R together with the reference spectrum \TOCSY1D\001001.1R into the **Multiple Display** window of WIN-NMR. Each of the 1D TOCSY spectra represents one sugar fragment of the trisaccharide compound.

The analysis of the three TOCSY experiments can be performed in basically two different ways. The first way is to analyze each spectrum using the WIN-NMR Multiplets mode and optimize the parameters after exporting the data to WIN-DAISY. The second way is to use the pre-analyzed β-D-glucose data, connect the WIN-DAISY input with the TOCSY spectrum, adapt the resonance frequencies and optimize the data.

Although slower than the multiplet analysis, the second method guarantees that the chemical shifts in all the TOCSY spectra are in the same order as in β-D-glucose. Changes in the chemical shifts relative to β-D-glucose can give information about the glycosidic bonding. The following two "Check its" described the two different methods. Analyze the TOCSY data using first one and then the other method and draw conclusions regarding the applicability of the two different methods. Note that due to the number of parameters that require optimizing calculation times will take several minutes, the exact times depending on the type of computer being used.

Check it in WIN-NMR and WIN-DAISY:

Open... the first 1D TOCSY spectrum ...\TOCSY1D\002001.1R in WIN-NMR. Determine a **Linewidth** and then switch into the **Analysis | Multiplets** mode. Perform a complete multiplet analysis including the connection of the coupling constants. Enter the IUPAC nomenclature with a pre-placed alphabetical letter to identify the TOCSY experiment (e.g. A for the present spectrum, these letter will be used in the following) using the **Define Identifier** option in the context menu, e.g. enter the string <AC1-H> for the anomeric proton of TOCSY experiment ...\002001.1R. Build up the coupling network. Mark one signal with the **maximum cursor** and then **Export...** the data to **WIN-DAISY**, run a **Simulation** and use the Dual Display to check whether the connection are build up correctly. In case the signals between 5-5.3 ppm differ in the calculation significantly from the experiment the network was incorrect. If that is the case either exchange the corresponding chemical shifts in the **Frequencies** dialog or **Close** the WIN-DAISY document and toggle back to WIN-NMR. Measure again a **Linewidth**, enter the **Multiplets** mode and **Disconnect** the couplings in the **Report...** box and build up manually a different network, mark one peak with the spectrum cursor and **Export...** the new analysis again to **WIN-DAISY**. The biggest problem is the assignment of C3-H and C4-H. Choose one possibility and go on with the analysis as follows.

Open the **Scalar Coupling Constants** dialog box and apply a negative sign <–> to the geminal coupling constant. This coupling constant is easily identified because the two nuclei involved have three coupling constants. Run a spectrum **Iteration** and display the experimental and calculated spectra using the **Dual Display** mode of WIN-NMR.

Toggle back to **WIN-DAISY** and **Export WIN-NMR exp.** to keep the result stored with the experimental spectrum. Do *not* close the WIN-DAISY document because it will be needed in other "Check its".

Check it in WIN-NMR and WIN-DAISY:

Open... the experimental spectrum of β-D-glucopyranose pentaacetate \GLUCOSE\001001.1R with WIN-NMR. Switch directly into the **Analysis | Multiplets** mode and press the **Windaisy** button to display the WIN-DAISY optimized parameters (you need not to measure a linewidth). Open the **Report...** dialog box and **Export...** the data to **WIN-DAISY**. Load the first 1D TOCSY experiment \TOCSY1D\002001.1R into WIN-NMR and select the **Analysis | Interval** mode. **Load...** the predefined **Regions...** file and **Export...** them to **WIN-DAISY** *without* sending any multiplet data.

In WIN-DAISY **Connect** the TOCSY experimental data to the input file as explained in detail for the α-glucose. Run a spectrum **Simulation** and display both spectra using the **Dual Display** mode of WIN-NMR. Use the command <Ctrl + Y> to align the Y-intensities of both spectra and <Ctrl + X> to switch the X-units into Hz. Click on the **Move Trace** button and holding down the <Shift> key drag the two spectra over each other until the appropriate signals match. Make a note of the difference in resonance frequencies displayed in the header line. Run the **Calculator** and open the **Frequencies** dialog in WIN-DAISY and add the frequency differences to the current frequency values. Press the >>> button and place a mark in front of the IUPAC **annotation**s to identify the number of the TOCSY experiment, e.g. insert an <A> in front of "C1-H" (leading to "AC1-H") for the anomeric proton and so on. This will be of use later. **Simulate** and display the spectra in the **Dual Display**. If the coincidence of signals between calculation and experiment is acceptable perform a spectrum **Iteration** otherwise return to **WIN-DAISY** and modify the chemical shift of the appropriate spin accordingly. If the signal overlap is not particularly good, choose some smoothing parameters in the **Control Parameters**. Display the result using the **Dual Display** mode of WIN-NMR. Toggle back to **WIN-DAISY** and **Export** the result to **WIN-NMR exp**.

The results obtained from a 1D TOCSY spectrum are not as good as that for normal experimental 1D spectra; intensity distortions and broader linewidths means that the iteration statistics are worse, e.g. 5 – 9 %. These differences are easily seen in the WIN-NMR Dual Display. However, a perfect fit is not the main purpose of the analysis, the aim of the analysis is to determine the parameters for the separate fragments in order to solve the complete system.

Check it in WIN-NMR and WIN-DAISY:

Perform the analysis of the remaining two 1D TOCSY experiments ...\TOCSY1D\003001.1R and ...\TOCSY1D\004001.1R using one of the methods described for the first TOCSY example.

To try out another assignment of C3-H and C4-H display first the **Iteration Protocol** and note the *R-Factor*. Exchange the chemical shifts in the **Frequencies** dialog box (copy one value into the clip board <Ctrl + C>, overwrite the value in the same edit field with the shift of the other nucleus and paste the value from the clip board <Ctrl + V> into the other edit field). Run an **Iteration** and compare the *R-Factor* value with the one determined previously.

Remember to keep the WIN-DAISY input documents open, to give a different annotation for every spin relating to the TOCSY experiment and **Export** the result to the **WIN-NMR exp**erimental spectrum. In addition, change the title line displayed in the input document main window in order to be able to distinguish between the 7-spin fragments (e.g. <B - ring of the trisaccharide for the solution of ...\TOCSY1D\003001.1R). To change the title, double click on the title line in the list box and enter a string into the **Title** edit field in the **Main Parameters** dialog box. In case not all three documents are available in WIN-DAISY you may toggle to **WIN-NMR,** load the experimental TOCSY spectrum, e.g. ...\TOCSY1D\002001.1R, enter the **Multiplets** mode, press the **Windaisy** button, open the **Report...** box and **Export...** the data to WIN-DAISY (ignore that there is no linewidth available because there is the optimized value transferred). In this way the optimized spin system data are restored in a WIN-DAISY document.

The result of the analysis of the individual 1D TOCSY experiments is three separate WIN-DAISY documents containing the NMR parameters of each sugar fragment. The remaining step is the combination of the three input files and connection to the trisaccharide spectrum.

Check it in WIN-DAISY:

When more than one input document is open the **Edit Copy fragment to...** button of the WIN-DAISY toolbar is enabled. (The corresponding menu entry is found in the **Edit** pull-down menu: **Copy fragment to**....) In your WIN-DAISY window the three TOCSY input files should be present. With the last analyzed TOCSY input file corresponding to the experiment ...\TOCSY1D\004001.1R as the active (highlighted) window. Press the **Copy fragment to ...**. A dialog box appears which offers in a combo box the filenames of all the other WIN-DAISY input files that are currently open. Select the input document corresponding to the first TOCSY experiment ...\TOCSY1D\002001.1R as the file to copy this fragments to. The default filenames are numbering consecutive e.g. DAISY1.MGS, DAISY2.MGS etc. as they are created. Press the **OK** button. Copy the remaining TOCSY data into the first TOCSY experiment file as well to combine all three fragments in one document. Take the destination WIN-DAISY document to the foreground (**Window I Cascade**) and **Close** the document windows relating to the second and third TOCSY experiments. The remaining input document main window should now contain the titles of all three fragments with the word (copy) at the end of the second and third title string. Toggle to **WIN-NMR**.

Check it in WIN-NMR and WIN-DAISY:

Open... the experimental trisaccharide spectrum, ...\TOCSY1D\001001.1R, in WIN-NMR. Call the **Analysis I Interval** mode mark one peak with the spectrum cursor, **Load...** the predefined regions and **Export...** them to **WIN-**

DAISY and accept the message with **Yes**. In WIN-DAISY **Connect** the data of the TOCSY reference spectrum just sent from WIN-NMR. Open the dialog box of the **Main Parameters** (double click on any title line in the main document window). Deselect any checked **Output Options**. Use the **Next** or **Previous** buttons to display the **Main Parameters** of the other two fragments and make sure that they also have no **Output Option** is selected. Run a spectrum **Simulation** and display the result together with the experimental spectrum using the **Dual Display** mode of WIN-NMR.

Export to WIN-NMR exp. the optimized NMR parameters. All the **Output Options** in the **Main Parameters** of the fragments have to be switched off because the number of calculated lines for all three fragments exceeds the maximum number of 1000 transitions that WIN-NMR is able to store and display. Zoom and shift the spectrum, each multiplet is labeled with the fragment type and spin number. In the **Report...** box the shifts are ordered according to their spin system.

Table 3.6 lists the changes in chemical shift in decreasing order of (absolute) magnitude for the three glucose fragments and Table 3.7 the differences in values of the scalar coupling constant.

Table 3.6: Approximate shift differences in comparison to the β-D-glucose data.

ring A			ring B			ring C		
nucleus		Δ [Hz]	nucleus		Δ [Hz]	nucleus		Δ [Hz]
C6-H$_A$*	−	315	C2-H	−	720	C1-H	−	507
C6-H$_B$*	−	46	C1-H	−	630	C6-H$_B$	+	90
C5-H	+	20	C5-H	−	87	C3-H	+	65
C2-H	+	18	C4-H	−	80	C2-H	−	61
C3-H	+	15	C3-H	−	41	C5-H	−	51
C4-H	+	9	C6-H$_A$	−	10	C6-H$_A$	+	34
C1-H	+	9	C6-H$_B$	+	4	C4-H	+	22

Comment: The * for the methylene protons at C6 in ...\TOCSY1D\002001.1R indicates that although the labeling has been kept in the same order in the table the shifts have interchanged so that C6-H$_B$ now represents a higher shift than C6-H$_A$. In this way the coupling constant indices in Table 3.7 do not change.

Table 3.7: Comparison of coupling constants with β-D-glucose data

3 bond constants, coupling partners		β-glucose value [Hz]	ring A Δ [Hz]	ring B Δ [Hz]	ring C Δ [Hz]
C1-H	C2-H	8.34	+ 0.06	− 0.81	− 0.26
C2-H	C3-H	9.58	+ 0.12	− 0.30	+ 0.10
C3-H	C4-H	9.42	− 0.06	+ 0.20	− 0.05
C4-H	C5-H	10.12	− 0.04	− 0.30	− 0.09
C5-H	C6-H$_A$	4.53	+ 1.09	+ 0.46	+ 1.28
C5-H	C6-H$_B$	2.22	− 0.10	+ 0.12	+ 0.60
C6-H$_A$	C6-H$_B$	−12.51	+ 1.63	+ 0.27	+ 0.34

What conclusions can be drawn from the analysis of the 1D TOCSY data?

Comparing the β-D-glucose pentaacetate spectrum with the TOCSY spectrum ...\TOCSY1D\002001.1R, it is very clear from the chemical shift differences and coupling constants, that the structural parts of C1, C2, C3 and C4 remains almost unchanged. There are significant changes in the chemical shift of C5-H and of the methylene protons C6-H_A and C6-H_B, the shifting of C6-H_B beyond C6-H_A to low frequency being particularly notable. These effects are easily observed by comparing the spectra as shown in Fig. 3.52. In addition, the large changes in the vicinal and geminal coupling constants of the these three spins leads to the conclusion that the glycosidic bonding must be at C6 (Fig. 3.53).

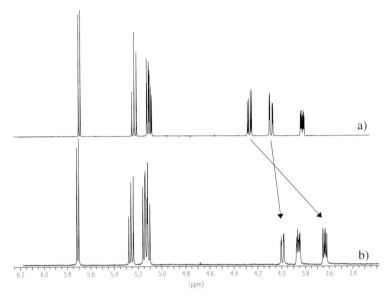

Fig. 3.52: a) β-D-glucose pentaacetate spectrum
b) ...\TOCSY1D\002001.1R spectrum,
the arrow indicates the frequency shifts.

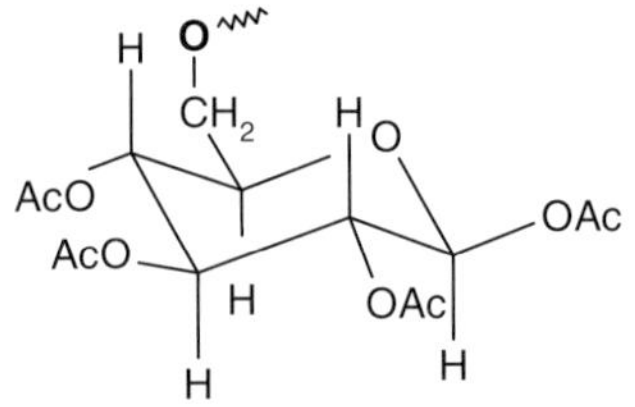

Fig. 3.53: Resulting structure fragment from ...\TOCSY1D\002001.1R.

In the second TOCSY experiment ...\TOCSY1D\003001.1R C1-H and C2-H are effected the most (Fig. 3.54). The changes in chemical shift of C1-H (ca. 5.7 ppm to ca. 4.4 ppm;) and C2-H (ca. 5.1 ppm to ca. 3.5 ppm) are larger than those found in

...\TOCSY1D\004001.1R. In addition the most significant changes in coupling constants are found in the coupling constants originating from these two protons. There is a small change in chemical shift for C5-H (comparable to that occurring in ...\TOCSY1D\002001.1R), C6-H_A and C6-H_B remain almost unchanged while the effect on C3-H and C4-H is quite small. The low-frequency shift of C1-H (ca. 1.3 ppm) and of C2-H (ca. 1.4 ppm) leads to the conclusion that this glucose fragment must be the central sugar unit and is bonded to both the other fragments from C1 and C2 (Fig. 3.54).

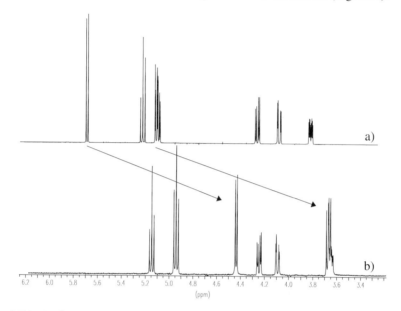

Fig. 3.54: a) β-D-glucose pentaacetate spectrum.
b) ...\TOCSY1D\003001.1R spectrum, the arrow indicates the shifting.

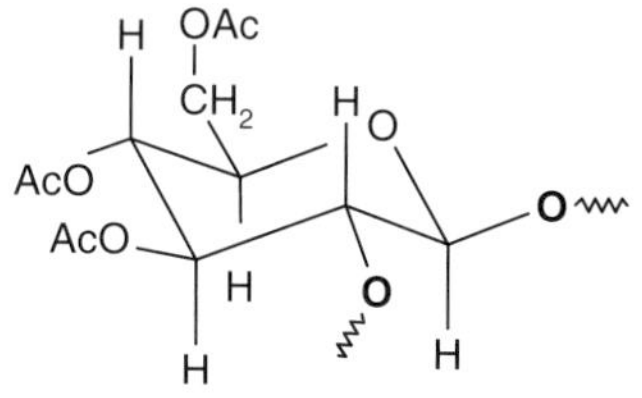

Fig. 3.55: Resulting structure fragment from ...\TOCSY1D\003001.1R.

A comparison of the β-D-glucose pentaacetate spectrum with the last TOCSY spectrum ...\TOCSY1D\004001.1R is shown in Fig. 3.56. The most significant change is the increase in shielding of the anomeric proton as indicated by its shift from ca. 5.7 ppm to ca. 4.7 ppm; all the other chemical shift changes are much smaller and decreases the further away from C1 the proton is located. Almost no difference is found in the shift of

C4-H which is furthest away from C1. The only decrease in shielding found in the spectrum is C6-H$_B$. The effect of this is to move C6-H$_A$ and C6-H$_B$ closer together and leads to a much stronger second order effect given by a perturbation parameter of ca. $\lambda_{AB} \approx 0.37$. The smallest change in coupling constant is C3-H to C4-H. For a direct glycosidic bonding at C6 the effect on the geminal coupling constant seems to be too small, so the detected effect might result from conformational rearrangements of this exocyclic spin system. To conclude, the most likely glycosidic bonding would be to C1 (Fig. 3.57).

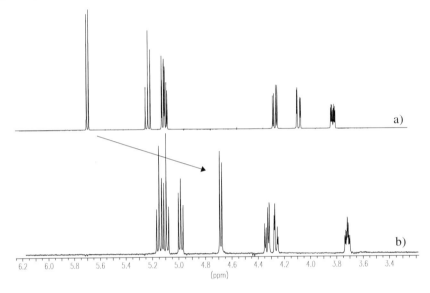

Fig. 3.56: a) β-D-glucose pentaacetate spectrum.
b) ...\TOCSY1D\004001.1R spectrum, the arrow indicates the C1-H shift.

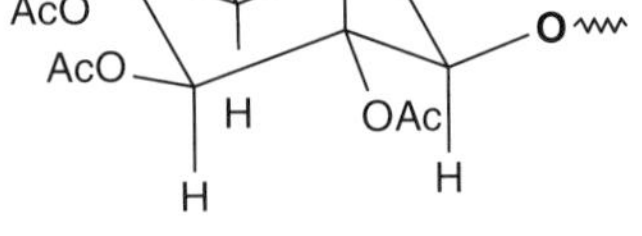

Fig. 3.57: Resulting structure fragment from TOCSY\004001.1R

Which sugar is bonded to which position cannot be determined from the data.

The possible glycosidic connections are either 1-6 and 2-1(Fig. 3.58a) or 1-1 and 2-6 (Fig. 3.58b); based on a knowledge of glycosidic bonding, an anomeric-anomeric linkage is highly unlikely so this would favor the first possibility. However, it is not possible to distinguish between the two possible structures with any degree of certainty basis on the results obtained from the spectrum analysis.

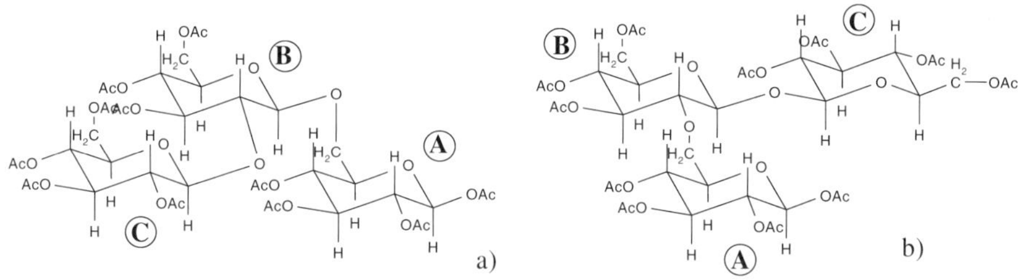

Fig. 3.58: a) peracetylated 2-O-(β-D-glucopyranosyl)-β-gentiobiose
gentiobiose = 6-O-(β-D-glucopyranosyl)- β-D-glucopyranose.
b) peracetylated 2-O-(β-D-glucopyranosyl-β D-glucopyronsid)-6-O-β-D-glucopyranose.

3.3.3.2 Piperine and Nicotine

The final examples in this chapter will illustrate the effect of long range couplings and field effects on spectra. It will also underline again that spectrum analysis is an interactive process. In any analysis it is necessary to have a clear understanding of the nature of the problem under investigation, this will not only speed up computational times but ensure that the final result is meaningful chemically.

Check it in WIN-NMR:

Open... the proton spectrum of piperine, ...\ABMX\004001.1R. The spectrum only displays the aromatic and olefinic protons. Switch to the **Analysis I Integration** mode and integrate the spectrum to determine the number of protons in each of the signals.

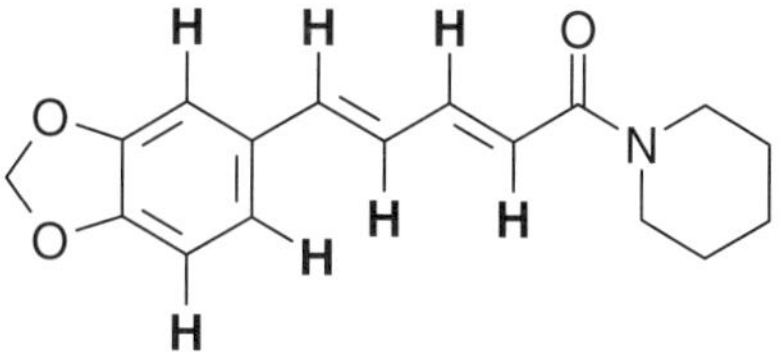

piperin

Measure a **Linewidth** and switch to the **Analysis I Multiplets** mode. The aromatic and olefinic protons printed in bold in the structure are of interest in this analysis. When considered as two separate fragments, the aromatic protons form a AMX type spin system and the four protons in the butadiene fragment a strongly coupled AMNX spin system. Try to identify the aromatic signals using the multiplet grids. Take into account the size of coupling constants you expect to find in a 1,3,4 trisubstituted benzene (section 2.2.3). Define the multiplet trees for these spins and connect the coupling constants.

One of the aromatic signals is found in the region where three signals overlap. Using the value of the coupling constants determined from the other two aromatic signals it is easy to determine the third aromatic signal.

Once the position of all the aromatic protons has been determined, the location of the remaining four olefinic protons is straight forward. Considering the four butadiene protons as an AMNX system; in the A part the weak outer lines are just visible, the MN part shows a very strong second order effect and the X part seems to contain just a doublet. Based on this information analyze the signal at 7.39 ppm ignoring the weak outer lines as a doublet of doublets of doublets, the signal at 6.43 ppm as a doublet and the intermediate shifts as two doublets of doublets with a very strong roof effect. The value of $^3J_{trans}$ for the two center protons of the two double bonds appears to be smaller than for the isolated double bonds. The analysis of the second order part is not very easy, but it can be done to a first order approximation.

Note: it is not necessary to assign the individual signals to the structure but it is helpful in the final analysis if **Identifier**s are assigned, e.g. <A1>, <M1>, <X1> for the aromatic ring and <A2>, <M2>, <N2> and <X2> for the butadiene.

Check it in WIN-NMR and WIN-DAISY:

When all coupling constants are connected mark one peak with the spectrum cursor and **Export...** the data to **WIN-DAISY** *with* fragmentation of the spin system. Open the **Control Parameters** and check the **Output Option Parameter Limit warning**. **Simulate** a spectrum and display the calculated lineshape together with the experimental spectrum using the **Dual Display** mode of WIN-NMR. The spectra differ significantly in the MN spectrum region of the butadiene spin system. There are two ways to proceed, either

- adapt the MN frequencies to get a better overlay of the simulated and experimental signals and run an iteration without smoothing
- or run an iteration with smoothing parameters

To adapt the frequencies proceed in a similar way to the α-D-glucose "Check it". Use the **Move Trace** Option in WIN-NMR to estimate the approximate frequency difference the resonance frequencies of the M and N nuclei must be shifted towards each other (about ±2.5 Hz) and modify the appropriate values in the **Frequencies** dialog box in WIN-DAISY. Rerun the **Simulation** and inspect the result in the **Dual Display** mode of WIN-NMR. Repeat this process until there is a reasonable agreement. Run a parameter **Iteration** without any smoothing parameters in WIN-DAISY.

To use smoothing, open the **Control Parameters** dialog box and select **low / medium Broadening** and **medium No. of cycles**. Run an **Iteration**.

Displaying the results using the Dual Display mode of WIN-NMR shows that the fit between experiment and simulation is not particularly good.

Check it in WIN-DAISY:

Load the Iteration Protocol list into an editor using the **Iteration protocol** command. Some parameters reached their limits during optimization and the value of *R-Factor* for the final result in this iteration is quite large. Check the standard deviations of the individual parameters, some values might be larger than others. Also check the *Best Parameter Vector* and compare this value with the **lower** and **upper** iteration limits. Compare both spectra again in the **Dual Display** of WIN-NMR and **zoom** in the on the appropriate regions and check the starting chemical shift frequencies. Up to now the sign of the coupling constant over four bonds in the double bond has been ignored.

Open the **Control Parameters** and click on the **Scalar Coupl.** button, in the butadiene fragment alter the limits for J_{12} e.g. from <−2> Hz to <+2> Hz and for J_{13} e.g. from <+6> Hz to <+14> Hz. Run one or more **Iteration**. Provided that the chemical shifts are correct an *R-Factor* of ca. 2.7% can be obtained.

The result obtained in this iteration is still not a particularly good fit, because only the large coupling constants leading to visible first order splitting in the spectrum have taken into account. The next step of analysis will be the consideration of long-range couplings within both fragments.

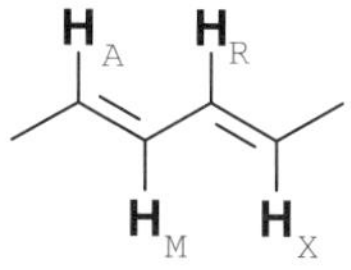

Fig. 3.59: Trans-trans arrangement of four protons in a butadiene fragment. Note: labeling is for identification purpose only.

Table 3.8: Coupling constants in conjugated double bond fragments.

Coupling Constant	Coupling value [Hz]
$^3J_{AM}$	ca. 17
$^3J_{MR}$	ca. 10
$^4J_{AR}$, $^4J_{MX}$	ca. −1
$^5J_{AX}$	ca. 0.5

Check it in WIN-DAISY:

Open the **Scalar Coupling Constants** dialog box and for both fragments click the **Iterate all** button to allow the 4- and 5-bond coupling constants to be varied. Using the values in Table 3.8 as a guide, select suitable starting values for J_{24} and J_{34} in the butadiene fragment. Select a suitable value for J_{13} in the aromatic fragment. Run another iteration either with or without smoothing parameters. Experiment with different signs for the long-range coupling constants in the butadiene fragment but keep the sign of the para coupling constant in the aromatic fragment positive since this sign has no influence on the spectrum. Continue until you cannot improve the result any more. Inspect the **Iteration Protocol** list. The value of *R-Factor* should be about 0.89% implying an good correlation between the experimental and calculated spectrum.

When this value is reached, click on the **Export** to **WIN-NMR exp.** button to transfer the WIN-DAISY result to WIN-NMR and store the analyzed NMR parameters with the experimental spectrum. **Close** the WIN-DAISY input file.

Another point to consider is the presence of long-range coupling constants *between* both fragments. There should be some coupling interaction between the aromatic protons ortho to the double-bond fragment and the nearest vinylic protons as shown in Fig. 3.60.

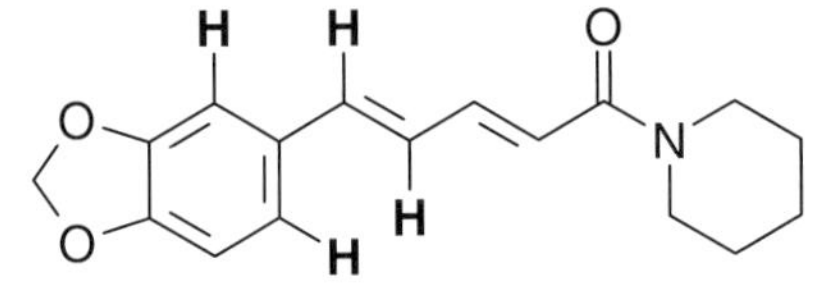

3 Structure and Spin System Parameters

Fig. 3.60: Protons displaying possible additional long-range coupling constant.

These coupling constants cannot be implemented while the total spin system is fragmented into two subsystems, it is necessary to recombine both fragments. The easiest way to do this is to export the previously determined result from WIN-NMR to WIN-DAISY *without* fragmentation. In this way the NMR parameters are arranged in the order they appear in the individual subsystems which simplifies the implementation of the additional long-range couplings.

Check it in WIN-NMR and WIN-DAISY:

Open... the experimental spectrum ...\ABMX\004001.1R in WIN-NMR. Switch into the **Analysis | Multiplets** mode. Press the **Windaisy** button. Open the **Report...** box and **Export...** the data to **WIN-DAISY** *without* fragmentation.

In WIN-DAISY a seven spin system is generated. Open the **Scalar Coupling Constants** dialog box, enter small values for the long-range coupling constants J_{25}, J_{26}, J_{35} and J_{36} and select them for iteration. Run an **Iteration** without smoothing or with weak smoothing parameters. Manually change the signs of the long-range coupling after every iteration. Check the value of R-Factor, the lowest value that can be obtained is 0.22%.

Check it in WIN-NMR:

Open... the 400 MHz proton spectrum of *nicotine*, ...\NICOTINE\001001.1R, in WIN-NMR. Use the **Analysis | Integration** mode to determine the number of protons in each signals. The aromatic signals are well-separated from the aliphatic region, but the N-methyl group does overlap with one of the aliphatic protons.

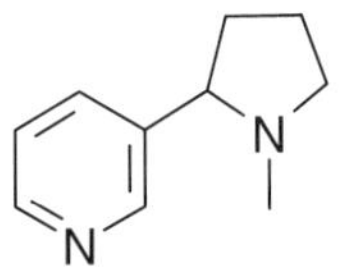

nicotine

The analysis of the aromatic region is relatively straight forward although there seem to be one long-range coupling present. The aliphatic region is probably the hardest example in this chapter.

Check it in WIN-NMR:

First measure a suitable **Linewidth** in the e.g. the N-methyl singlet and then switch to the **Analysis | Multiplets** mode.

Start with the analysis of the aromatic spectral region. Build up a coupling tree and assign the couplings. There might be some small long-range couplings to the pyrrolidine ring. Use the **Define Identifier** option to label the aromatic signals, e.g. C2'-H (prime to identify the pyridine carbons).

3.3 Conformation and NMR

Since the N-methyl signal overlaps with one of the aliphatic protons, a parameter optimization ignoring the N-methyl signal would be impossible. Mark the top of the signal with the cursor, click the right mouse button and select the **Define Singlet** option.

Before continuing with the multiplet analysis, the expected signal multiplicities should be considered. To simplify the following discussion the carbons of the pyrrolidine ring are labeled as shown in Fig. 3.61. Since the molecule is chiral, all methylene groups are diastereotopic and all multiplicities are based on 'simple' doublet splittings.

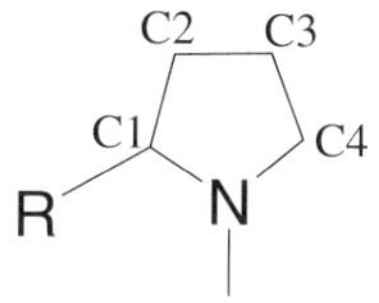

Fig. 3.61: Carbon labeling in the pyrrolidine fragment

It shall be assumed that only two and three-bond coupling constants are resolved (larger than the linewidth) and that first order rules can be applied. On this basis, for the proton(s) of every carbon atom labeled in Fig. 3.61 the expected number of coupling constants, the coupling pattern and the maximum number of lines can be calculated as shown in Table 3.9.

Table 3.9: Expected number of coupling constants and resolved lines.

carbon atom	no. of protons	expected number of coupling constants	pattern	number of lines = $2^{\text{number of couplings}}$
1	1	2	dd	4
2	2	4	dddd	16
3	2	5	ddddd	32
4	2	3	ddd	8

In addition to second order effects and long-range coupling, due to line overlap a doublet of doublets can appear as a pseudo-triplet (both coupling of a similar magnitude) and a doublet of doublets of doublets can appear as either a pseudo-quartet (three couplings of a similar magnitude) or a pseudo-doublet of triplets (only two couplings of a similar magnitude).

Check it in WIN-NMR:

Continue the **Analysis | Multiplets** with the signal at 3.23 ppm. As all the coupling patterns are based on doublet splittings the assignment of the multiplets to the protons in the pyrrolidine fragment is fairly obvious (refer to Table 3.9). The signal overlapping with the methyl singlet can also be analyzed starting with the well resolved part. The signals to the right of the methyl signal show quite wide spread patterns with two of them are overlapping slightly.

Four signals can be easily analyzed but one dddd and both the ddddd patterns are very difficult. The overlapping signals can be analyzed starting from either the right or the left hand side of the signal. The values obtained for the coupling constants are quite biased but the main thing at this stage of the analysis is the determination of the correct number of coupling constants to be able to fit the information given in Table 3.9. As an example one 32 line pattern is shown in Fig. 3.62.

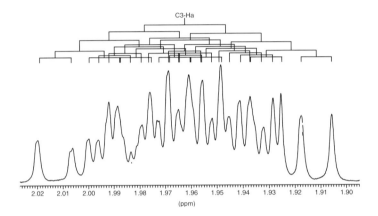

Fig. 3.62: Signal of C3-Ha at 1.96 ppm with ddddd (32 line) pattern.

When the first order analysis has been completed, the following splitting pattern order according to chemical shift has to be found (going from left to right).

pattern:	ddd	dd	ddd	dddd	ddddd	ddddd	dddd
proton at:	C4	C1	C4	C2	C3	C3	C2

Check it in WIN-NMR:

When this pattern has been successfully use the **Define Identifier** option again to label the multiplets based on Fig. 3.61 (e.g. C4-Ha and C4-Hb for the ddd). Manually connect the geminal coupling constants.

In the next step connect the remaining vicinal (three-bond) couplings to the neighboring protons. Do not worry if the size of the couplings are quite different, the multiplet pattern is just the basis for the construction of the spin system. The main thing is to accurately define, to a first order approximation, the spin system. If an error is made at this stage it will not be possible to optimizing the parameters satisfactorily. Do *not* connect any long-range couplings between the pyridine and pyrrolidine protons otherwise it will not be possible to fragment the spin system. Note that in the **Report...** box a question mark appears beside the large coupling constants, this implies second order effects and indicates that parameter optimization with WIN-DAISY is necessary.

If the setup of the multiplet trees for the signals causes problems, Table 3.10 contains hints about the spin system connections.

3.3 Conformation and NMR

Table 3.10: Approximate absolute values and connections of multiplets

Identifier	value of ^2J [Hz]	values of ^3J Hz]	connection to
C4-Ha	9	2, 8	C3-H
C1-H	none	8, 9	C2-H
C4-Hb	9	8, 9	C3-H
C2-Ha	12	5, 8, 9	C1-H, C3-H
C3-Ha	12	5, 8, 9, 11	C2-H, C4-H
C3-Hb	12	2, 5, 8, 9	C2-H, C4-H
C2-Hb	12	5, 9, 11	C1-H, C3-H

Resulting from this connection the following approximate coupling constants build up the spin system:

Table 3.11: Connections ordered according to the magnitude of coupling value

splitting [Hz]	connection			
2	C3-Hb ; C4-Ha			
5	C2-Ha ; C3-Ha	C2-Hb ; C3-Hb		
8	C1-H ; C2-Ha	C3-Ha ; C4-Ha	C3-Hb ; C4-Hb	
9	C4-Ha ; C4-Hb	C1-H ; C2-Hb	C2-Ha ; C3-Hb	C3-Ha ; C4-Hb
11	C2-Hb ; C3-Ha			
12	C3-Ha ; C3-Hb	C2-Ha ; C2-Hb		

Check it in WIN-DAISY:

When all couplings are connected **Export...** the data to **WIN-DAISY** with fragmentation of the spin system and *inclusion* of non-assigned couplings.

WIN-DAISY will generate three independent spin systems: the aromatic ring, the aliphatic ring and the N-methyl group. The aromatic system will contain five spins - the four aromatic protons and plus an additional spin to represent the long-range couplings. The aliphatic system only contains the seven aliphatic ring protons, no long-range couplings have been included so no additional nuclei is required. It is not possible to optimization the aromatic region without including the long-range couplings, however if long-range couplings are included for the aliphatic region the calculation time would be prohibitive. Initially the effect of the long range couplings can be neglected, later in the analysis they can either be included or the effect can be represented by selecting individual linewidth parameters.

Check it in WIN-DAISY:

If long-range couplings for the aliphatic region have been included in the analysis, the spin system will contain more than seven spins. To rectify this problem, open the **Main Parameters** dialog box and *decrease* the **Number of spins or groups with magnetical equivalent nuclei** to <7>. In the **Scalar**

226 *3 Structure and Spin System Parameters*

Coupling Constants dialog box apply a *negative* sign to the three geminal coupling constants J_{12}, J_{34} and J_{56}. Open the **Frequencies** dialog box for the one-spin, N-methyl, fragment and extend the number of **Spins in group** from <1> to <3> . Click the **Previous** or **Next** button in the header to switch to the aromatic fragment and check that Spin 5 is **disable**d and cannot be selected for iteration. Run a spectrum **Simulation** and display the results using the **Dual Display** mode of WIN-NMR.

The signals in the aromatic region and some of the signals in the aliphatic region are quite well reproduced, but the wide-spread aliphatic multiplets do not fit particularly well. However, the position of the signals and the intensity distribution over the spectral region is essentially correct. If the spectrum is simulated again but with the signs of the geminal coupling constants reversed, there is a difference in the appearance of the spectrum particularly in the signals at high-field. This is a clear indication that second order effects are present, although the effects are not very strong. Because there is already a reasonable fit between the simulated spectrum and the experimental data smoothing parameters should *not* be used in the optimization. Optimizing the parameters is very time consuming, it is vitally important that the initial multiplet analysis is correct. If there are large discrepancies between the experimental spectrum and the simulation spectrum repeat the multiplet analysis and simulate the spectrum again.

Check it in WIN-DAISY:

Open the **Lineshape Parameters** dialog box and *disable* the **iterate** linewidth option in *all* of the fragments. Run an **Iteration**, depending on the quality of the starting parameters one or more may be necessary. The final value of *R-Factor* should be ca. 1.5%.

As discussed earlier, in the pyridine spin system each proton will have its own individual linewidth. Open the **Lineshape Parameters** dialog box for the aromatic fragment, select the **Nuclei specific linewidths** option and press the **Iterate All** button. Spin 5, the X-spin, cannot be selected and remains disabled. In the aliphatic fragment and the N-methyl fragment select the **Global** Linewidth, for **iterat**ion. Run an **Iteration**. The lowest *R-Factor* value than can be obtain by iteration of ^2J and ^3J coupling constants and the linewidths is ca. 0.74%.

The different linewidths determined by the iteration is a result of neglecting the long-range coupling constants primarily *within* the pyrrolidine ring. The C1-H proton, couples to the pyridine protons yet its signal exhibits no fine structure. The main effect of these long-range coupling is line-broadening which is converted by the iteration algorithm into a broader linewidth parameter. A simultaneous iteration of long-range coupling and nuclei specific linewidths would fail because both types of parameters operate in a similar way on the lineshape and consequently the iteration might not lead to any error reduction.

Check it in WIN-DAISY:

Display the **Options** dialog and check the option **Allow sign change**. Open the **Scalar Coupling Constants** dialog box of the aliphatic seven spin fragment and press the **Iterate all** button. Run another **Iteration** and check the value of *R-Factor*. If no error reduction can be obtained (initial and best parameters identical) open the **Lineshape Parameters** dialog box, decrease the **Global Linewidth** in the second fragment and run the **Iteration** again. The value of *R-Factor* should fall to below 0.6%. Compare the result using the **Dual Display** mode of WIN-NMR.

There still might be some small differences in splitting visible and due to the sign of the long-range coupling constants. These could be checked out by an iteration including smoothing parameters, but this would be very time consuming. The easiest way to find the correct sign combination is to change the signs of some of the long-range coupling constants and run the iteration without any smoothing parameters.

Check it in WIN-DAISY:

Open the **Scalar Coupling Constants** dialog box of the aliphatic fragment and edit the sign of the small long-range coupling constants (their magnitude is in every case smaller than 1 Hz) and run another iteration. Afterwards check the result visually using the **Dual Display** and the examine the value of *R-Factor* in the **Iteration Protocol** list. The best *R-Factor* that can be obtained in this way is ca. *0.56%*. Finally open the **Lineshape Parameters** dialog box for the aliphatic fragment, select the **Nuclei specific linewidths** option and press the **Iterate All** button. Run a final **Iteration**. Afterwards examine the **Lineshape Parameters**; note that the linewidth parameter for the C1-H proton is much larger than the all others in the aliphatic fragment.

This increase in linewidth difference for C1-H cannot be related to the neighboring nitrogen, because then the two C4-H protons would also show a comparable increase in linewidth. The reason is found in neglecting the long-range coupling constants to the pyridine protons. Because of the lack of fine-structure in the signal we will have to be satisfied with that result. The best R-factor that can be obtained is ca. 0.55%.

In the final check we will compare the result obtained based on the 400 MHz proton spectrum with the experimental 250 MHz spectrum of the same sample.

Check it in WIN-NMR and WIN-DAISY:

Open... the 250 MHz ^{1}H spectrum of nicotine ...\NICOTINE\002001.1R in WIN-NMR. Enter the **Analysis | Interval** mode, press the **Regions...** button and **Load...** the regions file. **Export...** the data to **WIN-DAISY** *without* any multiplet data. Accept the message box WIN-DAISY displays and press the **Connect** button. A message box appears indicating that there is a difference in the spectrometer frequency of the WIN-DAISY document (400 MHz) and the new experimental data (250 MHz) and asking whether the WIN-DAISY spin system data should be adapted i.e. keep the same ppm values, but modifying the frequencies in Hz. Answer with **Yes**. In the status line for **Experimental Spectrum** the new spectrum name appears. Run a spectrum **Simulation** and compare the spectra in the WIN-NMR **Dual Display**. Zoom in in various regions in the spectrum.

The 250 MHz ^{1}H spectrum of nicotine is more second order that the 400 MHz spectrum. The high-field part in particular looks quite complicated, because the signals of the three protons overlap. A multiplet analysis of this spectrum is considerably more difficult than for the high-field spectra but the signs of coupling constants can be studied more easily in the low-field spectra. It is possible that the parameter values may be slightly different between the two spectra, because of temperature differences. An iteration of the 250 MHz spectrum will not give better results, because the overall lineshape is not as good as in the 400 MHz spectrum.

3.4 References

[3.1] Haigh, C.W., *J. Chem. Soc.*, 1970, *A*, 1682.
[3.2] Corio, P.L., Structure of High-Resolution NMR Spectra, New York: Academic Press, 1966.
[3.3] Hargittai, I., Hargittai, M., *Symmetry through the Eyes of a Chemist*, Weinheim: VCH, 1986.
[3.4] Weber, U., *Ph.D. Thesis*, University of Düsseldorf, 1994.
[3.5] Dischler, V.B., Englert, G., *Z. f. Naturforsch.*, 1961, *16A*, 1180.
[3.6] Govil, G., Whiffen, D.H., *Mol. Phys.*, 1967, *12*, 449.
[3.7] Freeman, R., Anderson, *J. Chem. Phys.*, 1962, *37*, 2053.
[3.8] Wakefield, R.H., Memory, J.D., *J. Chem. Phys.*, 1968, *48*, 2174.
[3.9] Karplus, M., *J. Chem. Phys.*, 1959, *30*, 11.
[3.10] Pachler, K.G.R., *J. Magn. Reson.*, 1972, *8*, 183 and
       Pachler, K.G.R., *J. Chem. Soc. Perkin II*, 1972, 1936.
[3.11] Abraham, R.J., Pachler, K.G.R., *Mol. Phys.*, 1964, *7*, 4 and 165.
[3.12] Wei-Chuwan, L., *J. Chem. Phys.*, 1969, *50*, 1890,
       Wei-Chuwan, L., *J. Chem. Phys.*, 1970, *52*, 2805 and
       Wei-Chuwan, L., *J. Chem. Phys.*, 1973, *58*, 4971.
[3.13] Durette, P.L., Horton, D., *Org. Magn. Reson.*, 1991, *3*, 417.
[3.14] Haasnoot, C.A.G., De Leeuw, F.A.A.M., Altona, C., *Tetrahedron*, 1980, *36*, 2783 and
       Huggins, M.L., *J. Amer. Chem. Soc.*, 1953, *75*, 4123.
[3.15] Hägele, G., Batz, M., Peters, R., Niemeyer, U., Engel, J., *Z. Arzneimitell Forsch / Drug Res.*, 1990, *40*, 599.
[3.16] Harris, R.K., Spragg, R.A., *J. Chem. Soc., Chem. Comm.*, 1966, 314.
[3.17] Finer, E.G., Harris, R.K., Bond, M.R., Keat, R., Shaw, R.A., *J. Mol. Spectroscopy*, 1970, *33*, 72.
[3.18] Hägele, G., Harris, R.K., *J. Chem. Soc., Daltron Trans.*, 1973, 79.
[3.19] Hägele, G., Lorberth, J., *Organ. Magn. Reson.*, 1977, *9*, 325.
[3.20] Braun, S., Kalinowski, H.-O., Berger, S., *100 and More Basic NMR Experiments*, Weinheim: VCH, 1996.
[3.21] Weber, U., Germanus, A., Thiele, H., *Fresenius J. Anal. Chem.*, 1997, *359*, 46.
[3.22] Bigler, P., *NMR Spectroscopy: Processing Strategies*, Weinheim: VCH Wiley, 1997.
[3.23] Pretsch, E., Clerc, J.T., *Spectra Interpretation of Organic Compounds*, Weinheim: Wiley-VCH, 1997.

4 Spin Systems and the Periodic Table

In section 2.1 some examples of spin active nuclei with their nuclear spin quantum number *I* were given (Table 2.3) and there was a brief discussion on the effect of the *spin quantum number I* on the spin orientation in a magnetic field. In chapters 2 and 3 the analysis of proton (the spin-½ nucleus ^1H) spin systems in organic molecules were discussed. Initially concentrating on the properties of the systems themselves, the influence of chemical structure on the magnitude of NMR parameters and consequently on the appearance of NMR spectra were also considered. The present chapter will be looking at the NMR spectra and spin systems of nuclei other than protons, the so-called "X-nuclei". Section 4.1 will deal with elements that have one spin-½ isotope in 100% natural abundance such as ^{19}F, ^{31}P and ^{103}Rh. Some of the examples will be purely homonuclear in nature while others will also contain heteronuclear coupling. In section 4.2, elements with spin-½ isotope(s) in less than 100% natural abundance will be discussed with particular emphasis on ^{13}C, ^{15}N, ^{29}Si, ^{195}Pt, ^{77}Se and $^{119}Sn/^{117}Sn/^{115}Sn$. The concept of main and satellite spin system(s) is introduced for the so-called dilute or rare isotopes The final section of this chapter 4.3 will deal with spin systems containing quadrupolar nuclei ($I > ½$) e.g. deuterium = ^2H), ^{14}N, ^{10}B and ^{11}B, ^{17}O and ^9Be. Some special applications will also be given such as the determination of hidden parameters using deuterium substitution or satellite spectra. Details of the NMR parameters of the various isotopes can be found in references [4.1]–[4.3].

4.1 Spin-½ Pure Elements

Only a few elements in the periodic table (20) exist as one naturally occurring isotope in 100% abundance, the so-called "pure elements". Five of these elements posses a nuclear spin quantum number I = ½: ^{19}F, ^{31}P, ^{89}Y, ^{103}Rh and ^{169}Tm. Hydrogen has two isotopes, the proton ^1H (I =½, natural abundant 99.98%) and the deuteron ^2H (I = 1, natural abundant 0.02%) therefore the proton is *not* a pure element. Nevertheless, it approximates to a "pure element" and all the analytical solutions derived for protons are equally applicable to the true pure elements. A common misconception is that other spin-½ nuclei are somehow different from protons, but the method of analyzing an AB spin system is identical irrespective if it occurs in a ^1H, ^{19}F or ^{31}P spectrum.

High natural abundance of the spin active nucleus in conjunction with high receptivity and sensitivity to the NMR experiment all help to make the observation of an NMR active isotope easier. The transmitter frequency used to detect a particular element can be calculated according to the magnetogyric ratio. A 500 MHz spectrometer means, that the approximate transmitter frequency for ^{1}H is 300 MHz; on the same instrument the transmitter must be set to 9.55 MHz for ^{103}Rh and to 121.44 MHz for ^{31}P.

Check it in WIN-DAISY:

Open the input document A.MGS. Call the **Spectrum Parameters** dialog box. Note that the **Isotope ^{1}H** is selected and both the **Frequencies adapted to Spectrometer Frequency** and the **Reference Transmitter Frequency** in the combo box are set to **300** MHz.

Click the **Isotope** button to open the **Element-Data** table. Check the option **Show Pure Elements Only**. The appearance of the periodic table changes and only a few elements are available for selection. Press the button **Rh** to display the data of the **isotope 103** of the element Rh and quit the dialog box with **Ok**. Automatically the data in the **Spectrum Parameters** dialog box is changed; the selected isotope is **^{103}Rh** and the **Reference Transmitter Frequency** is adapted to the current isotope data. Note that value of the **Frequencies adapted to Spectrometer Frequency**, which corresponds to the proton frequency, does not change.

Select other spin-½ pure elements from the **Element-Data** table and check the **Reference Transmitter Frequency**.

Because the reference transmitter frequency for the various isotopes is so different, the NMR resonances are detected separately. It is possible therefore to simplify the spectra of a heteronuclear spin system, e.g. containing both ^{1}H and ^{31}P, by measuring the ^{1}H spectrum while simultaneously decoupling the ^{31}P. The resulting ^{1}H{^{31}P}spectrum shows no influence of the phosphorous nuclei and is simpler to analyze. Conversely, the corresponding ^{31}P{^{1}H} spectrum can be used to extract only the phosphorous data. The normal ^{1}H and ^{31}P spectra will show heteronuclear ^{1}H-^{31}P interactions (J couplings). Fluorine, ^{19}F, presents its own experimental problems. Because of the similarity in the magnetogyric ratios of ^{1}H and ^{19}F (26.751 and 25.1665 respectively), the resonance frequencies are very close together and ^{19}F is not usually regarded as an X-nucleus. Consequently, spectrometers have to be specially equipped in order to be able to decouple ^{19}F while observing ^{1}H and vice versa.

Check it in WIN-DAISY:

Open the input document A.MGS and open the **Spectrum Parameters** dialog box. Press the **Isotope** button and click the **F** button in the **Element-Data** table. Quit the dialog box with **Ok** and check the **^{19}F Reference Transmitter Frequency**.

4.1.1 Spin Systems Containing ^{19}F

In this section a variety of homonuclear and heteronuclear spin systems in inorganic and organic compounds will be discussed [4.3]. The determination of the spin system based on the structure will be explained as well as the handling of heteronuclear coupling constants in the Multiplets mode of WIN-NMR. Finally, the special NMR characteristics of fluorine spin systems will also be considered.

4.1.1.1 All-Fluorine Spin Systems

The chemical shift range for ^{19}F spectra is much larger than for proton spectra, about 1200 ppm compared with about 15 ppm for ^1H (or 50 ppm if inorganic hydrides are included). $CFCl_3$ is usually used as reference and conversion tables are available for other compounds [4.4]. The spectrum of $CFCl_3$ will be discussed in section 4.1.1.3.

The inorganic compound F_3SSCl has the structure shown in Fig. 4.1. When the lone pairs on the sulfur are taken into consideration, the sulfur in the SF_3 fragment is in an essentially bipyrimidal environment while the sulfur in the SCl fragment is in a tetrahedral environment. In the SF_3 fragment two of the fluorines are axial with the remaining fluorine in a equatorial position. In the absence of any dynamic exchange all three fluorines are in different environments and at $-100°C$ a spin system with three different chemical shifts is observed [4.5]. At higher temperature pseudo-rotation takes place.

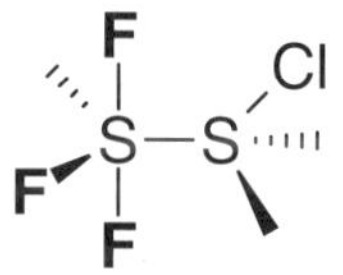

Fig. 4.1: Low temperature structure of F_3SSCl.
The free bonds symbolize the lone pairs.

Check it in WIN-DAISY:

Open the WIN-DAISY input document F3SSCL.MGS and examine the values in the **Frequencies** dialog box. Toggle the **Frequency** units between **Hz** and **ppm** to obtain an indication of the large chemical shift difference e.g. ca. 5000Hz. Spin 1 and 2 belong to the axial fluorines and spin 3 to the equatorial fluorine. Open the **Scalar Coupling Constants** dialog box and examine the values. Note the large differences in the magnitude of the coupling constants due to the different chemical environments. Run a spectrum **Simulation** and **Display** the **Calculated Spectrum** (red peak button) in WIN-NMR. Close the WIN-DAISY document.

Check it in WIN-NMR and WIN-DAISY:

Zoom in on the different parts of the spectrum. The spin system type is not easily classified but considering the chemical shift differences and the magnitude of the coupling constants it is best described as ABX rather than AMX. Enter the **Analysis | Multiplets** mode and analyze the signals (zoom in on each set of signals to ensure that the multiplet lines occur in the correct place). Connect the coupling constants in the **Report...** box. Note that a question mark indicating second order effects appears beside the large coupling constant. Quit the **Report...** box with **Ok** and leave the **Multiplets** mode storing the multiplet analysis (**Yes**). Measure a **Linewidth** and then enter the **Analysis | Interval** mode, define two suitable spectral regions and

232   4 Spin Systems and the Periodic Table

Export... the intervals and the multiplet analysis (**Yes**) to **WIN-DAISY**. Run a spectrum **Simulation** and display both spectra in the **Dual Display** mode of WIN-NMR.

Again, **zoom** in on the different signals. Switch the X-axis units to Hz <Ctrl + X> and holding down the <Shift> key, use the **Move Trace** button to overlay the individual signals. Make a note, in Hz, of the frequency shift of the second trace displayed in the status line.

The current example appears to be an ABX spin system where the AB spins do not exhibit a strong second order effect. To overlay the first-order based simulation with the original spectrum it is necessary to shift the spectrum by more than 20 Hz, indicating that the default values used for the analysis of proton spectra are not necessarily suitable for other nuclei. (This should be no surprise since the values in the Frequencies and Scalar Coupling Constants dialog boxes are a lot larger than any of the proton only systems examined.) To enable the parameters to be optimize, either the parameter limits have to be modified or the default limits changed. Changing the default limits is only valid for the present document. In the full program version the new default limits can be stored in the initialization file for use in other calculations by checking the option **Save and update all settings** in the program **Options** dialog box but this feature is not available in this teaching version.

Check it in WIN-DAISY:

In the program **Options** dialog box change the **Chemical Shift** iteration limits to <30> Hz. Open the **Control Parameters** and press the **Frequencies** button. To apply the new click one on the **Select none** button, then it changes to **Select all**, press it again. Quit the dialog box with **Ok** to accept the limits. Select smoothing parameters, e.g. **medium Broadening** and **medium No. of Cycles**. Run an **Iteration** and display the result in the **Dual Display** mode of WIN-NMR.

In the next example the spectrum of F_3SSF will analyzed by modifying the F_3SSCl input document.

Check it in WIN-DAISY:

Open the input document F3SSCL.MGS. In the **Main Parameters** dialog box. Change the **Number of spins or groups with magnetical equivalent nuclei** to <4> to represent the four fluorine nuclei. Open the **Frequencies** dialog box, change the display units to **Hz** and enter the resonance frequencies given in Table 4.1.

F_3SSF

Table 4.1: Spin System parameters of F$_3$SSF [4.5].

parameter	value [ppm]
ν_1	+ 53.2
ν_2	+ 5.7
ν_3	− 26.3
ν_4	− 204.1
J_{12}	86.3
J_{13}	32.8
J_{14}	40.2
J_{23}	32.2
J_{24}	156.0
J_{34}	63.5

Check it in WIN-DAISY:

Open the **Scalar Coupling Constants** dialog box and enter the appropriate coupling constants (Table 4.1). In the **Spectrum Parameters** dialog box, press the **Auto Limits** button. Note the large **Number of spectral points** required to give adequate digital resolution even at this low spectrometer frequency 60 MHz! Quit the dialog box with **Ok** and run a spectrum **Simulation**. Display the **Calculated Spectrum** in WIN-NMR (red peak button).

Check it in WIN-NMR:

Using the calculated spectrum, measure a suitable **Linewidth** and then switch to the **Analysis I Multiplets** mode. Perform an analysis (**zoom**ing the signals) and connect the coupling constants automatically (**Auto Connect**) in the **Report...** box. Note that no question mark indicating second order effects appears. **Export...** the spin system data to WIN-DAISY (no interval data is necessary because the data will *not* be used for iteration). Run a **Simulation** and display both spectra using the **Dual Display** mode of WIN-NMR.

Calculated chemical shift values using the IGLO method [4.6] reported in [4.7] are in good agreement with the experimental data. The present example illustrates how the substitution of Cl for F can alter the spin system data drastically and change the order of the spectra. It is not necessary to iterate the parameters because the spectrum is a simple first order spin system called AMRX.

Very strong second order effects can also occur in all-fluorine spin systems.

Check it in WIN-DAISY:

Open the input document C4CL5F3.MGS relating to the derivative shown opposite. The molecule has no symmetry and no magnetic equivalence. Examine the values in the **Frequencies** and **Scalar Coupling Constants** dialog boxes.

1,1,2-trifluoro-2,3,3,4,4,5,5-heptachloro cyclobutane

234 4 Spin Systems and the Periodic Table

The chemical shifts reported here [4.8] are referenced to C_6F_6 as 0 ppm, which resonates at -162.9 ppm according to $CFCl_3$ [4.3].

Check it in WIN-DAISY:

Run a spectrum **Simulation** and **Display** the **Calculated Spectrum** in WIN-NMR. **Zoom** in the various spectrum regions and determine the type of spin system.

The spectrum shows a AXY type spin system approaching ABC. Compared with other AXY type spectra discussed in previous chapters (refer to 2.3.3) the appearance of the A resonance (ca. 54.8 ppm) differences significantly with the weak ("outer") lines situated inside a strong "doublet of doublets". In this situation, a first order analysis is only a *starting point and will be quite wrong*!

Check it in WIN-NMR and WIN-DAISY:

Open the **Analysis | Integration** option and integrate all the signals; do not break the integral up in the region from 53.0 ppm to 56.0 ppm but integrate the remain "doublets" separately. From the value of the integrated areas it is obvious that the signals at 48.9 ppm and 50.9 ppm relate to the second fluorine and the signal at 49.5 ppm to the third fluorine. Analyze the spectrum as an first order system, ignoring the weak "inner" lines. The position of the X spin inside the lines of the M spin is confirmed by the Multiplet analysis. **Return** from the Multiplet mode saving the Multiplet Analysis. Measure a suitable a **Linewidth**. Enter the **Analysis | Interval** mode and define two spectral regions for iteration (48.5 ppm to 51.5 ppm and 53.0 ppm to 56.0 ppm). Do not break up the interval region because the iteration algorithm must be able to shift the peaks over the complete spectral range. **Export...** both the interval and the multiplet analysis to **WIN-DAISY**. Run a **Simulation** first and compare the spectra in the **Dual Display**. Switch the x-axis units into Hz <Ctrl + X> and use the **Move Trace** option to overlay the signals. Note the approximate magnitude of the shift. Due to the reason that this value is bigger than the default **iteration limits** for **Chemical Shifts** open the **Options** dialog and enter <30> Hz. Display the **Control Parameters** and select the **Frequencies** button. Therein click twice on the **Select none / Select all** button to apply the new limits. Select smoothing parameters (**Broadening** and **No. of Cycles**) and run one or more **Iteration**s. The minimal *R-Factor* obtainable based on these parameters is about *1.19%*. Display both spectra in the **Dual Display** mode of WIN-NMR and zoom in the signals. The A part of the AXY (low field, corresponds to X in ABX) looks reliable, but the XY part (AB of ABX) is different!

The wrong iteration result is determined, because no sign change has been allowed for the coupling constants.

Check it in WIN-DAISY:

Open the **Scalar Coupling Constants** dialog and change either the value of J_{12} or J_{23} to a negative value: Insert a <-> in front of the value (then automatically the iteration limits are adapted). Run another **Iteration**. In case the sign of J_{23} was changed, the correct result is obtained, the *R-Factor* in the **Iteration Protocol** is *0%*. If J_{12} was negative, a very large *R-Factor* will result and the value of J_{13} has reached the lower limit of *0 Hz*. Then change the value of J_{13} to a small negative value (e.g. <–0.1> Hz) and run another **Iteration** which will lead to the correct result.

Both correct solutions refer to the same relative sign combination of the scalar coupling constants.

4.1.1.2 Heteronuclear Spin Systems

An example of an AKMRX spin system is the fluorinated aromatic compound 3-fluoro-acetophenone [4.9], spins A to R refer to ^{1}H and X to the single ^{19}F nucleus. The fluorine chemical shifts are referenced to C_6F_6.

Check it in WIN-NMR:

Open... the ^{19}F spectrum of *3-fluoroacetophenone* ...\AKMRX\003001.1R in WIN-NMR. Perform a first order **Multiplets Analysis**. No assignment of coupling constants is possible because all the couplings are heteronuclear. Annotate the multiplet with the **Identifier** <C3-F> and store the data when you **Return** to the normal display mode.

Load the ^{1}H spectrum of *3-fluoroacetophenone* ...AKMRX\004001.1R into WIN-NMR. Analyze the spectrum in the **Multiplets** mode andin the **Report... Auto Connect** the proton-proton coupling constants. Annotate the multiplets <C2-H> etc.

3-fluoroaceto-phenone

To assign the $^nJ_{FH}$ couplings double click on any line in the **Report...** box and proceed as follows: In the new dialog box click on the **Displ. hetero nuclear** radio button and then the **Hetero List...** button. Another dialog box appears, click on the **Load...** button and select ...\AKMRX\003001.MLT, the file from the multiplet analysis of the corresponding fluorine spectrum. Set the **Iso** edit field to <19F> for fluorine. Click on the **Ok** button to close the various dialog boxes until you are back in the **Report...** box. Finally, click on the **Auto Connect** button to assign the $^nJ_{FH}$ couplings. **Export...** the data to WIN-DAISY.

Check it in WIN-DAISY and WIN-NMR:

In the **Frequencies** dialog box all chemical shifts except the fluorine resonance frequency, spin 5, are selected for iteration. The coupling constant J_{15} which does not lead to a detectable splitting in the spectrum should also be included in the **iteration**. Open the **Lineshape Parameters**. dialog box. Because the linewidth parameters might be different in the ^{1}H and the ^{19}F spectra choose the option **Nuclei specific linewidths**. Select the first four linewidths for **iteration** (do *not* select the fluorine nucleus Spin 5 for iteration). Run a spectrum **Iteration** and examine the results using the **Dual Display** mode in WIN-NMR.

Load the experimental ^{19}F spectrum ...\AKMRX\003001.1R into WIN-NMR. Switch to the **Analysis I Interval** mode and either define the region for iteration or **Load,,,** the predefine iteration region. Press the **Export...** button

236   *4 Spin Systems and the Periodic Table*

and transfer only the interval data and *not* the multiplet data to WIN-DAISY. A message box appears in WIN-DAISY, press the **OK** button and then press the **Connect** button in the current WIN-DAISY document main window. Open the **Spectrum Parameters** dialog box, click the **Isotope** button and select in the **Element-Data** table F. Close all the dialog boxes with **OK**. Automatically the **Experimental Spectrum** and **Calculated Spectrum** names in the status lines are updated to the appropriate fluorine spectrum. Run a spectrum **Simulation** and display both spectra using the **Dual Display** mode in WIN-NMR. It may be necessary to adapt the Y-scale of both spectra (<Ctrl + Y> and **ALL** button).

Open the **Linewidth Parameters** dialog box. Automatically the proton linewidth parameters (Spin 1 to 4) are disabled for iteration, but the fifth spin (the linewidth of the fluorine nucleus) is enabled.

When changing the Isotope in the Spectrum Parameters dialog box while an iteration mode is still selected in the Control Parameters, the shifts and nuclei specific linewidths (if this option is selected) for the particular isotope are automatically selected and all others deselected. Similarly in the Scalar Coupling Constants dialog box all the coupling constants involving the isotope are selected for iteration, all others are deselected. This automatic selection method is always applicable for first order spin systems but may have to be modified for special second order cases.

Check it in WIN-DAISY:

Open the **Control Parameters** dialog box and press the **Frequencies** button. As expected only Spin 5 is selected for **iterat**ion. Click the **Next** button to open the **Scalar Coupling Constants** dialog box, only the proton-fluorine coupling constants J_{i5} (i = 1, 2, 3, 4) in the last column are selected for **iterat**ion.

Run a spectrum **Simulation** to compare the experimental and calculated fluorine spectrum using the $^nJ_{FH}$ coupling constants determined from the ^1H spectrum. The coupling constants fit quite well, but the chemical shift is a little inaccurate. Run an **Iteration** and display the results in the **Dual Display** mode of WIN-NMR (use <Ctrl + Y> and **ALL** to overlay the spectra).

Switch back to the ^1H spectrum by selecting the **^1H Isotope** in the **Spectrum Parameters** dialog box. Run a spectrum **Simulation** and display the calculated and experimental spectra using the **Dual Display** mode of WIN-NMR.

In the present case the values of $^nJ_{FH}$ follow the rule $^3J_{ortho} > {}^4J_{meta} > {}^5J_{para}$. However there are examples in which $^4J_{meta}$ is bigger than the $^3J_{ortho}$ the latter having a positive sign. The magnitude of $^5J_{para}$ is normally less than 3 Hz and both positive and negative signs are possible. In addition through-space coupling interaction between proton and fluorine nuclei that are separated by many bonds but are close spatially is also possible.

The present example is relatively simple with only one signal in the fluorine spectrum. The method will be developed in the next example which contains three different fluorine nuclei and exhibits both homonuclear coupling (proton-proton and fluorine-fluorine) and heteronuclear proton-fluorine coupling interactions. In aromatic systems, the values of $^3J_{ortho}$ is ca. −20 Hz while the size of the $^4J_{meta}$ and $^5J_{para}$ can vary widely with $^4J_{meta}$ displaying both positive and negative signs. $^5J_{para}$ is usually larger than $^4J_{meta}$ and is generally positive. The values of $^nJ_{FH}$ and $^nJ_{FF}$ can be calculated quite accurately using increment systems [4.3].

Check it in 1D WIN-NMR:

Open... the ^1H spectrum of *2,3,6-trifluoro-acetophenone* [4.9], ...\AKMRSX3\001001.1R, with WIN-NMR. Perform a first order **Multiplets Analysis** of the three signals; all splittings are based on doublets. There is also a through-space coupling [4.10],[4.11] between the methyl protons and the ortho-fluorine. Only $^3J_{HH}$ can be connected in the multiplet table. **Define** at least for the methyl group an **Identifier**, e.g. <CH3>.

2,3,6-trifluoroaceto-phenone

Open... the corresponding ^{19}F spectrum ...\AKMRSX3\002001.1R. Measure a suitable **Linewidth**. Switch to the **Analysis | Multiplets** mode and analyze the three fluorine signals. Think about the multiplicities required keeping in mind that two signals show a coupling to the CH_3 group.

Each fluorine signals features at least four coupling constants. The very small coupling need not to be taken into consideration as the splitting is not resolved for all the lines and it can be selected later on in WIN-DAISY for iteration.

Check it in WIN-NMR:

Click on the **Report...** button. First of all connect the two large fluorine-fluorine coupling constants which are easily identified. If you have measured the small splitting between fluorine 1 and 2 connect this as well. All other coupling constants are heteronuclear. Double click on any entry line in the **Report...** box, preferably one of the quartet multiplicities: In the appearing **Define Multiplet Identifier and Connection** dialog box click on the **Displ. hetero nuclear** radio button and then the **Hetero List...** button. Another dialog box appears, click on the **Load...** button and select the file ...\AKMRSX3\001001.MLT, from the multiplet analysis of the corresponding proton spectrum. Enter the correct isotope in the **Iso** edit field <1H>. In the **Hetero Multiplet** list the three proton multiplets appear with the indices 4, 5 and 6. Press **Ok** to return to the connection definition. Starting with the two quartet splitting (Multiplicity 4) connect identifiers 1 and 2 to the methyl group 6, identifier "CH3" pressing the **Displ. hetero nuclear** option to display the heteronuclear coupling constants. The remaining heteronuclear couplings may be connected either automatically (**Auto Connect**) or manually using the **Hetero List...** option in the **Define Multiplet Identifier and Connection** dialog box. When all couplings are connected use the **Export...** button to transfer the multiplet and interval data to **WIN-DAISY**.

Check it in WIN-DAISY:

WIN-DAISY will generate an 8-spin system. Open the **Control Parameters** dialog box and press the **Frequencies** button. The three fluorine shifts are selected for iteration. Click on the **Next** button to open the **Scalar Couplings Constants** window. If the coupling J$_{12}$ was not determined in the multiplet analysis select it for iteration. Open the **Lineshape Parameters** dialog box and check the **Nuclei specific linewidths** option; select the first three linewidths for iteration. Run a spectrum **Simulation** and compare the calculated and the experimental spectra using the **Dual Display** mode of WIN-NMR. There is good fit between the sets of data. Run a spectrum **Iteration** and display the result in the WIN-NMR **Dual Display**. The final *R-Factor* in the **Iteration Protocol** should be about 0.22%. Toggle to **WIN-NMR** and load the ^{1}H spectrum ...\AKMRSX3\001001.1R. Enter the **Analysis I Interval** mode and **Load...** the **Regions...** file. Press the **Export...** button and transfer only the interval data and *not* the multiplet data to WIN-DAISY. A message box appears in WIN-DAISY, press the **Ok** button and then press the **Connect** button in the current WIN-DAISY document main window. Open the **Spectrum Parameters** dialog box, click the **Isotope** button currently displaying ^{19}F and select in the **Element-Data** table **H**. The **Spectrum Parameters** are automatically altered to the ^{1}H spectrum. Quit the dialog box with **Ok**. Open the **Control Parameters** dialog box; the **Frequencies** and **Scalar Coupl.** data has been automatically adapted to the ^{1}H isotope. In the **Scalar Coupling Constants** dialog box both the homonuclear and heteronuclear couplings involving ^{1}H are selected. Run a spectrum **Iteration** using these default values and display the results using the **Dual Display** mode of WIN-NMR. Toggle back to **WIN-DAISY** and define again a ^{19}F **Isotope** in the **Spectrum Parameters**.

It is now necessary to consider the signs of the coupling constants; ortho ^{3}J$_{FF}$ are negative and meta ^{4}J$_{FF}$ can be either positive or negative. For the heteronuclear coupling, both the ortho ^{3}J$_{FH}$ and meta ^{4}J$_{FH}$ are positive while the para ^{5}J$_{FH}$ can be either positive and negative.

Check it in WIN-DAISY:

Open the **Scalar Coupling Constants** dialog box and apply a negative sign to ^{3}J$_{FF}$ (J$_{23}$). To determine which sign might be correct run an **Iteration** and examine the value of *R-Factor*. It can be possible to detect even the sign of the first-order heteronuclear coupling constant, because the two protons contained in the spin system have a second order relation! Change the sign of ^{5}J$_{FH}$ (J$_{35}$) in the **Scalar Coupling Constants**, run another **Iteration** and compare the value of *R-Factor* with the previous value. Now change the sign of the meta coupling constant ^{4}J$_{FF}$ (J$_{13}$) and run again an **Iteration**. Display the protocol list and compare the value of *R-Factor* with the other list. Decide which sign might be the correct one. Final *R-Factor* is about *0.197%*.

As expected from previous "Check its", there is very little difference in the R-Factor values and a more reliable way to determine the relative signs of the coupling constants is to perform a tickling experiment and compare it with a corresponding simulation.

Check it in WIN-DR and WIN-NMR:

Press the **Export WIN-DR** button to transfer the spin system data into the double resonance simulation program. In WIN-DR the **Perturbation Parameters** dialog box is automatically opened and the **Double Resonance** option is selected. Leave the **Experiment** as **Free**. Set the **Frequency** to

<1468.8> Hz, **Field in B2** to ca. <12> Hz and check all **Effected Spin(s)** (1, 2, 3, 4, 5 and 8). Set a **Min. Absolute Linewidth** of <0.2> Hz. Run a **Simulation** and **Display** the **Calculated Spectrum** in WIN-NMR. In WIN-NMR use the **Display | Dual Display** to load experimental ^{19}F spectrum ...\AKMRSX3\003001.1R as the **second file**. In this spectrum the line at 7.338 ppm (1468.80 Hz) in the ^{1}H spectrum has been tickled.

Compare the spectra and toggle back to **WIN-DR**. Change the sign combinations in the **Scalar Coupling Constants** dialog box. Run a spectrum **Simulation** and inspect the result as done before in **WIN-NMR Dual Display**. Try out as well different signs combinations for J_{13} and J_{35} and compare the simulation with the experimental spectrum. Were the signs determined with WIN-DAISY correct?

4.1.1.3 Symmetry

An example of a symmetrical spin system is the $[ABX]_2$ spin system of *1,2-difluorobenzene* [4.12]–[4.15]. Compared to the $[AB]_2$ spin system *1,2-dichlorobenzene* the NMR inactive chlorine nuclei are replaced by the pure element and spin–½ nuclei ^{19}F. The labels A and B refer to the protons and X to the fluorine.

Check it in WIN-NMR:

Open... the 200 MHz proton spectrum of *1,2-difluorobenzene* ...\SYMMETRY\011001.1R into WIN-NMR. Inspect the spectrum and compare it with the characteristics of an $[AB]_2$ spin system.

Measure a suitable **Linewidth** and then switch into the **Analysis | Multiplets** mode.

1,2-difluorobenzene

The spectrum looks very different compared to the previously analyzed spectra. The symmetrical appearance of the spectrum has disappeared due to the different H-F coupling constants of protons A and B.

Check it in WIN-NMR:

Define two **Singlets** as starting values for the chemical shifts in both signal groups e.g. at about 7.473 and 7.365ppm and **Export...** the data to **WIN-DAISY**.

Do *not* fragment the spin system (**No**). In the **Main Parameters** dialog box extend the spin system to <4> spins. Select **C2** symmetry and in the **Symmetry Description** dialog box enter, as usual, the values <4>, <3>, <2>, <1>. Open the **Scalar Coupling Constants** dialog box and enter approximate values, e.g. J_{12} = <7> Hz, J_{13} =<2> Hz, J_{14} = <1> Hz and J_{23} = <8> Hz. Run a **Simulation** and compare the results using the **Dual Display** mode of WIN-NMR.

The calculated spectrum differs significantly from the experimental spectrum because the fluorine spins have not yet been included in the WIN-DAISY input document. But some characteristics of the simulated spectrum can be found at least in the left-hand side A-spectrum part. Toggle back to **WIN-DAISY**. Open the **Main Parameters** dialog box and expand the system to <6> spins.

Keep the selected **Symmetry Group** but modify the **Symmetry Description**. The new indices 5 and 6 appear in the second row. Enter in the edit field below index 5 a <6> and vice versa. This means that the symmetry operation converts spin 5 into spin 6 and vice versa. Open the **Frequencies** dialog box and press the **PSE** button next to spin 5 and select **F** from the **Element-Data** table. Press the **Next** button to open the **Scalar Coupling Constant** dialog box. Enter some values for the proton-fluorine coupling constant: e.g. J_{15} = <4> Hz, J_{16} =<3> Hz, J_{25} = <2> Hz and J_{26} = <–1> Hz. J_{26} *must* be negative if a reliable result is to be obtained. Later you can experiment with different values to observe the variations on the spectrum, but the values suggested here will give the correct result in the least number of steps. It is very important that the fluorine-fluorine coupling J_{56} is not left at zero, otherwise it will not be possible to iterate the proton spectrum successfully. Enter a value of <–22.5>Hz for J_{56}. Quit the dialog box with **Ok**.

Open the **Lineshape Parameters** dialog box, select **Nuclei specific linewidths** and check the spins **1** and **2** for **iteration**. Open the **Control Parameters** dialog box and select **Advanced Iteration Mode** with high **Broadening** and high **No. of Cycles**. Check the option **Starting Frequencies poor**. Press the **Scalar Coupl.** button and click on the **Select all** button. Deselect the fluorine-fluorine coupling constant J_{56}. Modify the iteration limits for the proton-fluorine coupling constants; J_{15}, J_{16}, J_{25} to be from <0> up to <15> Hz and for J_{26} from <–15> up to <0> Hz. Run a spectrum **Iteration** and examine the results using the **Dual Display** mode of WIN-NMR Fig. 4.2b and e (*R-Factor 0.86%*).

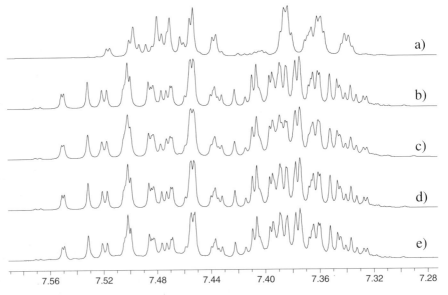

Fig. 4.2:   a) Initial simulation of *1,2-difluorobenzene* before iteration.
b) First iteration result without optimizing the F-F coupling.
c) Second iteration result including F-F-coupling adjustment.
d) Result of the iterated fluorine spectrum.
e) Experimental proton spectrum.

Check it in WIN-DAISY:

The results do not look too bad. To optimize the parameters further, open the **Control Parameters** set **Broadening** to **none**, deselect the **Starting Frequencies poor**. Press the **Scalar Coupl.** button and select the fluorine-fluorine coupling constant J_{56} for iteration over its default iteration limits. Because the biased lineshape results in a too large linewidth parameter, deselect it for iteration and fix the **Global Linewidth** parameter to <0.35> Hz. Run another **Iteration** and inspect the results using the **Dual Display** mode of WIN-NMR Fig. 4.2c and e (*R-Factor 1.04%* - this value is higher because of the fixed linewidth parameters).

It would now be interesting to see how well these values fit the experimental fluorine spectrum.

Check it in WIN-NMR and WIN-DAISY:

Load the ^{19}F spectrum ...\SYMMETRY\012001.1R into WIN-NMR. Measure a suitable **Linewidth** and make a note of the value (ca. 0.1 Hz). Switch into the **Analysis I Multiplets** mode and define a doublet using the very intense lines bounding the signal in order to determine the center shift of the signal. Note this value (2.6449 ppm) as well. Quit the Multiplets mode and enter the **Analysis I Interval** mode. Open the **Report...** box and **Load...** the predefined iteration region. Quit the dialog with **OK** and press the **Export...** interval data to **WIN-DAISY**. Do *not* export any multiplet data.

When WIN-DAISY is moved to the WINDOWS foreground accept the dialog with **OK** and press the **Connect** button. To change the experimental spectrum from proton to fluorine open the **Spectrum Parameters** dialog box. Press the **Isotope** button and select **F** from the **Element-Data** table. While an iteration mode is selected the spectrum limits of the selected isotope are automatically chosen. This is also shown in the Experimental and Calculated spectra filenames. Enter the fluorine chemical shift of <2.6449> ppm in the **Frequencies** dialog box. and enter the fluorine linewidth in the corresponding edit field <0.1> Hz in the **Lineshape Parameters** dialog box and select it for **iteration**. Execute a **Simulation** and display the results in the **Dual Display** of WIN-NMR **Zoom** in the baseline and look at the inaccuracy of the weak lines. Execute an **Iteration** and look at the result in WIN-NMR. **Zoom** in the baseline to examine the weak outer lines.

Change the **Isotope** in the **Spectrum Parameters** dialog box back to ^1H. Display the proton spectra with the final optimized parameters in WIN-NMR **Dual Display** mode. This is the most accurate result that can be obtained for this spectrum, because there are some baseline perturbations in the signal region. Zoom in the outer border regions of the signals, the weak lines are well reproduced.

The present example illustrates quite clearly how, in symmetrical heteronuclear spin systems, the appearance of the ^1H spectrum depends not only on the proton-proton and proton-fluorine coupling constants but also upon the fluorine-fluorine coupling constant. Similarly, the ^{19}F spectrum depends upon the proton-proton coupling constants. Consequently, it would not be possible to iterate the ^1H spectrum successfully if the fluorine-fluorine coupling constant had been set to zero!

The following example is designed to further the understanding of heteronuclear spin systems. Study the structure of methylene sulfur tetra fluoride given in Fig. 4.3, assuming a rigid structure [4.16], think about the complete spin system involving both the ^{19}F and ^{1}H nuclei keeping in mind the definition of isochronous spins and magnetic equivalence [4.17],[4.18].

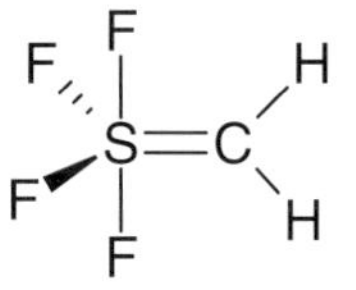

Fig. 4.3: Structure of methylene sulfur tetra fluoride.

Check it in WIN-DAISY:

Open... the input document SF4CH2.MGS and examine the **Main Parameters**, **Symmetry Group**, **Symmetry Description** and **Frequencies** dialog boxes. Think about the various parameters and the how they relate to the structure of methylene sulfur tetra fluoride. Run a spectrum **Simulation** and **Display** the **Calculated Spectrum**.

Toggle back to **WIN-DAISY** and **Change** the calculated spectrum name e.g. ...\999998.1R. Consider the ^{19}F{^{1}H} spin system. Alter the appropriate parameters in the **Main Parameters**, **Symmetry Group**, **Symmetry Description** dialog boxes in the current input document SF4CH2.MGS to describe the ^{19}F{^{1}H} spin system. Run a spectrum **Simulation** and **Display** the **Calculated Spectrum** in WIN-NMR. Using the **Dual Display** mode, load the previous calculated spectrum (...\999999.1R) as the second trace and compare both spectra.

The simplification of the ^{19}F{^{1}H} spectrum with respect to the ^{19}F spectrum is obvious and arises because the axial fluorine atoms which are isochronous in the ^{19}F spectrum become magnetically equivalent in the ^{19}F{^{1}H} spectrum. This observation may be verified by loading the input document SF4_CH2.MGS.

4.1.1.4 Isotope Effect

Generally the reference compound used in fluorine NMR is CFCl$_3$. It would be expected that the ^{19}F spectrum of CFCl$_3$ should exhibit a singlet, what is not the case.

Check it in WIN-NMR:

Open... the experimental spectrum ...\CFCL3\001001.1R in WIN-NMR. **Zoom** in the region at 0 ppm. Inspect the signal pattern. Open the **Analysis | Deconvolution 1** mode and define with the cursor the **Region** around 0 ppm. Open the **Options...** dialog box and set the **Number of peaks** to <4> and **Execute** the fitting procedure. Press the **Report...** button to display the relative intensities of the four peaks.

The four peaks show an intensity ratio of ca. 44.7:40.2:13.6:1.7 from low field to high field. The signals are separated by approximately 2.4 Hz at 376.5 MHz (400 MHz ^{1}H frequency). The peak for reference is the second from left [4.3].

Chlorine has two naturally occurring isotopes ^{35}Cl and ^{37}Cl, both have a nuclear spin of $\frac{3}{2}$ and possess a nuclear quadrupole moment. Quadrupolar relaxation is a very effective relaxation mechanism which in most cases prevents the detection of coupling interaction with quadrupolar nuclei. Chlorine-fluorine coupling cannot be observed and the strange appearance of the ^{19}F spectrum depends, not on coupling interactions, but on the statistical isotopic composition of the chlorine atoms present in CFCl$_3$. The natural abundance of ^{35}Cl is 75.77% and 24.23% for ^{37}Cl. (The isotope ^{36}Cl is radioactive.)

^{19}F–C^{35}Cl$_3$ ^{19}F–C^{35}Cl$_2$^{37}Cl ^{19}F–C^{35}Cl^{37}Cl$_2$ ^{19}F–C^{37}Cl$_3$
    a)   b)   c)   d)

Fig. 4.4: Chlorine isotopomers of CFCl$_3$.

The substitution of one atom by one of its heavier isotopes alters the vibrational energies in the bonds. As a consequence the electron distribution and the nuclear screening is changed by the isotopic substitution giving rise to small changes in the chemical shifts of the neighboring spin active nuclei. The changes are very small and usually reported in parts per billion (ppb). In most cases the heavier isotope leads to an improved screening at the nearby nuclei [4.19],[4.20].

In the present example the isotope shift for each chlorine ^{35}Cl substituted by a ^{37}Cl is ca. –6.4 ppb. The different isotopic compositions of a molecule are called isotopomers. The statistical probability for the four chlorine isotopomers CF^{35}Cl$_n$^{37}Cl$_{3-n}$ (n = 0, 1, 2, 3) is given by the binomial coefficient shown below. For CFCl$_3$ the probability is 1:3:3:1; the ^{35}Cl / ^{37}Cl mixed isotopomers are statistically three times more abundant than the all ^{35}Cl and all ^{37}Cl isotopomers, because there are three possible places available for the substitution. Taking into account the natural abundance of the chlorine isotopes, the relative probabilities of the isotopomers are as follows (refer also to the appendix, section 6.2):

$$\binom{3}{3-n} \cdot a(^{35}Cl)^n \cdot a(^{37}Cl)^{3-n} \cdot 100 \ [\%]$$

with a(^{35}Cl) = natural abundance of ^{35}Cl isotope (100% = 1.0), and n = 0, 1, 2, 3.

Check it with the Calculator:

Calculate the relative probabilities for the four chlorine isotopomers of CFCl$_3$ and compare the values with the WIN-NMR deconvolution results.

Calculated and experimental results should be comparable with the following data:
n = 3: $1 \cdot 0.7553^3 \cdot 0.2447^0 \cdot 100$ = 43.09 % (experimentally obtained ≅ 44.3 %)
n = 2: $3 \cdot 0.7553^2 \cdot 0.2447^1 \cdot 100$ = 41.88 % (experimentally obtained ≅ 40.4 %)
n = 1: $3 \cdot 0.7553^1 \cdot 0.2447^2 \cdot 100$ = 13.57 % (experimentally obtained ≅ 13.6 %)
n = 0: $1 \cdot 0.7553^0 \cdot 0.2447^3 \cdot 100$ = 1.47 % (experimentally obtained ≅ 1.7 %)

244 4 Spin Systems and the Periodic Table

The isotope effect discussed here is referred as to the *secondary isotope effect on nuclear shielding*, which occurs because of the difference in nuclear shielding of nuclei *caused* by two different isotopes. (The primary isotope effect, which has been only sparingly investigated is obtained by comparing the nuclear shielding of different isotopes). [4.19][4.20].

4.1.2 Spin Systems Containing ^{31}P

For compounds containing phosphorous, coupling interactions of the pure element ^{31}P have to be considered. Because the magnetogyric ratio of ^{31}P is very different from that of ^{1}H, the technique of phosphorous decoupling can be used to remove ^{31}P–^{1}H coupling and simply ^{1}H spectra.

^{31}P is an important element in organic and inorganic chemistry and biochemistry. 85% ortho phosphoric acid is normally used as an external reference standard [4.3]. The chemical shift range of phosphorous compounds is quite large covering ca. 4000 ppm.

Proton decoupled ^{31}P{^{1}H} NMR spectra of polyphosphorus compound represent interesting all–^{31}P spin systems. The complex homo and heteronuclear spin systems of phosphonic and phosphinic acids and their derivatives will be discussed in the following.

As a first example of this type of compound consider the spectrum of *tris (1,1-dimethylethyl) phosphonic acid anhydride* [4.21]. This molecule has a six-membered ring structure of three [–O–P(O)(C(CH$_3$)$_3$)–] units. The very bulky tertiary butyl groups prefer the equatorial position in the cyclohexane type skeleton and there are two possible conformers as shown in Fig. 4.5. Determine the symmetry and the spin system of the two compounds shown in Fig. 4.5 and hence the appearance of the ^{31}P{^{1}H} spectrum.

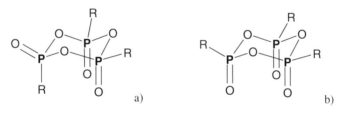

Fig. 4.5: Structures of *tris (1,1-dimethylethyl) phosphonic acid anhydride*, R = tert. Butyl.

Check it in WIN-NMR:

Open... the experimental ^{31}P{^{1}H} spectrum, ...\ANHYDRID\001001.1R, in WIN-NMR. In the **Analysis I Integration** mode integrate all the signals and split the integral at 23.6 ppm. Deduce which of the two possible spin systems (structures are given in Fig. 4.5) is present.

In Fig. 4.5a the spin system is A$_2$B or AB$_2$ and in Fig. 4.5b A$_3$. The experimental ^{31}P{^{1}H} spectrum corresponds to either an A$_2$B or AB$_2$ spin system; of these two possibilities, the integration confirms that the spin system is AB$_2$. The spectrum analysis of AB$_2$ spin systems has already been discussed in depth in section 2.4.6.1 and consequently only a brief outline will be given in the following "Check it".

Check it in WIN-NMR and WIN-DAISY:

Measure a suitable **Linewidth** and then switch to the **Analysis | Multiplets** mode. Define the third signal representing v_A as a **Singlet** and define peaks 5 and 7 as a **Doublet**, because the arithmetic mean corresponds to v_B. **Export...** the data to **WIN-DAISY** *without* fragmentation (**No**) of the spin system and *do not* include any non-assigned couplings (**No**). In this way only the chemical shifts are exported to WIN-DAISY. Open the **Frequencies** dialog box and for Spin 2 increase the number of **Spins in group** to <2>. In the **Scalar Coupling Constants** dialog box enter a value for J_{12}. e.g. <50> Hz. Open the **Control Parameters** dialog box. Press the **Scalar Coupl.** button, press the **Select all** button. Run an **Iteration** and inspect the result using the **Dual Display** mode of WIN-NMR. The agreement is very good and even the 9th transition at 21.7 ppm is clearly visible.

The next spin system corresponds to the three spin $^{31}P\{^1H\}$ spin system of the ethane trisphosphinic acid ester shown in Fig. 4.6 [4.22].

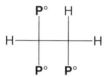

Fig. 4.6: Structure of *1,2,2-ethane tris (P-methyl phosphinic acid isopropylester)*, **P°**= $CH_3P(O)OiC_3H_7$, iC_3H_7= isopropyl.

The symbol **P°** in Fig. 4.6 refers to an asymmetric phosphorous atom. Assuming that there is no hindered rotation, 4 (2^{n-1} where n is the number **P°** units) diastereomers are possible and distinguishable by NMR. The spectra of isomeric mixtures are very complex, all the phosphorous atoms are non isochronous and often only the $^{31}P\{^1H\}$ spectra can be analyzed. A Karplus-like dihedral and substituent dependent vicinal $^3J_{PP}$ has been suggested to correlate with the P–C–C–P fragments.

Check it in WIN-NMR:

Open... the experimental spectrum ...\AMX\011001.1R in WIN-NMR. The spectrum relates to one of the isomer of the compound shown in Fig. 4.6 and is obtained by fractional crystallization [4.22]. Measure a suitable **Linewidth** and then switch to the **Analysis | Multiplets** mode. Perform a first order analysis and **Export...** the data to **WIN-DAISY**.

Check it in WIN-DAISY:

Because all the coupling constants in the spin system cannot be determined directly from the spectrum, click the **Iterate all** button in the **Scalar Coupling Constants** dialog box. In the **Lineshape Parameters** dialog box switch to **Nuclei specific linewidths** and select all spins for iteration. Run an **Iteration**. Examine the different values in the **Scalar Coupling Constants** dialog box. The smallest value corresponds to the geminal $^2J_{PP}$ and the other to the vicinal $^3J_{PP}$'s. In the **Lineshape Parameters** dialog box inspect the individual linewidths and note the smaller value for the third nucleus. Display the experimental and calculated spectra in the **Dual Display** of WIN-NMR. Keep the WIN-DAISY document *opened* and toggle back to **WIN-NMR**.

Check it in WIN-NMR:

Open... the experimental spectrum ...\ABX\009001.1R in WIN-NMR. This spectrum belongs to another isomer of the compound shown in Fig. 4.6. Measure a **Linewidth** - in comparison with the former spectrum, the half-height linewidth is much larger, about 6.5 Hz. Moreover two of the shifts show a strong second order effect (spin system AXY). The weak lines in the spectrum are caused by a small amount of another isomer and are not part of the AXY spin system.

Switch to the **Analysis | Multiplets** mode and for the low-field signal define a doublet of doublet. The two nuclei in the high-field spectral region form a strongly second order XY system; two of lines are degenerate and only three resolved lines are visible. Define two doublets for the coupling constant to the A nucleus based on the first and second line and the first and third line. The XY coupling constant cannot be determined in this first order approximation - but it must be included in the iteration (see next "Check it")! With the spectrum cursor in **maximum cursor** mode mark one of the small lines at about 51 ppm. Connect the coupling constants and **Export...** the data to **WIN-DAISY**.

Check it in WIN-DAISY:

Open the **Scalar Coupling Constant** dialog box and select the XY coupling constant, J_{23}, for **iterat**ion. The default iteration limits should be large enough because the geminal coupling constant was very small in the previous example. Run a spectrum **Simulation** and display experimental and calculated spectra using the **Dual Display** mode of WIN-NMR. An iteration using the present experimental spectrum would fail because of the additional signals not belonging to the spin system.

To check if the impurities in the current spectrum originate from the already analyzed isomer toggle back to **WIN-DAISY**. Two documents should be open, the first relates to the previous spectrum ...\AMX\011001.1R and the second to the current spectrum ...\ABX\009001.1R. Switch to the input document of the other isomer ...\AMX\011001.1R (use the **Window** pull down menu) and using either the pull-down menu **Edit | Copy fragment to...** or the toolbar button copy this information into the actual document belonging to ...\ABX\009001.1R document. Press the **Ok** button in the dialog box that appears. Toggle back to the ...\ABX\009001.1R document, this now contains two fragment relating to both of the isomers. Open the **Main Parameters** of the second fragment and enter a **Statistical weight** of <0.03>. Display the **Lineshape Parameters**; deselect the **Nuclei specific linewidth** option and enter a bigger value for the **Global Linewidth**, e.g. <8> Hz. Do *not* check the iterate option. Because of the low concentration of the second isomer it is not possible to optimize the associated parameters. However, the minor isomer does contribute to the overall lineshape and the final *R-Factor* value and if it was ignored it would not be possible to optimize the parameters of the major isomer correctly. Open the **Control Parameters** dialog box and press the

Frequencies button. Display the shifts for the second fragment and press the **Select none** button. Use the **Next** button in the bottom line to open the **Scalar Coupling Constants** dialog box and click for the second fragment the **Select none** option. Run a spectrum **Simulation** and use the **Dual Display** option to inspect the result. Run an **Iteration** and display the result in the WIN-NMR **Dual Display** mode.

Check it in WIN-NMR and WIN-DAISY:

With the right mouse button click the **ALL** button in WIN-NMR to zoom in the second order region. There is still a visible difference in the XY spectrum part. Remember that the other isomer had different linewidths. Toggle back to **WIN-DAISY**. Open the **Lineshape Parameters**, check the option **Nuclei specific linewidths** option and click the **Iterate all** button. Run another **Iteration** and display the result together with the experimental spectrum in the WIN-NMR **Dual Display**. Toggle back to **WIN-DAISY**. Click the blue FID button (**Export to WIN-NMR exp.**) to export the WIN-DAISY result to the experimental lineshape in WIN-NMR. WIN-NMR will automatically change into the **Multiplets** mode with the **Windaisy** option selected. The transitions calculated by WIN-DAISY are displayed below the experimental lineshape (normalized to the cursor position when the multiplet data was exported to WIN-DAISY). **Zoom** in the low-field region of the spectrum where the two combination lines in the A spectrum part - as weak "inner" lines are visible. The *standard deviation* in of the geminal coupling constant is very large, so it cannot be determined in any accuracy from this kind of spectra.

Continuing with the isomers of ethane oligophosphinic compounds, the ethane 1,1,2,2 tetrakis phosphinic acids and the corresponding esters (Fig. 4.7) result in 8 enantiomeric pairs of which only 6 can be distinguished by NMR [4.22].

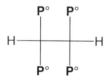

Fig. 4.7: Structure of *ethane-1,1,2,2-tetrakis (methyl phosphinic acid Na salt)*, $P°= CH_3P(O)ONa$.

The isomers display a variety of spin systems in the $^{31}P\{^1H\}$ spectrum; A_4 (all P magnetically equivalent), ABCD up to AMRX (all P non-isochronous) and $[AB]_2$ up to $[AX]_2$ (with point group symmetries C_2, C_s or C_i) [4.23].

Check it in WIN-NMR:

Open... the experimental $^{31}P\{^1H\}$ spectrum ...\SYMMETRY\013001.1R in WIN-NMR. From the appearance, the spectrum belongs to one of the isomers featuring a twofold symmetry. Set up a symmetrical spin system. First measure a suitable **Linewidth** and then switch to the **Analysis | Multiplets** mode. For each signal group define a doublet based on the two most intense lines. Connect the "coupling" and **Export...** the data to **WIN-DAISY**.

Check it in WIN-DAISY and WIN-NMR:

Open the **Main Parameters** dialog box and extend the **Number of spins or groups with magnetical equivalent nuclei** to <4>. Click the **Next** button and in the **Symmetry Group** window select either **C2** or **Cs** or **Ci** symmetry. Press **OK**. In the **Symmetry Description** dialog box enter the values <4> <3> <2> and <1> in the edit fields. Quit the dialog with **Ok**. In the **Options** dialog box set new default limits for the **Scalar Coupling Constants** of ± <70> Hz and select the option **Allow sign change**. Open the **Control Parameters** dialog box and select **Advanced Iteration mode** with high **Broadening** and high **No. of Cycles**, check the option **Starting Frequencies poor**. Press the **Scalar Coupl.** button and enter <0> for both the **Lower limit** and **Upper limit** of the selected coupling constants; press the **Select all** button. Quit all the dialog boxes with **Ok**. Open the **Scalar Coupling Constants** dialog box and enter some values for J_{13}, J_{14} and J_{23} e.g. <5>, <10> and <15>. Run an **Iteration** and display the iteration result together with the experimental lineshape in the **Dual Display** mode of WIN-NMR. The fit is not particularly good, but in the **Dual Display** the different linewidths in the A and B parts of the spectrum is obvious. Toggle back to **WIN-DAISY**, open the **Lineshape Parameters**, select the option **Nuclei specific linewidths** and click on the **Iterate all** button. Open the **Control Parameters** and select **Standard Iteration Mode** with **none Broadening**. Optimize the parameters via **Iteration**.

When a satisfactory result has been obtained open the **Scalar Coupling Constants** dialog box and examine the values. One coupling constant has a *negative* sign. Interchange the *signs* of J_{12} and J_{13}, run a spectrum **Simulation** and compare the calculated and experimental spectra in the **Dual Display** mode of WIN-NMR. There is no visible difference in the lineshape. Open the **Scalar Coupling Constants** dialog box again and interchange the *values* J_{12} and J_{13}. Copy the first coupling constant value into the clipboard, <Ctrl + C>, and then enter the value of J_{13} into the edit field for J_{12}. Double click with the mouse in the edit field of J_{13} and paste in the original value of J_{12}, <Ctrl + V>. Run another **Simulation** and compare the spectra in the WIN-NMR **Dual Display**. The spectrum appearance is not altered by interchanging the values and only the absolute value of the sum $J_{AB} + J_{AB'}$ can be obtained. Open the **Scalar Coupling Constants** dialog box and change the sign(s) of J_{14} and / or J_{23}. Run a Simulations and inspect the spectra in the WIN-NMR **Dual Display**.

Note: The results obtained in this "Check it" are exactly the same as those obtained in chapter 3 on the analysis of symmetrical proton spin systems and reinforces the argument that the analysis of the spectra of "pure" elements follows the same rules as for the analysis of proton spectra.

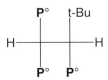

Fig. 4.8: Structure of 1-tert. Butyl-ethane 1,2,2-tris (phosphonic acid) trisaniline salt
P°= HOP(O)O⁻ ⁺NH$_3$C$_6$H$_5$ [4.23].

Fig. 4.8 shows the structure of a oligophosphonic acid. Due to the bulky nature of the tertiary butyl group free rotation around the C–C single bond is hindered and different rotamers can occur in varying amounts.

Check it in WIN-NMR:

The files ...\AKMRX\005001.1R (^{31}P), ...\AKMRX\006001.1R (^{1}H), ...\AKMRX\007001.1R (^{31}P{^{1}H}) and ...\AKMRX\008001.1R (^{1}H{^{31}P}) represent the complete set of proton and phosphorous spectra available for this compound. Load each spectrum into WIN-NMR and decide which type of spectrum (coupled or decoupled) would be the best to start the analysis with.

Think about how many phosphorous nuclei define one spin system in the ^{31}P{^{1}H} spectrum and how many protons define the spin system in the ^{1}H{^{31}P} spectrum. **Open...** the ^{31}P{^{1}H} spectrum ...\AKMRX\007001.1R in WIN-NMR and use the **Analysis | Integration** mode to decide which signals belong to the same spin system.

The number of phosphorous nuclei per spin system and the overall integral values makes it clear, that there are two rotamers present in a ratio of about 1 : 10. The minor rotamer also shows less coupling interactions than the major rotamer.

Check it in WIN-NMR:

Measure a suitable **Linewidth**. Enter the **Analysis | Multiplets** mode and analysis the spectrum keeping in mind that all nuclei are non-isochronous, so that only doublet splittings are present. Define the peak of the low abundant rotamer that shows no splitting as a singlet; mark the peak with the arrow in maximum spectrum cursor, press the right mouse button and select **Define Singlet**. Open the **Report...** box and connect the coupling constants. **Export...** the data to **WIN-DAISY** with *fragmentation* of the spin system.

The automatic fragmentation algorithm always forms the smallest fragment sizes. Because a singlet was defined, three fragments are obtained; a three spin system for the major rotamer, and a two and one spin system for the minor rotamer. To define two three-spin systems, both the spin systems for the minor rotamer have to be merged.

Check it in WIN-DAISY:

First of all it is necessary to merge the two fragments for the minor rotamer into one. Highlight the first fragment title line (**... 2-spin system...**) in the list box. Using either the **Merge fragment** command or button in the tool bar. Select in the **Fragment Handling** dialog box the fragment with the title **... 1-spin system...** as the destination. After merging the fragments delete the superfluous fragment (**Yes**). Edit the **Title** entries in the **Main Parameters** of both fragments to make the identification of the content clear, e.g. <major rotamer 31P(1H)>. To take into account the different abundance of the rotamers enter a **Statistical weight** of <10> for the major rotamer in the **Main Parameters** dialog box. Open the **Scalar Coupling Constants** dialog box and **Iterate all** parameters of both rotamers. Run a spectrum **Simulation** and compare the spectra in the **Dual Display** mode of WIN-NMR. Then run an **Iteration** with default values and inspect the result again in the **Dual Display** of WIN-NMR. A comparison of the lineshapes clearly shows, that the linewidth of the phosphorous nuclei in the main rotamer are very different (*R-Factor ca. 13.2%*). Open the **Lineshape Parameter** dialog box; for both fragments select the option **Nuclei specific linewidths** and click on the **Iterate all** button. Depending on the value of the selected linewidth it might be necessary to increase the **Upper limit** of the linewidth parameters, e.g. to <20> Hz. Run a spectrum **Iteration** and examine the results in the WIN-NMR **Dual Display** (*R-Factor ca. 1.13%*). Click the blue FID button (**Export WIN-NMR exp.**) to export the WIN-DAISY result to the experimental spectrum. Keep the WIN-DAISY document open.

Check it in WIN-NMR:

Load the ^{1}H{^{31}P} spectrum ...\AKMRX\008001.1R into WIN-NMR and use the **Analysis I Integration** mode to decide how many protons are in each signal and to which spin system they belong. Measure a suitable linewidth. Enter the **Analysis I Multiplets** mode and analyze the data. Open the **Report...** box and connect the coupling constants. **Export...** the data to WIN-DAISY with fragmentation of the spin system.

Check it in WIN-DAISY:

Merge the two one spin systems for the minor rotamer. In the **Main Parameters** dialog box enter a suitable **Title** for each fragment e.g. <minor rotamer 1H(31P)>. For the major rotamer enter a **Statistical weight** of <10>. Select the **Scalar Coupling Constant** J_{12} of the minor rotamer for **iterat**ion. Run an **Iteration** using default values and inspect the result in the WIN-NMR **Dual Display**.

A comparison of experimental and calculated lineshape shows that the linewidths for each proton may be different in both rotamers and for the minor rotamer there may also be a small unresolved coupling.

Check it in WIN-DAISY:

Open the **Scalar Coupling Constants** dialog box for the minor rotamer, enter a small value for the coupling constant, e.g. <0.1> Hz and check the **Iterate** option. Open the **Lineshape Parameters** dialog box, for each fragment select the option **Nuclei specific linewidths** and **Iterate all**. Run an **Iteration** and check the result in the WIN-NMR **Dual Display**. Open the **Scalar Coupling Constants** dialog box of the minor rotamer and check the sign of coupling constant J_{12}. If the sign is negative change it to <+>; the sign has no effect on the spectrum and $^3J_{HH}$ should be positive. **Export** to **WIN-NMR exp.** the WIN-DAISY result. Keep the WIN-DAISY document open.

The next stage is to prepare the WIN-DAISY document for the iteration of the ^{1}H and ^{31}P spectra. Three steps are required: the combination of the ^{1}H{^{31}P} and ^{31}P{^{1}H} data into one WIN-DAISY document, the merging of the ^{1}H and ^{31}P spin system fragments for both the minor and major rotamers and finally the connection of the ^{1}H and ^{31}P experimental spectra and the modification of the iteration data.

Check it in WIN-DAISY:

Starting with the $^{31}P\{^1H\}$ WIN-DAISY document (exp. spectrum ...\AKMRX\007001.1R), highlight the first fragment title line in the list box. Using either the **Edit | Copy fragment to...** command or the corresponding button in the tool bar copy the **minor rotamer 31P(1H)** fragment into the $^1H\{^{31}P\}$ document. Repeat the process for the major rotamer fragment. **Close** the ^{31}P document. Note that the title has been modified and now includes the word (copy). Now merge the two minor rotamer fragments (select the 1h document on top of the list box), deleting the superfluous fragment, as described in detail in the previous "Check its". Repeat the process for the major rotamer fragment.

In the next step the coupled experimental spectra need to be connected to the new combination document. Load the ^1H spectrum ...\AKMRX\006001.1R into WIN-NMR. Switch to the **Analysis | Interval** mode. Click the **Regions...** button and **Load...** the predefined iteration region. Close the dialog box with **Ok** and **Export...** the interval data *only* to **WIN-DAISY**. In WIN-DAISY, accept the dialog box and then press the **Connect** button in the main window. Toggle back to **WIN-NMR**. Load the ^{31}P spectrum ...\AKMRX\005001.1R and then **Load...** and **Export...** the predefined iteration regions to **WIN-DAISY**. Accept the dialog box and press the **Connect** button on the main window. Notice that the **Experimental Spectrum** and **Calculated Spectrum** file names are not updated as was the case for the ^1H spectrum. To check the position of the signals in the ^1H spectrum, run a **Simulation** and display the coupled experimental spectrum with the calculated (still decoupled) spectrum using the **Dual Display** mode. If necessary use <Ctrl + Y> to adapt the Y-axis and the **ALL** button to overlay the spectra. Repeat this for the ^{31}P spectrum by changing the **Isotope** entry in the **Spectrum Parameters** dialog box to **31P**. Run a **Simulation** and display the results using the WIN-NMR **Dual Display**.

Using this approach the data for the decoupled spectra is taken as the basis for the coupled spectra analysis and only the heteronuclear coupling constants $^nJ_{PH}$ have to be optimized from scratch. Consequently smoothing parameters will need to be define for the spin system.

Check it in WIN-DAISY:

Open the **Spectrum Parameters** dialog box and select the **31P Isotope** lineshape data. Open the **Options** dialog and change the **Parameter limits for iteration** for the **Scalar Coupling Constants** to ± <50> Hz and **Allow sign changes**. Open the **Control Parameters** dialog box and choose **Advanced Iteration mode, high Broadening, high No. of Cycles** and **Starting Frequencies poor**. Change the **maximum number of iterations** to <150>. Press the **Frequencies** button and *deselect* all shifts from iteration (use **Select all** followed by **Select none**) for both fragments. Switch to the **Scalar Coupl.** window, press the **Select all** button; deselect the $^3J_{HH}$ for both rotamers and for the minor rotamer the two zero-value $^nJ_{PP}$.

Note: The exact indices for the coupling constants will depend on whether the ^1H data has been merged into the ^{31}P data or vice versa. The heteronuclear coupling constants are arranged in the same place in both fragments either as *two rows of three columns* (^1H merged into ^{31}P) or *three rows of two columns* (^{31}P merged into ^1H). In the former case $^3J_{HH}$ corresponds to J_{12}, the three

252   4 Spin Systems and the Periodic Table

$^nJ_{PP}$ couplings to J_{34}, J_{35} and J_{45} and $^nJ_{PH}$ the six remaining couplings. In the second case J_{45} now corresponds to $^3J_{HH}$ and J_{12}, J_{13} and J_{23} to the three $^nJ_{PP}$ couplings. The value of $^3J_{HH}$ is ca. 0.8 Hz and ca. 7 Hz in the minor and major rotamer respectively.

Open the **Lineshape Parameters** dialog box deselect the linewidth for iteration in both fragments (use **Iterate all** followed by **Iterate none**). Open the **Scalar Coupling Constants** dialog box and for each fragment set some arbitrary values for $^nJ_{PH}$, e.g. <5>, <6>, <7>, <8>, <9> and <10>. Run an **Iteration** and examine the results in the **Dual Display** mode of WIN-NMR. Although the overall lineshape is not too bad, the *R-Factor* is over *10%* mainly because of incorrect shifts. Open the **Control Parameters** dialog box, remove the smoothing parameters (**none Broadening**) and deselect **Starting Frequencies poor**. In the **Frequencies** dialog box select the three phosphorous shifts of both fragments for **iterat**ion. Run another **Iteration** (*R-Factor* ca. 5%). Finally in the **Lineshape Parameters** include the phosphorous nuclei linewidths in both fragments for optimization. Run another **Iteration**, the final *R-factor* should be ca. *0.13 %*.

When this *R-factor* value has been achieved, change the **Isotope** in the **Spectrum Parameters** to **¹H**. Run a spectrum **Simulation** and look at the result in the WIN-NMR **Dual Display**. If necessary align the Y-axis with <Ctrl + Y> and the **ALL** button. It is most likely that the simulated spectrum does not coincide with the experimental proton spectrum.

The reason for this difference is as follows. In the iteration of the ^1H coupled ^{31}P spectrum the heteronuclear coupling constants $^nJ_{PH}$ were obtained from an arbitrarily selected set of starting values. Because of the large difference in the resonance frequencies of ^1H and ^{31}P, the coupling interaction is strictly first order (refer to heteronuclear AX spin systems, section 2.2.2.1). Consequently, the appearance of the ^{31}P spectrum does not depend on whether the coupling constant $^nJ_{PH}$ belongs to either the high or the low field proton. However, the origin of the $^nJ_{PH}$ interaction does influence the ^1H spectrum and a comparison of the experimental and calculated ^1H spectra indicates which, if any, of the coupling have been interchanged. If all the $^nJ_{PH}$ couplings have been interchanged for a rotamer, the calculated spectrum will be the mirror image of the experimental spectrum.

Check it in WIN-DAISY:

Toggle to **WIN-DAISY**. Open the **Scalar Coupling Constants** dialog box. Starting with the major rotamer, interchange the pairs of $^nJ_{PH}$ in a stepwise fashion systematically. The fastest way to interchange the *pairs* (in columns when index 1,2 are protons, otherwise when index 3, 4 and 5 are protons in rows) is to highlight one value and copy it to the clipboard, <Ctrl + C>, overwrite it with the neighboring value, highlight the neighboring value (double click with the mouse) and paste the clipboard content, <Ctrl + V>. After each exchange run a **Simulation** and compare the pattern in WIN-NMR. Repeat the process for all the pairs of $^nJ_{PH}$ until the simulated and experimental spectra match. Do the same optimization for the minor rotamer. Finally run an **Iteration** without smoothing parameters (*R-Factor* 0.09%). In the **Spectrum Parameters** change the **Isotope** to ^{31}P. Simulate the ^{31}P spectrum and compare the result in the **Dual Display** mode of WIN-NMR. **Export** to **WIN-NMR exp.** the WIN-DAISY results.

Check it in WIN-NMR:

Open... the 200 MHz $^1H\{^{31}P\}$ spectrum of *benzene-1,2-diphosphonic acid* in $D_2O/KOD+$ 10% MeOD [4.24] ...\SYMMETRY\014001.1R in WIN-NMR. Inspect the spectrum and decide which type of spin system is present.

Due to the alkaline conditions, the acidic protons are removed and the compound exists in anionic form. Deuteromethanol is added to keep the outer lines resolved. Due to the ^{31}P decoupling, all the phosphorous interactions are removed resulting in a proton only spin system.

benzene-1,2-diphosphonic acid

Measure a suitable **Linewidth** and switch into the **Multiplet I Analysis** mode. Analyze this type of spectra in the usual way and **Export...** the data to WIN-DAISY. Set up a spin system with two fold symmetry, use typical values for the proton-proton coupling constants e.g. J_{13} = <2> Hz, J_{14} = <0.5> Hz and J_{23} = <8> Hz. Optimize the parameters and store the result as multiplet data with the experimental spectrum **Export WIN-NMR exp..**

Check it in WIN-NMR:

Open... the 200 MHz ^1H spectrum of *benzene-1,2-diphosphonic acid anion* ...\SYMMETRY\015001.1R, in WIN-NMR. Enter the **Interval** mode, **Load...** the **Regions...** and **Export...** them to **WIN-DAISY**. **Connect** the data, run a **Simulation** and display both spectra in the **Dual Display** (refer to Fig. 4.9).

The two sets of signals are still well-separated. To analyze this spectrum, the best way is to modify and extend the parameters obtained for the phosphorus decoupled spectrum. If the data are no longer present in WIN-DAISY, use the $^1H\{^{31}P\}$ spectrum, ...\SYMMETRY\014001.1R, to switch into the **Analysis I Multiplet** Mode. Press the **Windaisy** button and **Export...** the data to **WIN-DAISY**.

Check it in WIN-DAISY:

Open the **Symmetry Group** dialog box and select a twofold symmetry, e.g. **Cs**. In the **Symmetry Description** window define the permutation operators <4>, <3>, <2> and <1>.

In the **Main Parameters** dialog box extend the spin system to <6> nuclei. Return to the **Symmetry Description** and enter the value <6> below the index 5 and vice versa. Open the **Frequencies** dialog box and use the **PSE** button to select **31P** for spin 5.

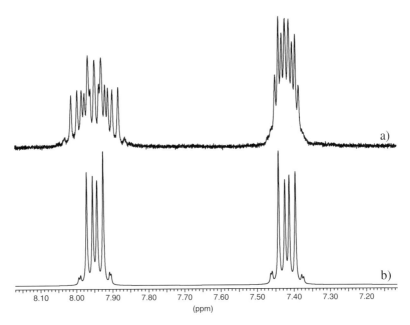

Fig. 4.9: a) experimental ^{1}H NMR spectrum of *benzene-1,2-diphosphonic acid*
b) experimental ^{1}H{^{31}P} NMR spectrum of the compound

Open the **Program Options and Settings** dialog box and change the default limits for **Scalar Coupling Constants** to ±<15> Hz and make sure that the option **Allow sign change** remains *unchecked*. Then open the **Control Parameters** dialog box and select smoothing parameters. Open the **Scalar Coupl.** box. *Deselect* all the proton-proton coupling constants and select all the proton-phosphorous coupling constants. Quit the **Control Parameters** with **OK**.

Open the **Scalar Coupling Constants** dialog box and enter some values for the H-P coupling constants J_{15}, J_{16}, J_{25} and J_{26}, e.g. <4.0>, <3.0>, <2.0> and <1.0> (all values positive). The P-P coupling constant, not selected for iteration, should be set to ca. <10> Hz. Execute one or more spectrum **Iteration**s till a good result is obtained.

When the proton spectrum is fitted satisfactory, toggle back to **WIN-NMR** and **Open...** the proton coupled ^{31}P spectrum ...\SYMMETRY\016001.1R. Switch into the **Analysis I Interval** mode, **Load...** the predefined iteration **Regions...** file and **Export...** the interval data only to **WIN-DAISY**.

Accept the dialog box and press the **Connect** button in the actual WIN-DAISY document main window. Open the **Spectrum Parameters** dialog box, press the **Isotope** button and select the **P** nucleus from the **Element-Data** table. Notice that the **Experimental Spectrum** and **Calculated Spectrum** file names are now updated for the ^{31}P data. Open the **Frequencies** dialog box and in the edit field for Spin 5 enter the ^{31}P chemical shift <1014.9> Hz. Run a spectrum **Simulation** and compare the experimental and calculated spectra

4.1 Spin-½ Pure Elements

in the **WIN-NMR Dual Display**. If necessary use <Ctrl + Y> to adapt the Y-axis and the **ALL** button to overlay the spectra.

Now optimize the ^{31}P spectrum: In the **Lineshape Parameters** dialog box select **Nuclei specific linewidths** and Spin 5 for iteration (take care that the upper iteration limit is ca. <2> Hz). Open the **Control Parameters** and press the **Frequencies** button. Modify the lower limit for Spin 5 (the ^{31}P) to <1000> Hz. Run a **Standard Iteration** without using any smoothing parameters.

When a satisfactory result is obtained, open the **Spectrum Parameters** dialog box, click the **Isotope** button and select **H** from the **Element-Data** table. Run a spectrum **Simulation** to calculate the ^{31}P coupled ^{1}H spectrum using the parameters optimized in the ^{31}P spectrum.

Many examples are known of organophosphorous compound of phosphonic acids and their esters. In the ethyl ester the methylene protons of the ethoxy groups are often diastereotopic.

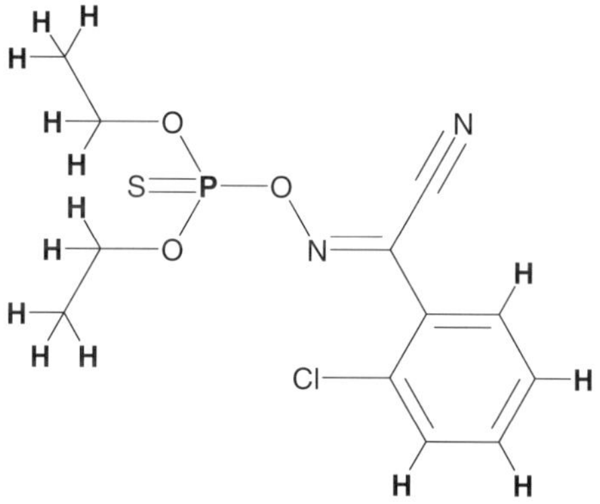

Fig. 4.10: Structure of *O,O-Diethyl-2-chloro(alpha-cyanobenzyliden amino) phosphorothioate*.

Check it in WIN-NMR:

Load the ^{1}H spectrum, ...\ABM3X\001001.1R, of the phosphorothioate shown in Fig. 4.10 into WIN-NMR. **Zoom** in and inspect the various signals. Use the **Analysis I Integration** mode to assign the signals.

The aromatic region is straight forward and is quite similar to the *9-fluorenone* example discussed in section 2.3.4. Considering only the aliphatic spin system in detail, the methyl group is at 1.40 ppm and the methylene protons at 4.35 ppm. Because the methylene protons are diastereotopic, the methylene signal is quite complex and corresponds to the AB part of an ABM$_3$X spin system (X=^{31}P). In a similar manner to the first order analysis of the AB part of an ABX spectrum, the coupling constant values can be determined quite accurately unlike the chemical shifts of A and B.

Check it in WIN-NMR:

Measure a suitable **Linewidth**. Switch to the **Analysis | Multiplets** mode. In the methylene region there are two signals; in a first order analysis each signal is split into a doublet (J_{AB}) of doublets (J_{AX}) of quartets (J_{AM}). But this is only a very rough starting guess! The composition of this signal is shown in Fig. 4.11.

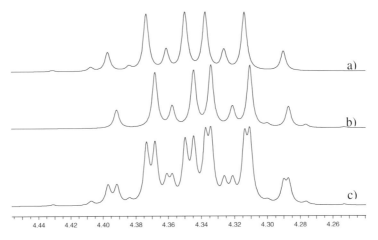

Fig. 4.11: Superposition of A and B in AB spectrum part of $ABMX_3$ spin system.
        a) A signal part   b) B signal part   c) sum = AB part.

Check it in WIN-NMR:

Analyze the aliphatic spectrum region. Start with the methyl signal and keep in mind, that it contains only doublet splittings including vicinal couplings to *both* the methylene protons. For the A signal start from the left hand side of the methylene signal and for the B signal start from the right. Open the **Report...** box and use the **Auto Connect** function. Three heteronuclear coupling constants, $^nJ_{PH}$, remain unconnected, one in each multiplet. Press the **Export...** button and transfer the data to **WIN-DAISY**.

Check it in WIN-DAISY:

Include the non-assigned coupling constants in the analysis (**Yes**). Open the **Frequencies** dialog box and change the fourth spin from ^1H to ^{31}P using the **PSE** button. Switch into the **Scalar Coupling Constants** dialog box and press the **Iterate none** button. Apply a negative sign to the geminal coupling constant J_{12}. Open the **Spectrum Parameters**, press the **Iter. Regions** button and *deselect* the first line (the aromatic spectral region) from the iteration, because the data is not included in the WIN-DAISY document. Open the **Control Parameters** and select **Standard Iteration mode, medium Broadening, medium No. of Cycles**. Run an **Iteration** and display the result in the WIN-NMR **Dual Display**. There is already good agreement between the calculated and experimental lineshape.

4.1 Spin-½ Pure Elements 257

Open the **Control Parameters** and select **none Broadening**. Press the **Scalar Coupl** button; press the **Select all** button. Run another **Iteration** and look at the result in the **Dual Display** mode of WIN-NMR.

Check it in WIN-NMR and WIN-DAISY:

You may also analyze the aromatic spectrum part in an adequate manner. This spin system is quite similar to a type already analyzed in section 2.3.4. Remember to *deselect* the **Iter. Regions** in the **Spectrum Parameters** of the aliphatic part! When this spin system is optimized, merge the two WIN-DAISY documents and calculate the complete spectrum in one lineshape.

Another type of heteronuclear phosphorous spin system but this time involving ^{31}P and ^{19}F is the cation of the complex $[Au(PF_3)_2][SO_3F]$ (Fig. 4.12). The cation is linear, the PF_3 groups undergo free rotation and the spin system is labeled $[AX_3]_2$ with twofold symmetry [4.25].

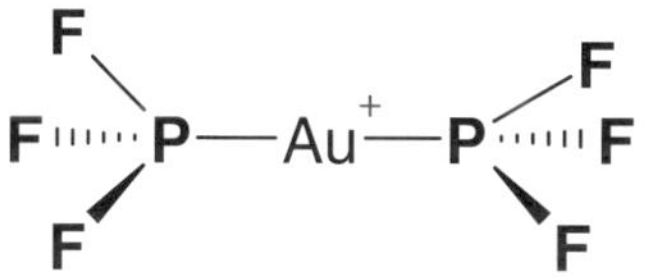

$$F\diagdown_{F\cdots\cdots}P-\overset{+}{Au}-P_{\cdots\cdots F}\diagup^F_F$$

Fig. 4.12: Structure of $[Au(PF_3)_2]^+$ cation.

Check it in WIN-DAISY:

Load the input document AU(PF3)2.MGS. Look at the entries in the various dialog boxes; **Main Parameters** - the **Number of spins or groups with magnetical equivalent nuclei, Symmetry Group, Symmetry Description, Frequencies** - the isotopes and **Spins in group, Scalar Coupling Constants** - the magnitude and sign of the coupling constants, **Spectrum Parameters** - the lineshape isotope to be calculated. Open the **Control Parameters** dialog box; the parameters have been selected for a simulation sequence of the ^{31}P spectrum. Press the **Scalar Coupl.** button, $^4J_{FF}$ (J_{14}) is selected to be varied in 0.2 Hz steps starting at −1.8 Hz. Run a **Simulation Sequence** and then click the **Display the Calculated Spectrum** (red peak button) to display the reference spectrum (original data set with calculated spectrum name ...\999999.1R) in WIN-NMR. Use the **Display | Multiple Display** command and select the nineteen spectra (...\999999.1R down to ...\999981.1R). Switch to the **Display | Display Options...** and select the **Multicolor** option. Click on the **Overlay** option on the button panel. Zoom in the different peak regions to examine the differences in the spectra.

Toggle back to **WIN-DAISY** and open the **Spectrum Parameters**. Press the **Isotope** button and select the **F** isotope from the **Element-Data** table and return with **Ok** to the **Spectrum Parameters**. Press the **Auto Limits** option to select the appropriate limits for the calculated lineshape. Quit the dialog box with **Ok**. Run a **Simulation Sequence** and then click the **Display** the **Calculated Spectrum** (red peak button). Select the **Display | Multiple Display** option, where the previous (^{31}P spectra) filenames are still selected.

Press **Ok** to display the nineteen ^{19}F spectra. Select the **Overlay** option to see the differences between the simulations.

Although the differences are small, the calculated spectrum sequence clearly shows that only the magnitude and not the sign of the long range coupling $^4J_{FF}$ has any effect on the spectrum. Furthermore, $^4J_{FF}$ has an effect on both the ^{19}F and ^{31}P spectra due to the symmetry.

4.1.3 Spin Systems Containing ^{103}Rh

The pure element Rhodium is widely used in organometallic chemistry. The sensitivity of ^{103}Rh to the NMR experiment is only 0.003% of that of ^{1}H and consequently the direct observation of ^{103}Rh requires very long measurement times. Coupling involving ^{103}Rh is very common in ^{1}H, ^{13}C, ^{31}P spectra and because rhodium is a pure element complicated spin systems can occur.

The complex [Rh$_4$(η^5-C$_5$Me$_5$)$_4$H$_4$][BF$_4$]$_2$ [4.26] contains a central Rh$_4$ cluster as shown in Fig. 4.13.

Fig. 4.13: [Rh$_4$(η^5-C$_5$Me$_5$)$_4$H$_4$]$^{2+}$ cation showing the Rh$_4$ central cluster.

In the solid state the Rh$_4$ cluster has S$_4$ symmetry with four long Rh–Rh bonds (normal lines) and two short ones (bold lines). The four hydrogens bridge the four long Rh-Rh bonds of the Rh$_4$ cluster. However, in solution the four Rh atoms have tetrahedral symmetry; both the hydrogens and the Rh are isochronous leading to a [AX]$_4$ T_d spin system, for detailed inofrmation about spin system symmetry refer to [4.27].

Check it in WIN-DAISY:

Load the input document RH.MGS. The file contains the symmetry description for an 8-spin-system consisting of four rhodiums and four hydrogens. Inspection of the **Symmetry Group** and the **Symmetry Description** shows the quite complicated arrangement of the symmetry operations. Open the **Main Parameters** dialog box; the **Output Options Output of symmetry data**, **Output of subspectra** and **Size of submatrices** have been selected. Inspection of the **Frequencies** and **Scalar Coupling Constants** dialog boxes shows the result of the symmetry: there are only two chemical shifts one ^{1}H and one ^{103}Rh and four independent coupling constants (instead of the 27 without symmetry). Run a spectrum **Simulation** and **Display** the **Calculated** ^{1}H **Spectrum** in WIN-NMR.

Check it in WIN-DAISY and WIN-NMR:

Generate subspectra to obtain the five subspectra corresponding to the five irreducible components of T_d. Toggle to **WIN-NMR**. In the **Display | Multiple Display** option, click the **Delete All** button if necessary, and then manually load the subspectra (...\999998.1r up to ...\999995.1r). Click the **Stacked Plot** button. It obvious, that the second subspectrum contains no transitions. Toggle back to **WIN-DAISY** and examine the **Simulation Protocol** list. In the section *Dimension of Submatrices* the first column contains the m_T-values and the following 5 columns the dimensions of the submatrices according to the irreducible components label given in the column header. The *A2* column shows zero's except for $m_T = 0$, where one element is given. Because transitions between different irreducible components are not allowed, there cannot be any transitions in this subspectrum.

Open the **Spectrum Parameters** and select a ^{103}Rh Isotope. Click on the **Auto Limits** button and run a **Simulation**. **Display** the **Calculated** ^{103}Rh **Spectrum** in WIN-NMR (red peak button). You may try your hands on the full tetrahedral symmetry definition (or a subgroup of T_d) using the input document RH_C1.MGS.

The ^{13}C spectrum of this compound will be discussed in section 4.2.1.2.

4.2 Spin-½ Rare Isotopes

Most of the elements of the periodic table are not pure elements, but exist as a number of different isotopes. Some do not have a nuclear spin whilst others may posses a nuclear spin of ½ or greater. In this section we will discuss the spectra of spin-½ nuclei that are not 100% abundant. The most commonly studied nucleus of this type has been the basis of chapters 2 and 3 - the proton although with a natural abundance of 99.98 %) the ^{1}H isotope is almost a pure element.

The majority of NMR active nuclei are of low natural abundance and this makes the measurement of such elements difficult. Another factor to consider is the sensitivity of a nucleus to the NMR experiment. This sensitivity depends on the magnetic moment and the spin quantum number and is often defined with respect to the sensitivity of ^{1}H (=1). When comparing the sensitivity of different isotopes, the absolute sensitivity, defined as the product of the natural abundance and the relative sensitivity should be used. One method to overcome the decrease in absolute sensitivity caused by a low natural abundance is to used isotopically enriched samples.

First the ^{13}C nucleus will be discussed first and in the second sections the focus of discussion will lie upon ^{195}Pt, ^{29}Si, ^{77}Se, ^{119}Sn, ^{117}Sn and ^{115}Sn.

4.2.1 ^{13}C Spectra and Spin Systems with ^{13}C

^{13}C NMR measurements are frequently used in structure elucidation, primarily because most molecules contain carbon and hydrogen. As discussed in chapters 2 and 3, ^{1}H NMR is a very useful tool for obtaining structural information about a compound. In section 4.1, the use of heteronuclear coupling in structural determination and the simplification of ^{1}H spectra using heteronuclear spin decoupling have been outlined.

Carbon occurs naturally as two isotopes: ^{12}C (I = 0, natural abundance 98.9%) and ^{13}C (I = ½, natural abundance 1.1%). The ^{13}C nucleus has a low relative sensitivity to the NMR experiment compared to ^{1}H ($1.59 \cdot 10^{-2}$) and taking into account the natural abundance it's absolute sensitivity is extremely low ($1.75 \cdot 10^{-4}$). Consequently, compared to ^{1}H, ^{13}C spectra need much longer measurement times and / or sample quantities and the resulting signal to noise ratio is often much worse [4.28]–[4.30].

4.2.1.1 Double Resonance Methods

Coupled spectra of pure or nearly pure elements do not display any obvious sign of coupling with ^{13}C in natural abundance, because ca. 98.9 % of the carbon atoms are the spin inactive ^{12}C isotope. The effects introduced into other isotope spectra by the small amount of spin active ^{13}C is discussed in subsection 4.2.1.3.

However, when measuring a ^{13}C spectrum, every ^{13}C directly bonded to a proton (99.98% spin active ^{1}H), is strongly affected by the coupling constant $^1J_{CH}$. ^{13}C–^{13}C interactions are not normally detected in natural abundance ^{13}C spectra, because the probability of having two neighboring ^{13}C isotopes is extremely low (1.1% of 1.1%).

The standard chemical shift range for ^{13}C spectra is ca. 300 ppm (compared to 15 ppm for ^{1}H) and the digital resolution is usually lower. In a ^{1}H coupled ^{13}C spectrum each signal is a multiplet due to coupling with the protons in the molecule and since ^{1}H-^{13}C coupling constants can be up to 250 Hz, signal patterns can overlap badly. More important however, is the reduction in signal to noise caused by these multiplet signals.

Consequently, routine ^{13}C NMR spectra are usually recorded using broadband ^{1}H decoupling. This means, that all couplings to the protons are eliminated and each carbon appears as a singlet (Coupling to other X-nuclei such as ^{31}P are preserved.) These spectra contain only chemical shift information and no coupling connectivity's. ^{13}C chemical shifts been intensively studied and for aromatic carbon atoms in particular, well defined increment systems exist. The amount of ^{13}C chemical shift data is huge and in data bases the bond connections of the carbon atoms are usually coded (using mainly the HOSE-codes). Using this information shifts can be easily predicted to support structural assignments in new compounds. For an overview of the ^{13}C chemical shifts of selected functional groups and structural fragments refer to the literature [4.28].

To establish C-H connectivity's, C-H correlation spectroscopy is used. The direct C-C connections can be determined using INADEQUATE spectroscopy. For detailed description of both these methods refer to the literature [4.31].

Although broadband proton decoupling simplifies the signals in a ^{13}C spectrum, information relating to the number of neighboring protons and the carbon-proton coupling constants is lost. A compromise between coupled and broadband decoupled spectra is the off-resonance decoupling method. In this technique, the decoupler frequency is set at ca. -5.0 ppm and the protons irradiated using continuous wave irradiation. The residual one-bond carbon-proton coupling pattern remain, but with a reduced value, enabling the number of *directly* attached protons to be determined. (The exact reduction in the values of $^1J_{CH}$ depends on the power of the decoupling field and the frequency offset). All long range couplings $^nJ_{CH}$ disappear. A quaternary carbon (C) appears as a singlet, a methine carbon (CH) as a doublet, a methylene carbon (CH$_2$) as a triplet and a methyl group (CH$_3$) as a quartet. Although the relative signal intensities are correct, e.g. 1:1, 1:2:1 and 1:3:3:1, the "inner" and "outer" lines of the multiplets have different linewidths and this can make it difficult to distinguish between C / CH$_2$ and CH / CH$_3$.

DEPT spectra

Another way to determine the number of directly attached protons is spectrum editing. In spectrum editing, C, CH, CH$_2$ and CH$_3$ carbons are detected separately by suppression of all the other signals using techniques such as DEPT spectroscopy (Distortionless Enhancement by Polarisation Transfer) [4.32]. DEPT spectra are not quite distortionless, because the method is based on the value of a single coupling constant. Typically $^1J_{CH}$ can vary over the range 120 - 200 Hz and a compromise value of ca. 135 Hz is used in the pulse sequence. In cases where $^1J_{CH}$ is substantially different from the set value, phase distortions can occur and this is particularly noticeable in proton-coupled DEPT spectra. Normally broadband proton decoupling is used during detection, so that multiplet patterns are not recorded. By altering the pulse angle of the final proton pulse in the pulse sequence, different subspectra are recorded:

Table 4.2: Subspectra content according to final proton pulse angle.

Carbon Type	DEPT angle 45°	90°	135°
C			
CH	X	X	
CH$_2$	X		−X
CH$_3$	X		X

The negative sign of CH$_2$ in the last column means, that the signals are negative. The DEPT 90 spectrum contains only the CH signals. The CH$_2$ signals can be determined by subtracting the DEPT 135 spectrum from the DEPT 45 spectrum and finally the CH$_3$ signals by subtraction of $\sqrt{2}$ DEPT 90 spectrum from the sum of DEPT 45 spectrum and DEPT 135* spectrum. The signals of the quaternary carbon atoms C do not appear in any of these DEPT spectra, they can be determined by comparing with the normal ^{13}C spectrum. This File Algebra can be done using either the FIDs or the transformed data. The first method is described in detail in the first volume of this series [4.33]. Here only the finished job files will be executed to obtain the spectra containing only the signals of one of the carbon types.

---

* Some authors use the factor $\frac{\sqrt{2}}{2}$ when the 90° DEPT is acquired with twice the accumulation time.

Check it in WIN-NMR:

From the **File** pull-down menu chose the **Serial Processing** option. In the **Automatic Serial Processing** dialog box and select the files ...\GCDP\004001.FID, ...\005001.FID and ...\006001.FID. These files are copies of the original DEPT spectra ...\GCDP\001001.FID, ...\005001.FID and ...\006001.FID. Select the **use different Jobs** option and press the **Define diff. Jobs...** button. The **Define Processing Jobs** dialog box opens. Highlight the first line 004 001 FID, press the **Select Job...** button and in the new dialog box select DEPTCH.JOB. Close the dialog box with **OK**. Repeat this process for the other two FID's selecting DEPTCH2.JOB for 005 001 FID and DEPTCH3.JOB for 006 001 FID. **Execute** the serial processing.

The final edited spectra ...\GCDP\004002.1R, ...\005002.1R and ...\006002.1R contain only CH carbons, CH_2 carbons and CH_3 carbons signals respectively. Load the normal $^{13}C\{^1H\}$ spectrum ...\GC\001001.1R into WIN-NMR. Switch to the **Display | Dual Display** mode and load the CH only DEPT spectrum, ...\GCDP\004002.1R as the second trace. Use <Ctrl + Y> to align the Y-scale of both spectra. Press the **Separate** button and zoom in the individual peak regions to work out which signals in the upper $^{13}C\{^1H\}$ spectrum correspond to the CH carbons. When all 5 carbons are identified click the **Return** button.

Switch to the **Analysis | Peak Picking** mode and in the **Peak Picking Options** dialog box select the options **Peak Labels ppm**, **Interpolation** and **Multiplicity**. Modify the PC value to <2.0> to avoid picking the impurity signals. Press the **Execute** button. When the **Multiplicity** option is selected the peak labels (in ppm) are reduced from four to two decimal places so that there is space to enter the multiplicities. To enter the multiplicity into the peak label, click the **Edit Cursor** button and select one of the CH signals with the left mouse button, e.g. the very left of the signals (which is the anomeric CH). When the spectrum cursor is positioned at the top of the peak enter <d> for doublet. Due to the attached proton, the carbon signal would be a doublet in the proton coupled ^{13}C spectrum. The letter "d" appears in the peak label. Click the **Report...** button, the letter "d" also appears in the chemical shift table. Mark the four remaining CH carbons with the doublet label using the arrow cursors <←> and <→> to switch to the neighboring peak. **Return** to the standard display mode.

Repeat the spectrum comparison and multiplicity assignment for the CH_2 DEPT spectrum, ...\GCDP\005002.1R, but this time label the single methylene carbon peak with the symbol <t> for "triplet". Continue with the CH_3 DEPT spectrum ...\GCDP\006002.1R. These signals belong to the acetyl methyl group in the glucose compound. Identify the signals and label the peaks with a <q> for "quartet" in the $^{13}C\{^1H\}$ spectrum. Think about how many signals should be present. In case one peak is not listed, click the **Edit Cursor** button again, mark the peak with the missing label with the spectrum cursor and press the right mouse button to insert a peak. Press the **Edit Cursor** button again and continue with the labeling.

Although the small peak at ca. 20.55 ppm is picked, there are still two peaks not resolved at about 20.57 ppm. Altogether there are five CH_3 groups present. The remaining signals in the $^{13}C\{^1H\}$ spectrum are either quaternary carbon atoms or solvent peaks. The three peaks around 77 ppm belong to $CDCl_3$ which does not contain any protons (see section 4.3.2.1). The remaining signals at ca. 170 ppm belong to the acetyl carbonyl groups.

Check it in WIN-NMR:

In the **Analysis | Peak Picking** mode, label the five peaks at ca. 170 ppm in the $^{13}C\{^1H\}$ spectrum ...\GC\001001.1R, with <s> for "singlet".

Repeat the procedure for the oligosaccharide DEPT spectra:

Check it in WIN-NMR:

Choose the **File | Serial Processing** option in WIN-NMR. Select the three DEPT FIDs ...\OCDP\004001.FID, ...\005001.FID and ...\006001.FID. Use the **use different Jobs** option and press the **Define diff. Jobs** button. Select the job files DEPTCH.JOB, DEPTCH2.JOB and DEPTCH3.JOB for 004 001 FID, 005 001 FID and 006 001 FID respectively. Press the **Execute** button to start the process.

Repeat the analysis of the $^{13}C\{^1H\}$ spectrum using the three DEPT edited spectra:

Check it in WIN-NMR:

Load into WIN-NMR the $^{13}C\{^1H\}$ spectrum of the oligosaccharide, ...\OC\001001.1R. Switch to the **Analysis | Peak Picking** mode and perform a peak picking with the **Interpolation** and **Multiplicity** options checked.

Switch to the **Display | Multiple Display** mode, select the three DEPT spectra and the $^{13}C\{^1H\}$ spectrum and compare the different spectra.

In the **Display | Dual Display** mode load the CH only DEPT spectrum, ...\OCDP\004002.1R as the second trace. Use <Ctrl + Y> to align the Y-scale of both spectra. Press the **Separate** button and zoom in the individual peak regions to work out which signals in the upper $^{13}C\{^1H\}$ spectrum correspond to the CH carbons. Return to the single spectrum display and enter the **Peak Picking** again. Press **OK** in the dialog box, because the Peak Picking has already been performed. Click the **Edit Cursor** button and select one of the CH carbon peaks with a left mouse click. Enter a <d> for doublet. Use the arrow cursors <←> and <→> to switch to the neighboring peaks and mark the remaining methine carbon peaks. Repeat the process for the CH$_2$, DEPT spectrum ...\OCDP\005002.1R, and the CH$_3$ signals, DEPT spectrum ...\OCDP\006002.1R. Keep in mind, that not all signals might be resolved. Finally label the singlet (quaternary) carbons with an <s>.

The signal at 29.8 ppm is an impurity signal containing a CH$_2$ group, because it appears in the corresponding DEPT spectrum.

4.2.1.2 Satellite Spectra

Isotope distribution leads to a statistically mixture of molecules containing ^{12}C and ^{13}C, the so-called ^{12}C and ^{13}C isotopomers. Molecules containing only one carbon atom consist of 99.89% ^{12}C isotopomer and 1.1% ^{13}C isotopomer. Compounds with two different carbon atoms consist of ca. 1% of each ^{13}C isotopomer, 0.01% of the two-^{13}C isotopomer and the rest all-^{12}C isotopomer. Because the probability of finding two ^{13}C

264   4 Spin Systems and the Periodic Table

isotopes in one molecule is very low, ^{13}C spectra do not usually exhibit ^{13}C-^{13}C scalar coupling and normally only chemical shift information is obtained. Heteronuclear coupling information appears in both the ^{13}C spectrum and the heteronucleus spectrum where the isotopically dilute ^{13}C nucleus gives rise to satellite lines in the spectrum around the ^{12}C isotopomer signal, each satellite line having ca. 0.5% of the intensity of the main signal. Depending on the magnitude of the ^{13}C-X coupling constant the satellite peaks may be masked by the base of the main signal.

The simplest example of a ^{12}C / ^{13}C isotopomer is chloroform which contains one proton and one carbon atom. The residual proteo chloroform signal is often observed when proton spectra are recorded using deutero chloroform as solvent and occurs as a singlet at about 7.26 ppm. Fig. 4.14 shows the ^{12}C and ^{13}C isotopomers of CHCl$_3$.

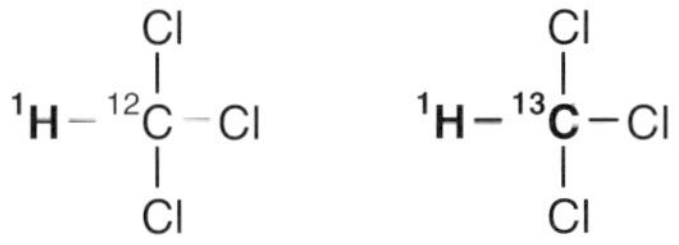

$$\begin{array}{cc} \text{Cl} & \text{Cl} \\ | & | \\ ^1\text{H}-{}^{12}\text{C}-\text{Cl} & {}^1\text{H}-{}^{13}\text{C}-\text{Cl} \\ | & | \\ \text{Cl} & \text{Cl} \\ 98.9\% & : \quad 1.1\% \end{array}$$

Spin System: A   AX

Fig. 4.14: Natural ^{12}C and ^{13}C isotopomer distribution in chloroform.

The ^{12}CHCl$_3$ (spin system A) represents about 98.9% of the total intensity in the proton spectrum with the ^{13}CHCl$_3$ (spin system AX) representing the remaining 1.1%.

Check it in WIN-NMR:

Load the ^{1}H spectrum of chloroform, ...\CHCL3\001001.1R into WIN-NMR. Call the **Analysis | Integration** and **zoom** in the baseline. Integrate main signal and the two small peaks. Click the **Options...** button and normalize the integral to 100 by setting the **Manual Calibration All Integrals** to <100.0>.

$$\begin{array}{cc} \text{Cl} & \text{Cl} \\ | & | \\ \text{H}-\text{C}-\text{Cl} \; + \; \text{H}-\text{C}-\text{Cl} \\ | & | \\ \text{Cl} & \text{Cl} \end{array}$$

Open the **Report...** box and inspect the values.   *Chloroform*

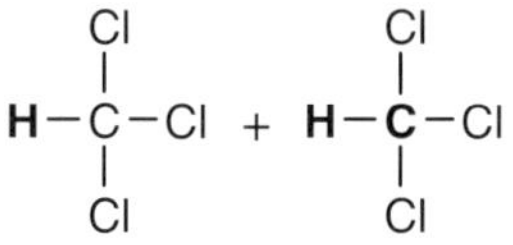

The chloroform signal is due to the residual proton-isotopomer in deutero chloroform and shows that the spin system of the ^{13}C isotopomer of chloroform is always present. To examine the satellite peaks (the A part of the AX spin system) zoom in on the baseline in this region just integrated.

Measure a **Linewidth** of the main signal and switch to the **Analysis | Multiplets** mode. Mark the main singlet of the ^{12}C isotopomer with the spectrum cursor and select **Define Singlet**. Define a doublet for the ^{13}C isotopomer satellite peaks. Open the **Report...** box and examine the shift values of the isotopomers.

4.2 Spin-½ Rare Isotopes

It is obvious that the chemical shift of both spin systems is not identical - the proton chemical shift of the ^{13}C isotopomer is slightly more shielded than the ^{12}C isotopomer. This phenomenon is called the isotope effect; the substitution of one isotope (^{12}C) by a heavier isotope (^{13}C) causes a shielding effect on the other spins.

Check it in WIN-DAISY:

Export... the spin system data *with* fragmentation of the spin systems to WIN-DAISY. *Include* the non-assigned coupling constant (the ^1H-^{13}C coupling constant). Two fragments are generated, one for each isotopomer. Open the **Frequencies** dialog box for the second fragment, click on the **PSE** button for the second spin and select **C** from the **Element-Data** table. Change the **Frequency** of the carbon atom to <77.0> ppm.

Open the **Main Parameters** dialog box, for the one-spin fragment enter a **Statistical weight** of <99.89> and for the AX spin system, the second fragment, enter <1.1>. The absolute values are not important, just the relative ratio. Execute a spectrum **Simulation** and display the calculated and experimental spectra in the mode **Dual Display** of WIN-NMR.

The examination of satellite lines is a way to extract not only the heteronuclear coupling constant, e.g. ^{13}C-^1H but also hidden coupling constants, e.g. ^1H-^1H which cannot normally be determined.

Check it in WIN-NMR:

Open... the experimental 300 MHz ^1H spectrum of *1,2,3,4,-tetrachloro benzene* ...\ABX\010001.1R into WIN-NMR. Examine the structure of the compound and decide which type of spin system is present. Study the spectrum, **zoom** in on the base line and search for the ^{13}C satellite lines.

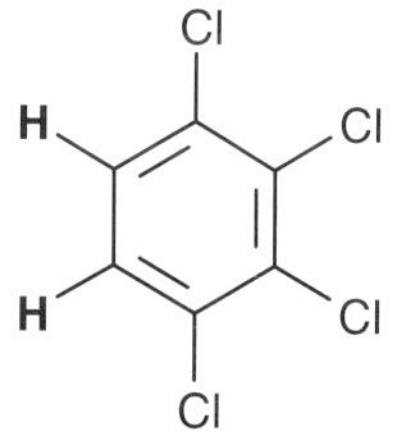

1,2,3,4 Tetrachloro benzene

The main proton spin system is named A_2, because there are no other spin active isotopes and the protons are magnetically equivalent. However, taking into account the ^{12}C / ^{13}C isotopes, the following isotopomers are present in the sample (Fig. 4.15).

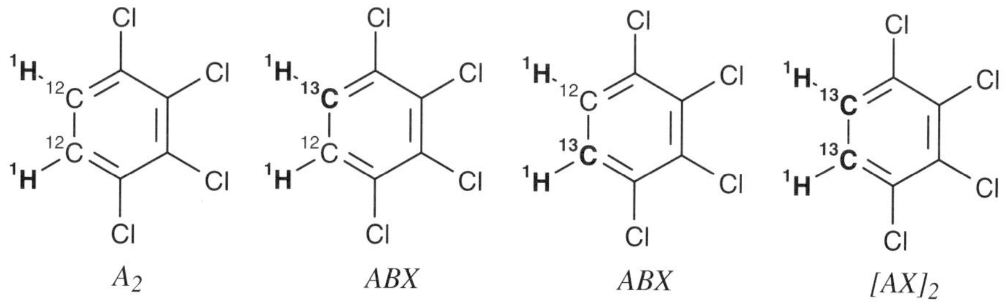

A_2 ABX ABX $[AX]_2$

Fig. 4.15: Spin Systems of carbon isotopomers of *1,2,3,4-tetrachloro benzene*. Spin active nuclei are printed bold; A, B= ^1H, X = ^{13}C.

$$\binom{2}{2-n} \cdot a(^{12}C)^n \cdot a(^{13}C)^{2-n} \cdot 100 \quad [\%]$$

with $a(^{13}C)$ = natural abundance of ^{13}C isotope,
and n = 0, 1, 2.

Check it in the Calculator:

Calculate the probability of the A_2, ABX and $[AX]_2$ spin systems with the assumption that ^1H is a 100% natural abundant isotope.

n = 2: $1 \cdot 0.9890^2 \cdot 0.011^0 \cdot 100$ = 97.81 % A_2
n = 1: $2 \cdot 0.9890^1 \cdot 0.011^1 \cdot 100$ = 2.18 % ABX
n = 0: $1 \cdot 0.9890^0 \cdot 0.011^2 \cdot 100$ = $1.21 \cdot 10^{-2}$ % $[AX]_2$

Hydrogen is not a pure element, consequently the individual probabilities for the three possible ^1H - ^{13}C spin systems have to be multiplied by the probability of the molecule containing ^1H or ^2H. There are two positions for hydrogen in the molecule:

$$\binom{2}{2-n} \cdot a(^1H)^n \cdot a(^2H)^{2-n} \cdot 100 \quad [\%]$$

with $a(^1H)$ = 0.99985 and $a(^2H)$=0.00015.

The resulting probabilities are as follows.

n = 2: $1 \cdot 0.99985^2 \cdot 0.00015^0 \cdot 100$ = 99.97 % all-^1H
n = 1: $2 \cdot 0.99985^1 \cdot 0.00015^1 \cdot 100$ = 0.03 % $^1H, ^2H$
n = 0: $1 \cdot 0.99985^0 \cdot 0.00015^2 \cdot 100$ = $2.25 \cdot 10^{-6}$ % all-^2H

Only 97.78% (99.97% of 97.81%) of the total number of molecules contribute to the A_2 spin system and not the expected 97.81% based on the $^{12}C/^{13}C$ probabilities. The same argument is also valid for both the ABX and the $[AX]_2$ spin systems, but here the original probabilities are so small, that such small differences are not detectable. When probabilities are considered, hydrogen is often regarded as a pure element.

Because the isotopomer with two neighboring ^{13}C isotopomers has vanishing probability (less than 0.1 %), only the first three isotopomers in Fig. 4.15 need to be considered. The two ABX spin systems are of course identical. Quite often this type of ABX spin system, where the chemical shift difference between A and B is a result of the isotope effect caused by X, is called a AA'X spin system.

Inspection of the baseline around the singlet of the all-^{12}C isotopomer shows, beside some impurity signals, two doublets symmetrically placed about the main signal. The ^{13}C satellites in the proton spectrum represent the AB part of an ABX spin system. In a ABX spectrum there should be two (ab) subspectra with the coupling $|J_{AB}|$ appearing four times. The coupling constants $|J_{AX}|$ ($^2J_{CH}$) and $|J_{BX}|$ ($^1J_{CH}$) can normally be determined from the spectrum. However in the present case, only one half of each (ab) subspectrum can be seen. Although $^1J_{CH}$ is about 170 Hz, the value of $^2J_{CH}$ is very small, typically less than 10 Hz, and hence the "other" half of each (ab) subspectrum is covered by the main ^{12}C isotopomer signal. Consequently, only |J_{AB}| and |J_{BX}| can be determined from the multiplet analysis.

Check it in WIN-NMR and WIN-DAISY:

Measure the **Linewidth** of the main signal and switch into the **Analysis | Multiplet** mode. **Define** a **Singlet** for the main signal and a doublet of doublet for the satellite lines. Position the mouse cursor a the top of one of the ^{13}C satellite lines. **Export...** the data to **WIN-DAISY** *with* fragmentation of the spin system and *inclusion* of non-assigned coupling constants.

In WIN-DAISY open the **Frequencies** dialog box for the single spin system fragment. Copy the value into the clipboard, <Ctrl + C>. Open the Scalar Couplings Constants dialog of the second fragment and check to which nucleus the smaller constant is defined. With **Previous** switch into the **Frequencies** dialog box for the second fragment and paste the clipboard entry into the **Frequencies** edit field of the appropriate shift, <Ctrl + V>. Click on the **PSE** button of the second spin and select **C** from the **Element-Data** table.

In this way the chemical shift of the protons in all-^{12}C isotopomer is taken as the shift of A in the ABX spin system. Select the **1-spin system** in the document main window and using the **Delete fragment** button, delete the fragment containing the main signal. Open the **Spectrum Parameters** dialog box and inspect the **Iter. Regions** limits. Only the spectral regions of the observable satellite lines are included, the other satellite lines are obscured but the main signal and cannot be taken into account in the iteration. Run a spectrum **Iteration** and optimize v_B and the two coupling constants J_{AB} and J_{BX}. Examine the experimental and calculated spectrum using the **Dual Display** mode of WIN-NMR. Toggle back to **WIN-DAISY** and remove the **Disable** flag from the second proton shift in the **Frequencies** dialog. Run a **Simulation** and **zoom** in on the base line, the "inner" satellite lines (the other half of the AB spectrum part) hidden in the experimental spectrum are clearly visible in the calculated spectrum. The coupling constant we are interested in is the J_{AB}.

The value of $^3J_{HH}$, 8.72 Hz, determined using the ^{13}C satellite lines is a typical value for an aromatic ortho coupling constant. In the next example, the analysis of the ^{13}C satellite lines is used to determine whether the spectrum refers to cis-butenedioic acid (maleic acid, Fig. 4.16a) or trans-butenedioic acid (fumaric acid, Fig. 4.16b)

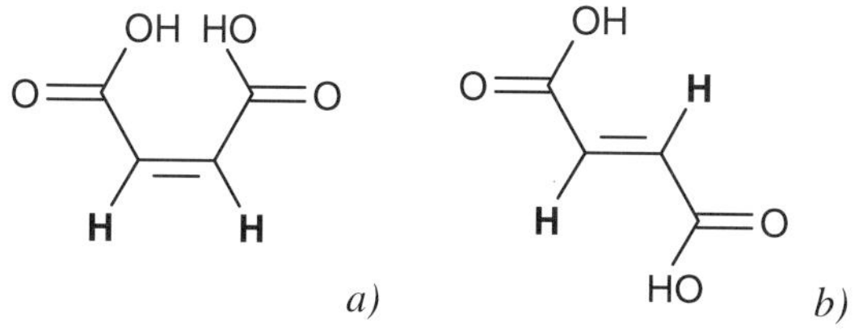

Fig. 4.16: a) Structure of maleic acid, b) Structure of fumaric acid.

Both the cis and the trans isomer exhibit twofold symmetry: the cis isomer has a mirror plane or a twofold axis and the trans isomer an inversion center. Because of the symmetry and the absence of any another coupling partner, the olefinic protons in both isomers are magnetically equivalent and appear as a singlet (A_2 spin system). However, in ^{13}C satellite spectrum the olefinic protons form the AB part of the ABX spin system.

Check it in WIN-NMR and WIN-DAISY:

Open... the experimental spectrum ...\ABX\011001.1R into WIN-NMR. Measure the **Linewidth** and switch to the **Analysis | Multiplet** mode. Proceed as in the previous "Check it". Define a singlet for the A_2 main spin system and a doublet of doublets for half of the AB part of the ABX (X=^{13}C) spin system. **Export...** the data to WIN-DAISY *with* fragmentation of the spin system and *inclusion* of non-assigned couplings. Copy the value of the **Frequency** in the single spin fragment into the edit field of the third spin of the three spin system. Again press the **PSE** button of the second spin and select C from the **Element-Data** table. Enter a chemical shift of about <133> ppm for the carbon. Delete the first fragment from the document. Run a spectrum **Simulation** and display the experimental and calculated spectra in the WIN-NMR **Dual Display**. Optimize the parameters by running an **Iteration**. Display the result in the **Dual Display** mode of **WIN-NMR**. Taking into account the values in Table 4.3 decide if the spectrum refers to fumaric acid or maleic acid.

Table 4.3: Comparison of NMR parameters of cis and trans butenedioic acid.

	δ_H [ppm]	J_{HH} [Hz]
Maleic acid	6.28	12
Fumaric acid	6.83	15

There are many examples in which the lines of the AB part of the ^{13}C satellite ABX spin system are obscured by the lineshape of the main signal. In such cases the X spectrum can be used to extract the coupling constant J_{AB}. For the phosphinic acid shown in Fig. 4.17, the X-part of the ^{13}C satellite ABX spin systems display a characteristic five line pattern (Fig. 4.18). The molecule contains two ABX spin systems, one involving the methyl group (highlighted in bold) and the second involving the bridging methylenes.

$$H_7C_3-O-\underset{\underset{O}{\|}}{P}-[CH_2]_2-\underset{\underset{^{13}CH_3}{|}}{\overset{\overset{O}{\|}}{P}}-O-C_3H_7$$
$$\overset{^{12}CH_3}{|}$$

Fig. 4.17: Structure *of P,P'-dimethyl-1,2-ethanebis (phosphinic acid diisopropyl ester)* with one ABX spin system highlighted in bold letters [4.22].

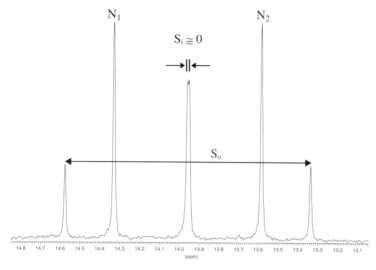

Fig. 4.18: $^{13}C\{^{1}H\}$ X-part of ABX spin system (X=^{13}C of CH$_3$) of P,P'-dimethyl-1,2-ethanebis (phosphinic acid diisopropyl ester).

The coupling constant J_{AB} (in the present case it is $^3J_{PP}$) can be determined using the following equation [4.34]:

$$J_{AB} = \frac{S_o}{2}\sqrt{\frac{I_i}{I_i + I_o}}$$

The distance between the N-lines (in Hz) corresponds to the sum of the first order coupling constants:

$$N = N_1 - N_2 = |J_{AX} + J_{BX}|$$

$$L = S_o\sqrt{\frac{I_o}{I_i + I_o}} = |J_{AX} - J_{BX}|$$

The degree of coincidence of the inner lines S_i depends upon the magnitude of the phosphorous isotope shift caused by the ^{13}C nucleus. Because the central line is composed of two transitions, the integral values and not the peak-heights have to be used for determining the auxiliary parameters I_i and I_o.

Check it in WIN-NMR and WIN-DAISY:

Load the spectrum ...\ABX\012001.1R into WIN-NMR. Measure a suitable **Linewidth** and then switch into the **Analysis | Multiplets** mode. Ignore the signals at 24.9 and 25.0 ppm and focus on the two multiplets at 13.9 and 23.4 ppm. For both multiplets define a *doublet* using the N lines. Leave the coupling unassigned. **Export...** the data to **WIN-DAISY** *with* fragmentation of the spin system and *inclusion* of non-assigned coupling constants. In WIN-DAISY the parameters for *both* fragments must be modified in the *same* way. In the **Main Parameters** dialog box increase the **Number of spins or groups with magnetical equivalent nuclei** from to 2 to <3>. Open the **Frequencies** window and change the isotope of the second and third spin to ^{31}P using the **PSE** button. In the frequency edit window enter the value <0.0> **Hz** for both phosphorous spins. Use the **Next** button in the bottom line to open the **Scalar Coupling Constants** window. Enter values for the coupling constants J_{13}, and J_{23} e.g. <0.1> Hz and <20.0> Hz respectively. Quit the dialog box with **OK**. Open the **Control Parameters** window and select the **Advanced Iteration mode** with **high Broadening** and **high No. of Cycles**. Also check the option **Starting Frequencies poor**. Click on the **Scalar Coupl.** button. **Select all** coupling constants for optimization and change the iteration limits: **Lower limit** <–110.0> Hz, **Upper limit** <110.0> Hz. Quit the dialog box with **Ok**. Run a spectrum **Iteration** and display the result using the **Dual Display** mode of WIN-NMR. The overall lineshape is reasonable but the intensity ratio of the inner and outer the lines is incorrect.

The iteration result shows a five line pattern for both signals. If the experimental signals are expanded the central line is actually split into two. Due to the isotope effect, the chemical shift of the ^{31}P nucleus attached to the ^{13}C atom is shifted slightly to higher field (lower frequency) and the two phosphorous shifts are no longer identical.

Check it in WIN-DAISY:

Open the **Frequencies** dialog box and switch the display units into **Hz**. Enter a very small value, <–0.1> **Hz** (not ppm!), for the first phosphorous **Frequency** (second line) to mimic the isotope effect on the chemical shift. Check the **Iterate** option for this shift. Repeat this procedure for the second fragment. Run an **Iteration** (without smoothing) and display the result in the WIN-NMR **Dual Display**. The experimental spectrum is reproduced exactly. Open the **Frequencies** dialog box and switch the display units into **Hz** again. Note that the magnitude of the isotope shift is different for both fragments. Open **Scalar Coupling Constants** dialog boxes, the sign of the long range P-C coupling (J_{13}) is negative.

The relative signs of J_{AX} and J_{BX} can be determined from the analysis of ABX spectra and the negative sign is real and not an aberration of the iteration process.

Check it in WIN-NMR and WIN-DAISY:

Another way to analyze the spectrum is to determine the coupling constant $^3J_{PP}$ beforehand using the intensities and frequencies obtained from the spectrum and the equations given earlier. If necessary, load, the spectrum …\ABX\012001.1R back into WIN-NMR. Open the **Analysis I Peak Picking** mode and perform a peak picking, in **Hz**, over the two multiplets at 13.9 and 23.4 ppm. Switch to the **Analysis I Integration** mode and carefully integrate the five lines in each multiplet. **Calculate** $^3J_{PP}$ using the appropriate equations and the parameters obtained from the spectrum.

Measure a **Linewidth** and then switch to the **Analysis I Multiplets** mode. **Export…** the data to **WIN-DAISY** with fragmentation of the spin system and *inclusion* of non-assigned coupling constants. Alter the parameters in both fragments in the same manner as in the previous "Check it". In the **Options**

4.2 Spin-½ Rare Isotopes

dialog enable **Allow sign change**. However in the **Scalar Coupling Constants** dialog box enter the value of the calculated $^3J_{PP}$ for J_{23}. Run a spectrum **Simulation** and display both the spectra in the WIN-NMR **Dual Display**. There is already good agreement between the experimental and calculated spectra. Run a spectrum **Iteration** using the **Standard Iteration mode** and the default iteration parameters.

The default iteration values are large enough to enable the correct result to be obtained because in the N parameter, $^1J_{PC}$ is many times larger than the long range P-C coupling.

Check it with the WINDOWS Calculator:

Determine the value of L using the intensities, frequencies and equations given earlier. From the values of N and L calculate J_{AX} and J_{BX} and compare the results with those obtained using the iteration method.

Another example of the analysis of the X-part of an ABX spin system is the spectrum of the diphosphonic acid (Fig. 4.19). In addition to the highlighted ABX system, two other carbon in the cyclohexane structure form ABX spin systems.

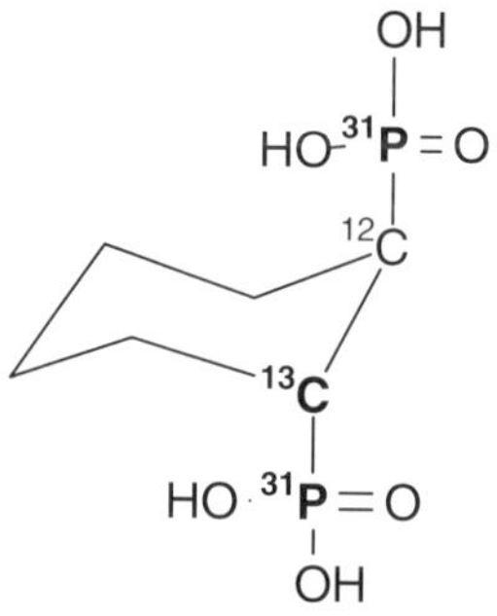

Fig. 4.19: Structure of *cyclohexane-1,2-diphosphonic acid* with one ABX system highlighted in bold letters [4.35].

Check it in WIN-NMR:

Open... the spectrum ...\ABX\013001.1R of the diphosphonic acid (Fig. 4.19) in WIN-NMR. The spectrum shows three multiplets corresponding to the X-part of three different ABX spin systems. **Zoom** in the different signals and think about the multiplet pattern.

Generally the five line pattern of the previous examples is reproduced but the width and the intensity distribution in the multiplets are different indicating that either the magnitude and / or signs of coupling constants have altered significantly. The low-field signal shows very intense outer lines, the inner lines are still resolved but are very weak In contrast, the high-field signals appear to be narrow triplets, the inner lines have collapsed to a single peak and are not resolved and the outer lines are very weak.

Check it in WIN-NMR:

Using the multiplet at 36.7 ppm, measure a **Linewidth**. Switch to the **Analysis | Multiplets** mode. For each signal define the N-lines as a doublet. Open the **Report...** box and **Export...** the data to **WIN-DAISY** *with* fragmentation of the spin system and *inclusion* of non-assigned coupling constants.

Check it in WIN-DAISY:

The automatic fragmentation produces three fragments. Open the **Main Parameters** of the first fragment and enter an appropriate **Title** e.g. P(a), and extend the **Number of spins or groups with magnetical equivalent nuclei** from to 2 to <3>. Repeat this process for the other fragments, giving each fragment a different **Title**. Open the **Frequencies** dialog box; for each fragment change the **PSE** isotope for Spin 2 and Spin 3 to ^{31}P with a **Frequency** of <0> ppm. Use the **Options** dialog to **Allow sign change** and modify the default limits for the **Scalar Coupling Constants** to ± <40> Hz. Open the **Control Parameters** and select **Advanced Iteration mode, high Broadening** and **No. of Cycles**, check the option **Starting frequencies poor**. Click on the **Scalar Coupl.** button and **Select all** for iteration, to automatically update the iteration limits set the **Lower limit** and **Upper limit** for all the coupling constants. Enter <0> for the displayed limits for the coupling J_{12} to apply new default limits as well. Repeat for all fragments. Open the **Scalar Coupling Constants** dialog box and, for every fragment, enter a value for J_{13} and J_{23} e.g. <0.1> Hz and <5> Hz respectively. Run an **Iteration** and examine the results in the **Dual Display** mode of WIN-NMR. The second fragment did not reach the correct result. Toggle back to **WIN-DAISY** and open the **Scalar Coupling Constants** and compare the signs and values of the fragments. Change something (e.g. the sign of J_{23} in the second fragment) and run another **Iteration** until the patter match in all spectrum parts (*R-Factor* about *0.67%*). Now introduce an isotope effect; in the **Frequencies** dialog box change the **Frequency** of Spin 2 to <-0.1> Hz and select for Iteration (first remove the **Disable** flag). Repeat for all fragments. Deselect the option **Starting frequencies poor** and run another **Iteration** with **none broadening**. The final value of *R-Factor* should be ca. *0.58%*. Open the **Scalar Coupling Constants** dialog box and examine the signs and magnitudes of the couplings. The *sign* of J_{23} ($^3J_{PP}$) has no effect on the spectrum, if J_{23} is negative change the sign. The magnitude of $^3J_{PP}$ has to be comparable in all three fragments, because it refers to the same parameter!

So far, all the examples of ^{13}C{^{1}H} spectra containing ABX type spin systems have been of compounds that exhibit twofold symmetry. Consider the complex [Rh$_4$(η^5-C$_5$Me$_5$)$_4$H$_4$][BF$_4$]$_2$ discussed in Section 4.1.3, refer to Fig. 4.13 In the ^{13}C spin system, the presence of one ^{13}C in the cyclopentadienyl ring will lower, not completely destroy, the symmetry of the Rh-C spin system [4.26]. Think about the symmetry of the Rh-C spin system.

Check it in WIN-DAISY:

Open the input document RH_13C.MGS. Display the **Symmetry Group** dialog box and check if your ideas about the symmetry where correct. Open . the **Symmetry Description** window and look at the permutations - Spins 1 to

4 are ^{103}Rh and Spin 5 ^{13}C. Use the **Next** button to change to the **Frequencies** dialog box - the four rhodiums still have the same chemical shift. Open the **Scalar Coupling Constants** window, the only coupling that is different is J$_{13}$ the one bond ^{1}J$_{RhC}$. Open the **Control Parameters** and look at **Mode**. Click on the **Frequencies** and **Scalar Coupl.** buttons and determine which parameter is to be varied in the Sequence Simulation. Close the dialog box and run the **Sequence Simulation**. **Display** the **Calculated Spectrum** (red peak button) in WIN-NMR. Choose the **Display | Multiple Display** option and load the eleven spectra ...\999999.1R to ...\999989.1R. Click the **Overlay** button and examine the differences in the simulations.

4.2.2 Spin Systems Containing ^{15}N

Nitrogen has two spin active isotopes, ^{14}N (I = 1, natural abundance 99.63%) and ^{15}N (I = ½, natural abundance 0.37%). In early NMR experiments, ^{14}N [4.36] was studied in preference to ^{15}N because of its high natural abundance even though it was a quadropolar nucleus and the linewidths in ^{14}N spectra were often very large, e.g. several kHz which precluded any spin system analysis. ^{15}N spectra could be measured directly using isotopic enrichment but this approach was both expensive and time consuming. Advances in measurement techniques have now made the direct observation of natural abundance ^{15}N relatively straight forward. One of the most important developments has been in ^{15}N{^{1}H} polarization transfer experiments e.g. INEPT, which give an enhancement of ca. 10 times compared with the standard observation methods [4.31]. Unlike the Nuclear OVERHAUSER Enhancement discussed in chapter 5, polarization transfer experiments do not depend upon the observed nucleus relaxing exclusively by dipole-dipole interactions. Nitrogen is an important component in bioorganic molecules but even using the latest techniques the measurement of natural abundance ^{15}N spectra can be rather time consuming [4.3].

As an introduction to ^{15}N spin systems, the discussion will be based upon ^{1}H spin systems and how they are modified by the inclusion of ^{15}N. From the parameters obtained by analyzing the ^{1}H spin system the ^{15}N spectrum will be simulated and can be compared with experimental data in literature.

Check it in WIN-DAISY:

Open... the WIN-DAISY input file 2_ACPY.MGS [4.37]. Inspect the values in the **Main Parameters**, **Frequencies**, **Scalar Coupling Constants** and **Spectrum Parameters** dialog boxes. They correspond to an aromatic four-spin system similar to that discussed in section 2.2.4.1 and one methyl group [4.38]. Because the latter is not coupled to the aromatic protons, the acetyl group forms a separate fragment. Run a spectrum **Simulation** and **Display** the **Calculated Spectrum** (red peak button).

2-acety pyridine

The ^{15}N couples to both fragments, consequently to include ^{15}N in the spin system both ^{1}H fragments have to be merged. To merge the acetyl group into the aromatic spin system high light the first fragment and then either press the

Merge fragments button or use the **Edit I Merge fragment with...** command. Press the **OK** button in the **Fragment handling** dialog box and accept the appear message box with **OK** to delete the acetyl fragment. Open the **Frequencies** dialog box and check that the acetyl is Spin 5.

Open the **Main Parameters** dialog box and extend the spin system to <6> spins. In the **Frequencies** dialog define a ^{15}N isotope for Spin 6 by clicking on the **PSE** button. In the **Element-Data** table select **N** and then click on the **Nuclear Isotope** and select **15** from the pull-down menu. In the **Frequency** edit field enter <–150> ppm. In the **Scalar Coupling Constant** window enter the following values for $^nJ_{NH}$: $^3J_{16}$ = <0.92> Hz, $^4J_{26}$ = <0.25> Hz, $^3J_{36}$ = <1.67> Hz, $^2J_{46}$ = <11.13> Hz and for the long-range coupling constant to the acetyl protons: $^4J_{56}$ = <0.51> Hz. Run a spectrum **Simulation** and **Display** the **Calculated Spectrum** (red peak button) in WIN-NMR. The ^1H spectrum displays additional splittings due to interaction with the ^{15}N.

Open the **Spectrum Parameters** dialog box and select the ^{15}N isotope using the **Isotope** button and **Element-Data** table. Press the **Auto Limits** button to obtain appropriate spectrum limits for the ^{15}N lineshape. **Simulate** the spectrum and **Display** the **Calculated Spectrum** (red peak button) in WIN-NMR. Compare the result with that expected using a simple first order analysis of the spin system

The identification of the individual coupling constants in the ^{15}N multiplet pattern is not straightforward, the small quartet splitting caused by coupling to the acetyl group considerably complicates the basic pattern. The selective decoupling of the acetyl group would simplify the pattern leaving only the aromatic coupling interactions.

Check it in WIN-DAISY and WIN-DR:

Press the **Export to WIN-DR** button or use the **Export I WIN-DR** command. Automatically the WIN-DR program will be started and the spin system data read in as parameters. Open the **Perturbation Parameters** dialog box and select the **Double Resonance** option. Check the **Decoupling** experiment option and set the **Field in B2** to <38> Hz. For the **Effected Spin(s)** select spin **7** (spins 5-7 are the magnetically equivalent protons of the acetyl group). Set the **Min. Absol. Linewidth** to <0.1> Hz. Run a spectrum **Simulation**. Using the red and blue peak button display the reference spectrum (unperturbed) and the double resonance calculation spectrum in the **Dual Display** mode of WIN-NMR. In Reference [4.37] and [4.3] the experimental double resonance spectra are given.

In the next example the spin system of pyrimidine and the ^{15}N isotopically labeled analogue are discussed (Fig. 4.20). The ^1H spectrum of the unlabelled pyrimidine contains four protons. Due to the low natural abundance of ^{15}N, ^{15}N-^1H interactions cannot be observed and essentially the ^{15}N does not take part in the spin system. The two protons H_4 and H_6, are magnetically equivalent due to the absence of any coupling spin in positions 1 and 3 (Fig. 4.20a). The ^{14}N isotope does not lead to a splitting, merely to a broadening of the signals of the neighboring protons.

Measurement of the ^{15}N coupled ^1H spectrum is possible using INEPT although again, due to the low natural abundance of ^{15}N, the spin system contains only one ^{15}N nucleus per molecule [4.37]. Due to the symmetry of the compound the two possible ^{15}N

isotopomers are identical (Fig. 4.20b). The substitution of ^{14}N by ^{15}N in either position 1 or position 3 removes the magnetic equivalence of H$_4$ and H$_6$.

In the fully ^{15}N-labeled [1,3-^{15}N$_2$] pyrimidine [4.39], the spin system is symmetric again, but due to the different coupling paths for J_{AX} and $J_{AX'}$ ($^2J_{NH}$ and $^4J_{NH}$) H$_4$ and H$_6$ are no longer magnetically equivalence and are only isochronous (Fig. 4.20c).

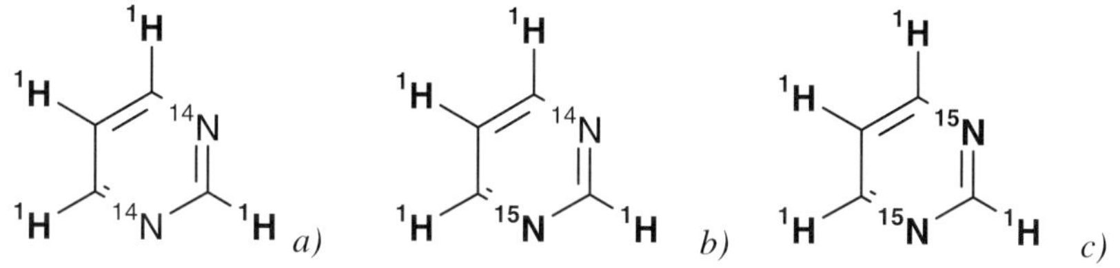

Fig. 4.20: Structure for the isotopomers of pyrimidine:
  a) Normal proton spectrum, symmetrical spin system AM$_2$R.
  b) One- ^{15}N isotopomers: unsymmetrical spin system AMNRX.
  c) ^{15}N labeled molecule, symmetrical spin system A[MX]$_2$R.

Check it in the WINDOWS Calculator:

Using the formula given in sections 4.1.1.3 and 4.2.1.3, calculate the probabilities of the spin systems shown in Fig. 4.20 using a(^{14}N) = 0.9963 and a(^{15}N) = 0.0037 and assuming that the natural abundance of ^{1}H is 100% (a(^{1}H)=1.0).

Check it in WIN-DAISY:

Open... the input document PYRIMIDI.MGS with WIN-DAISY. The 4-protons are therein defined as an A[M]$_2$R spin system with twofold symmetry. H$_4$ and H$_6$ are defined as being isochronous, not magnetically equivalent: Using the point group symmetry instead of the composite particle approach leads to the same spectrum but has the advantage that the spin system can be easily extended to incorporate the ^{15}N spin(s) which removes the magnetic equivalence.

Run a spectrum **Simulation** and **Display** the **Calculated Spectrum** in WIN-NMR. In the next step the spin system will be extended to include two ^{15}N nuclei to calculate the ^{15}N labeled [1,3-^{15}N$_2$] pyrimidine spin system Fig. 4.20c. Open the **Main Parameters** dialog box and extend the spin system from 4 to <6>. Open the **Symmetry Description** window and enter the permutations for spins 5 and 6, below the index 5 enter <6> and below index 6 enter <5>. Call the **Frequencies** dialog box; for Spin 5 click on the **PSE** button and select ^{15}N from the **Element-Data** table and then in the **Frequency** edit field enter <-84.5> ppm. Switch into the **Scalar Coupling Constants** dialog box and add the following ^{15}N-^{1}H coupling constants: J_{15} = <-10.64> Hz, J_{16} = <0.46> Hz, J_{35} = <-14.39> Hz and J_{45} = <-1.12> Hz. Enter a very small ^{15}N-^{15}N coupling constant for J_{56} = <0.1> Hz (this value is reported to be smaller than 0.3 Hz). Open the Lineshape Parameters and give the linewidth parameter for spin 5 to <0.4> Hz. Run a spectrum **Simulation** and **Display** the **Calculated Spectrum** in WIN-NMR. Open the **Spectrum**

276   4 Spin Systems and the Periodic Table

Parameters dialog box and change the lineshape **Isotope** from ^1H to ^{15}N by clicking on the Isotope button and selecting ^{15}N from the **Element-Data** table. Use the **Auto Limits** button to determine appropriate spectrum limits. Run a spectrum **Simulation** and **Display** the **Calculated Spectrum** in WIN-NMR. The WIN-DAISY input document PYRIM_N2.MGS can be used to check the spin system data or the experimental spectra are shown in reference [4.39].

Check it in WIN-DAISY:

To prepare the pyrimidine data for the unsymmetrical [1-^{15}N] pyrimidine spin system remove the symmetry: Open the **Symmetry Group** window and select the point group **C1**. The **Symmetry Description** window is now empty, close the window with **Ok**. Open the **Main Parameters** dialog box and reduce the size of the spin system from 6 to <5> to remove the second ^{15}N. Run a spectrum **Simulation** (still the ^{15}N spectrum!).

In this initial simulation H_4 and H_6 (spins M and M') still have identical chemical shift. In the unsymmetrical system H_4 and H_6 should have different resonance frequencies (spins M and N) due to the isotope effect of the ^{15}N. Check the entries for Spin 1 and Spin 2 in the **Frequencies** dialog box. To study the influence of Δ_{AB} a **Sequence Simulation** can be used. Open the **Control Parameters** and select **Sequence Simulation Mode**. Press the **Frequencies** click the check box in the second line (Spin 2). Enter **No. of Steps** <10> and **Step width** of <-0.1> Hz. Close all dialog boxes with **Ok**. Press the **Sequence Simulation** button on the toolbar. **Display** the **Calculated Spectrum** in WIN-NMR and then switch into the **Display | Multiple Display** mode. Select the eleven calculated spectra and compare them in the **Overlay** mode. If necessary choose the **Multicolor Color Mode** from the **Display | Display Options** pull-down menu to highlight the differences in the ^{15}N spectra more clearly: Only the combination lines show a different line pattern. The correct ^{15}N isotope shift determined by comparison with experimental data is $\Delta_{AB} = 0.8$ Hz. The reference document PYRIM_N1.MGS can be used to compare the spin system data. Reference spectra are given in the literature [4.39] as well.

The last example of ^{15}N spin systems is the ^{15}N labeled cis and trans isomers of difluoro diazine (Fig. 4.21) [4.40].

Fig. 4.21:   Structure of a) cis and b) trans difluoro diazine.

Both compounds are $[AX]_2$ spin systems and exhibit twofold symmetry; the cis isomer (Fig. 4.21a) has either C_s or C_2 symmetry and the trans isomer (Fig. 4.21b) C_i symmetry with the inversion center in the middle of the N=N bond.

Check it in WIN-DAISY:

Open... the input document N2F2_CIS.MGS. The values are taken from reference [4.40]. Examine the entries in the following dialog boxes; **Symmetry**

Group - the spin system definition, **Symmetry Description**, **Frequencies** and **Scalar Coupling Constants**. Run a spectrum **Simulation** and **Display** the **Calculated Spectrum** in WIN-NMR. To show that the selection of either Cs or C2 has no effect on the spin system data, open the **Symmetry Group** dialog box and change the **Symmetry Group** to **Cs**. Quit the dialog with **Ok**. Automatically the **Symmetry Description** window is opened for updating the permutation operators. In this case values need not to be changed, because the reflection interchanges the same spins as the rotation symmetry operation. Close the dialog box with **Ok**. The data in the **Frequencies** and **Scalar Coupling Constants** dialog box are still the same as for C2 symmetry. To confirm that the C2 and Cs give rise to the same spectrum run a **Simulation** and **Display** the **Calculated Spectrum** (^{19}F) in WIN-NMR. The experimental spectra is given in [4.40].

Table 4.4: Coupling Constant values for trans difluorodiazine-^{15}N$_2$.

Coupling Constant	value [Hz]
$^1J_{NN}$	18.5
$^1J_{FN}$	± 172.8
$^2J_{FN}$	∓ 62.8
$^3J_{FF}$	316.4

Check it in WIN-DAISY:

Open the **Symmetry Group** dialog box and change the **Symmetry Group** to **Ci**. Quit the dialog and examine the **Symmetry Description**. For the trans-isomer the same atoms are interchanged by the inversion and no change is necessary. Open the **Scalar Coupling Constants** dialog box and enter the $^nJ_{FN}$ values for the trans isomer as given in Table 4.4. Run a **Simulation** and **Display** the **Calculated Spectrum** in WIN-NMR. The WIN-DAISY document N2F2_TR.MGS can be used to check the spin system data and experimental and calculated spectra are given in reference [4.40].

4.2.3 Spin Systems Containing ^{29}Si

Silicon has three naturally occurring isotopes, ^{28}Si, ^{29}Si and ^{30}Si with a natural abundance of 92.23%, 4.67% and 3.10% respectively. Of these three isotopes only ^{29}Si is NMR active with I = ½. Due to its relatively high natural abundance ^{29}Si satellite often appear in the spectra of nuclei that are sensitive to the NMR experiment [4.1], [4.41].

As an example we will consider tetramethylsilane (TMS), the standard reference compound for ^{1}H, ^{13}C and ^{28}Si NMR experiments. TMS contains one silicon, four carbons and twelve protons and the most probable isotopomers are shown in Fig. 4.22.

$$\binom{r}{r-n} \cdot a(^iX)^n \cdot a(^0X)^{r-n} \cdot 100 \; [\%]$$

with a(^{i}X) = natural abundance of isotope ^{i}X,
and n = 0, 1 ... r   and r = number of places.

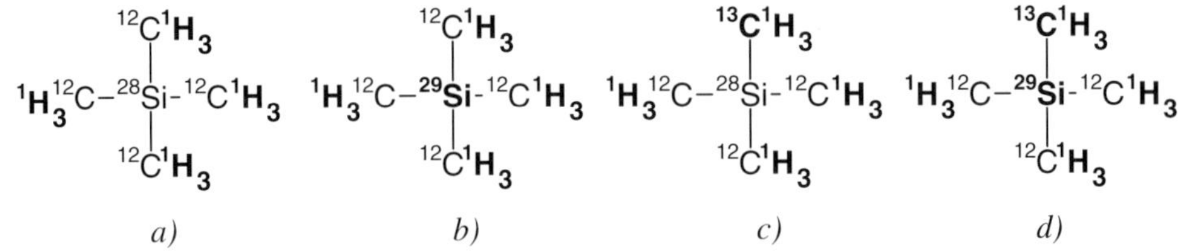

278   4 Spin Systems and the Periodic Table

$^1H_3^{12}C-^{28}Si(^{12}C^1H_3)_3$ $^1H_3^{12}C-^{29}Si(^{12}C^1H_3)_3$ $^1H_3^{12}C-^{28}Si(^{12}C^1H_3)_2(^{13}C^1H_3)$ $^1H_3^{12}C-^{29}Si(^{12}C^1H_3)_2(^{13}C^1H_3)$

a)   b)   c)   d)

Fig. 4.22: Most probable tetramethylsilane isotopomers:
a) $^1H^{28}Si^{12}C$ isotopomer,   b) $^1H^{29}Si^{12}C$ isotopomer,
c) mono-$^{13}C,^{28}Si^1H$ isotopomer,   c) mono-$^{13}C,^1H^{29}Si$ isotopomer.

Check it with the Calculator:

Calculate the probabilities of the four isotopomers of tetramethylsilane shown in Fig. 4.22 using the data given in Table 4.5. The ^{28}Si and ^{30}Si isotopes can be combined, because both have a nuclear spin quantum number $I = 0$, referred as ^0Si in the following. For the individual elements use the general formula given below (refer to sections 4.1.1.3 and 4.2.1.3 if necessary) and then multiply together the individual probabilities to calculate the overall isotopomer probability. Refer as well to the appendix, section 6.2.

Table 4.5: Natural abundance of isotopes contained in tetramethylsilane.

Isotope	a (^iX)
^1H	0.9998
^2H	0.0002
^{28}Si	0.9223
^{29}Si	0.0467
^{30}Si	0.0310
^0Si	0.9533
^{12}C	0.9890
^{13}C	0.0011

Check it in WIN-NMR:

The spin systems for the first three isotopomers in Fig. 4.22a-c can be found in the ^1H spectrum of TMS. Load the experimental spectrum ...\TMS\001001.1R into WIN-NMR. Measure a **Linewidth** and then switch into the **Analysis | Multiplets** mode. Define a **Singlet** for the main peak (most abundant isotopomer). Locate the ^{29}Si satellite lines ($^2J_{SiH}$ = 6.5 Hz) and define a doublet. Search for the ^{13}C satellite lines ($^1J_{CH}$ = 118 Hz) and define another doublet. Open the **Report...** box and **Export...** the data to WIN-DAISY with fragmentation of the spin system and *inclusion* of non-assigned coupling constants.

Check it in WIN-DAISY:

WIN-DAISY generates three fragments for the three isotopomeric spin systems. One fragment contains only ^1H as the spin active isotope, one other second fragment the ^{29}Si satellite lines and the remaining fragment contains the ^{13}C satellites. The order of the fragments depends on the chemical shifts. Open the **Main Parameters** dialog box; for each fragment enter a proper **Title** and **Statistical weight** based on the calculated probabilities. The isotopomer given in Fig. 4.22d has negligible natural abundance and cannot be observed. Open the **Frequencies** dialog box; using the **PSE** button define in the Si satellite spin system Spin 2 as ^{29}Si, and in the other two spin system with the large coupling as ^{13}C. Run a spectrum **Simulation** and compare the calculated and experimental TMS spectra using the **Dual Display** mode of WIN-NMR. If the calculated probabilities are correct there should be reasonable agreement between the intensities in both spectra. Run an **Iteration** using the default values and display the result in the WIN-NMR **Dual Display**. Compare the result with the values in the WIN-DAISY input document TMS.MGS where approximate values are given.

4.2.4 Spin Systems Containing ^{195}Pt

Platinum has only one NMR active isotope, ^{195}Pt with a natural abundance of 33.8% and I = ½ [4.1]. Many platinum organometallic compounds are known and spin systems involving more than one platinum atom are of special interest. Because only about one third of all platinum atoms in the structure have a nuclear spin, different spin systems are present in the same complex. Fig. 4.23 shows an unsymmetrical isomer of a Pt$_3$-complex [4.42].

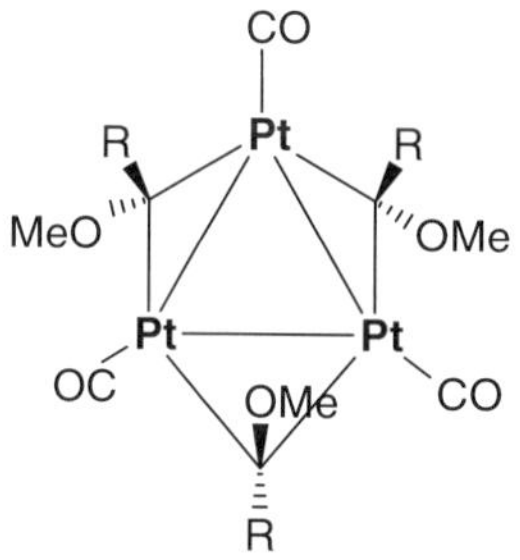

Fig. 4.23: Structure of unsymmetrical isomer of $Pt_3\{\mu\text{-C(OMe)}C_6H_4Me\text{-}4\}_3(CO)_3$, R = $C_6H_4Me\text{-}4$.

The composition of the various ^{195}Pt spin systems is as follows: the platinum atoms are in two different chemical environment and for a 100% ^{195}Pt isotopically labeled compound the spin system would be AB$_2$. It would be second order, because the platinum-platinum coupling constant ($^1J_{PtPt}$ = 3115 Hz) is very large compared with the chemical shift difference (Δ_{AB} = 1388 Hz at 21.46 MHz). But the natural abundance of ^{195}Pt is 33.8% so that the probability of a $^{195}Pt_3$ cluster is much lower. If the platinum isotopes with I = 0 are represented by ^0Pt, then all the isomers and isotopomers from $^0Pt_3$ to $^{195}Pt_3$ have reasonable probability.

Check it in the Calculator:

Calculate the probabilities for all the possible isotopomers, use $a(^{195}Pt) = 0.338$ and for all the other platinum isotopes use $a(^{0}Pt) = 0.662$. Think about the spin systems for the isotopomers that contain only one ^{195}Pt and two chemically different ^{195}Pt. According to the position of the ^{195}Pt nuclei in the structure, the spin systems are different. Tabulate the probabilities and spin system types and compare your results with the entries in Table 4.6.

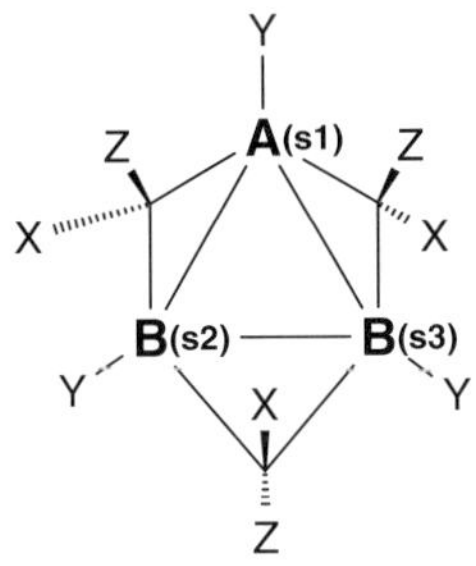

Fig. 4.24: Spin labels (A and B) of Pt positions (s1,s2,s3), all other substituents are labeled X,Y and Z.

Table 4.6: Isotopomers of Pt_3 cluster with probabilities and spin system type.

n	Number of isotopomers	Position(s) of ^{195}Pt	Corresponding Spin System	Probability [%]
0	1	–	–	29.0
1	3	A(s1)	A	14.8
		B(s2)	B	14.8
		B(s3)	B	14.8
2	3	A(s1)B(s2)	AB	7.6
		A(s1)B(s3)	AB	7.6
		B(s2)B(s3)	B_2	7.6
3	1	A(s1)B(s2)B(s3)	AB_2	3.9

A variety of spin systems are possible for the Pt_3 cluster and each has a significant probability. The $^{0}Pt_3$ isotopomer does not contribute to the spectrum, and will be ignored in the following discussion. It is not necessary to normalize the probabilities because the relative values can be taken into account using the statistical weight parameter. For identical spin systems the probabilities can be summed.

Check it in WIN-DAISY:

Open... the WIN-DAISY input document PT_AB2.MGS. This file contains the input data for the 100% ^{195}Pt isotopically labeled isotopomer, forming the AB_2 spin system. Run a **Simulation** and **Display** the **Calculated Spectrum** (red peak button) in WIN-NMR. The system is strongly second order.

4.2 Spin-½ Rare Isotopes

Now modify the input document in order to simulate all the isotopomers according to the data calculated in the previous "Check it". All the resonance frequencies needed to form the spin systems are contained in the AB_2 fragment. Five fragments are required to define the spin system. Always using the original AB_2 fragment, **Copy** the selected **fragment** (within the same document) and modify the parameters to correspond to the required spin system. In the **Main Parameters** dialog box edit the **Number of spins or groups with magnetical equivalent nuclei** and **Statistical weight** according to the desired spin system. In the **Frequencies** dialog box the **Spins in group** for the B and B_2 spin systems and the **Frequency** must be changed accordingly. Disregard any isotope effect on either coupling or chemical shifts. Run a **Simulation** and **Display** the **Calculated Spectrum** (red peak button) in WIN-NMR. Check your result by comparing the simulated spectrum with the WIN-DAISY input document PT3.MGS or the reference data in literature [4.42].

4.2.5 Spin Systems Containing ^{77}Se

Selenium has not less than six stable isotopes, but only one of them ^{77}Se, with a natural abundance of 7.6% and I = ½, is NMR active [4.1],[4.43]. Compared with ^{13}C the natural abundance is high, but in comparison with ^{195}Pt the natural abundance is rather low. In a spin system containing a pure element e.g. ^{31}P, and a single selenium atom, ^{77}Se satellites are clearly visible. However, if the spin system contains more than one selenium atom, how many different satellite spectra will be observed? This combinatorical problem is comparable with the unsymmetrical Pt_3-complex discussed in section 4.2.4. The structure of the anion $[P_2Se_6]^{4-}$ is shown in Fig. 4.25. The anionic structure under goes resonance hybridization so that it is not possible to distinguish between the three selenium atoms attached to the same phosphorus.

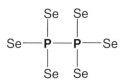

Fig. 4.25: Structure of the $[P_2Se_6]^{4-}$ anion;
all three selenium atoms of -PSe_3^{2-}-group are equivalent.

Check it with the Calculator:

Using the formula given below calculate the type and probabilities of the various spin systems including phosphorus and selenium. The selenium isotopes with I = 0 are represented by ^{0}Se. Assume that $a(^{77}Se) = 0.076$ and $a(^0Se) = 0.924$. Tabulate the probabilities, assign the spin system types and compare your results with the entries in Table 4.7.

$$\binom{6}{6-n} \cdot a(^{77}Se)^n \cdot a(^0Se)^{6-n} \cdot 100 \; [\%]$$

282 4 Spin Systems and the Periodic Table

When assigning spin systems to the probabilities connected with the number of n ^{77}Se in the compound a complication arises. For n = 2,3 and 4 more than one spin system is possible according to the position of the ^{77}Se atoms with respect to the ^{31}P nuclei. This fact also refers to a combinatorical problem in which the element n in "n over k" can be divided into two types of distinguishable items. The complete discussion of this problem is found in the appendix, section 6.2.

Table 4.7: Statistical probabilities for the various spin systems of ^{31}P- and ^{77}Se- nuclei in the isotopomers of $P_2Se_6^{4-}$ (A, B = ^{31}P; X,Y = ^{77}Se).

n	Probability [%]	Spin System(s)	Topology
0	1 · 0.076^0 · 0.924^6 · 100 = 62.23	A_2	—A—A—
1	6 · 0.076^1 · 0.924^5 · 100 = 30.71	ABX	X—A—B—
2	15 · 0.076^2 · 0.924^4 · 100 = 6.32 $\frac{9}{15}$ · 6.32 = 3.79	$[AX]_2$	X X'—A—A'—
2	$\frac{6}{15}$ · 6.32 = 2.53	ABX_2	X—A—B—X
3	20 · 0.076^3 · 0.924^3 · 100 = 0.69 $\frac{18}{20}$ · 0.69 = 0.62	ABX_2Y	X Y—A—B—X
3	$\frac{2}{20}$ · 0.69 = 0.07	ABX_3	X—A—B—X X
4	15 · 0.076^4 · 0.924^2 · 100 = 0.043 $\frac{9}{15}$ · 0.04 = 0.026	$[AX_2]_2$	X X'—A—A'—X X'
4	$\frac{6}{15}$ · 0.04 = 0.017	ABX_3Y	X Y—A—B—X X
5	6 · 0.076^5 · 0.924^1 · 100 = 0.0014	ABX_3Y_2	X Y—A—B—X Y
6	1 · 0.076^6 · 0.924^0 · 100 = 1.93·10^{-5}	$[AX_3]_2$	X X'—A—A'—X' X X'

For compounds containing two, three and four ^{77}Se atoms (n = 2 to 4 in Table 4.7) the binomial formula given above will only calculate the *total* probability of a compound containing ^{77}Se atoms. It does not take into account the *relative* position of the ^{77}Se atoms with respect to the ^{31}P nuclei. This fact also influences the spin system; for example for n = 2 the two ^{77}Se atoms could be bond to the same phosphorus atom giving an ABX$_2$ spin system or distributed equally between the two phosphorus atoms giving an [AX]$_2$ spin system. In addition the isotope shift caused by the ^{77}Se on the chemical shift of ^{31}P has to be considered. Thus for the ABX$_2$ spin system it is possible that no isotope shift can be detected in the ^{31}P NMR spectrum, in which case the two phosphorus atoms remain isochronous resulting in an AA'X$_2$ type spin system. In this case the prime notation is correct, because there is no symmetry operation converting A into A', a slight isotope shift might be expected however.

Table 4.7 shows that for values of n of three and greater, the probability of the isotopomers is too low for them to be detected in the ^{31}P NMR spectrum, Consequently the ^{31}P spectrum will show only the signals of *four* different spin systems and the ^{77}Se spectrum those of *three* (because the first spin system does not contain any ^{77}Se).

Starting with the ^{31}P spectrum, the NMR parameters will be optimized for the various phosphorus-selenium spin systems and the result checked by comparing the calculated ^{77}Se spectrum with the experimental spectrum. The analysis will start with the most abundant spin systems the A$_2$ which does not contain any ^{77}Se and the ABX which contains one ^{77}Se atom. In practice, the NMR parameters defining the various spin systems will be very similar and consequently the parameters determined for the ABX system may be used in the [AX]$_2$ and ABX$_2$ spin systems.

Check it in WIN-NMR and WIN-DAISY:

Open... the experimental ^{31}P spectrum of P$_2$Se$_6^{4-}$ ...\P2SE6\001001.1R [4.44]. Measure the **Linewidth** of the center signal and then switch into the **Analysis | Multiplets** mode. Mark the center signal with the spectrum cursor and use the right mouse button to **Define** a **Singlet** for the A$_2$ phosphorous spin system. The most abundant satellite spectrum will be the AB part of an ABX spectrum, although as discussed in the preceding section the spin system might be very close to an AA'X spin system. There must be two (ab) subspectra with the coupling constant J$_{ab}$ appearing four times in the spectrum. For the first order multiplet analysis define for A and B a doublet of doublets each *not* as an (ab) subspectrum (refer to section 2.3.3.2, first order solution of ABX spin systems). Try to identify the two signals keeping in mind that the arithmetic mean of the A- and B-multiplet midpoints (first order estimated) are almost identical (refer to Fig. 4.26).

When the four J$_{AB}$ distances are found (ca. 210 Hz) there are several ways to combine them to form two doublet of doublets but only one way will result in almost the same starting values for shifts A and B. In the **Report...** box connect the corresponding coupling constants and **Export...** the data to **WIN-DAISY** *with* fragmentation of the spin system (**Yes**) and *inclusion* of non-assigned couplings (**Yes**).

284   4 Spin Systems and the Periodic Table

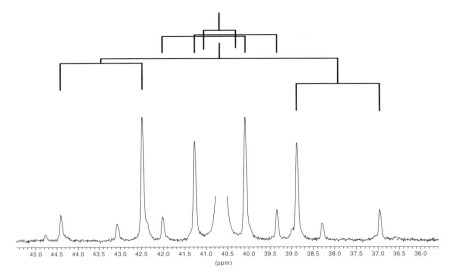

Fig. 4.26: First order Multiplets definition for the first satellite spin system in ...\P2SE6\001001.1R.

The spin system with the title **...1-spin system...** refers to the A_2 singlet and the **...3-spin system...** fragment to the ABX. Open the **Main Parameters** for both fragments and enter the appropriate **statistical weight** for the two different spin systems (refer to Table 4.7). In the **Lineshape Parameters** for the three-spin fragment check the **Nuclei specific linewidth** option and click on the **Iterate All** button. This will be necessary, because the selenium spectrum will feature a quite different linewidth parameter. Open the **Frequencies** dialog box the A_2 fragment and increase the number of **spins in group** to <2> according to the two magnetically equivalent nuclei. Click the **Next** button in the header to switch to the ABX fragment, press the **PSE** button of the third spin and select **Se** from the **Element-Data** table. Enter a chemical shift of about <120> ppm for the ^{77}Se. *Remove* the **disable** option for the selenium nucleus. It is not necessary here, because the resonance frequencies of phosphorous and selenium are quite apart. Open the **Control Parameters** and select **Advanced Iteration Mode, medium Broadening** and **medium No. of Cycles**, check the option **Starting frequencies poor**. Click on the **Scalar Coupl. button** and *deselect* J_{12} (the coupling between A and B, which is already well defined by the first order analysis) for iteration. For the other two couplings increase the **Lower limit** and **Upper limit** e.g. J_{13}: ±<700> and J_{23}: ±<150>. Remember that in a ABX spin system the relative signs of J_{AB} and J_{AX} can be determined but not the sign of J_{AB}. Quit the dialogs with **Ok** and run an **Iteration**. Display the result in using the **Dual Display** mode of WIN-NMR. Open the **Control Parameters** again and select the **Standard Iteration mode** *without* any smoothing parameters (**none Broadening**). Click on the **Scalar Coupl.** button and *select* J_{12} for optimization using its default iteration limits. Run another **Iteration** (which are sufficient). The final value of *R-Factor* should be ca. *2.58 %*. Open the **Scalar Coupling** Constants dialog box and inspect the optimized values, the X scalar coupling constants are of opposite sign. Assuming a negative sign for one-bond SeP coupling constants [4.45] interchange the signs, run a spectrum

Simulation and inspect the result. As expected, there is no difference in the appearance of the spectrum.

In the next step the WIN-DAISY document has to be enlarged to include the [AX]$_2$ and ABX$_2$ satellite spin systems with the appropriate weighting factors. In main window list box highlight the **...3-spin system...** and using either the **Copy fragment** command or toolbar button to copy the fragment *within the same document*. Automatically the **Main Parameters** dialog box of this new third fragment opens, enter a different **Title**, e.g. <[AX]2 spin system> and enter the appropriate **Statistical weight**. Increase the **Number of spins or groups with magnetical equivalent nuclei** to <4>. Open the **Symmetry Group** window and select a twofold symmetry, e.g. **Cs**. In the **Symmetry Description** dialog box enter the following line for the permutation: <2>, <1>, <4> and <3>. Spins 1 and 2 correspond to the ^{31}P while spin 3 and the new spin 4 correspond to the ^{77}Se. As expected, the **Frequencies** dialog box shows only one ^{31}P and one ^{77}Se chemical shift. In the original fragment J$_{14}$ and J$_{34}$ did not exist, consequently in the **Scalar Coupling Constants** dialog box there are only two non-zero coupling constants. Click the **Previous** or **Next** button in the header to switch to the three spin system, ABX, fragment. Copy J$_{23}$ (^{2}J$_{SeP}$) into the clipboard (<Ctrl + C>) return to the new fragment and paste the entry into J$_{14}$ (<Ctrl + V>). Run a spectrum **Simulation** and display the result in the WIN-NMR **Dual Display**. The comparison between experiment and calculation is getting closer, but a few peaks/shoulders are still missing.

To include the final satellite spin system copy the ABX fragment again into a new ABX$_2$ fragment. Proceed in a similar way to the [AX]$_2$ spin system. In the **Main Parameters** dialog box enter an appropriate **Title**, e.g. <ABX2 spin system>, and appropriate **Statistical weight** (Table 4.7). Open the **Frequencies** dialog box of the new fragment and increase the **Spins in group** for spin 3, the ^{77}Se, to <2>. Run another **Simulation** and compare the spectra in the WIN-NMR **Dual Display**. Keep the document open!

When all four satellite spin systems are included there is a very good agreement between the experimental ^{31}P spectrum and simulated spectrum. Small differences do still exist and are caused by very small isotope effects. In theory the parameters could be optimized but the intensities of the satellite spectra are too low to enable a satisfactory result to be obtained via lineshape iteration. In the next step the ^{77}Se spectrum will be calculated.

Check it in WIN-NMR and WIN-DAISY:

Open... the ^{77}Se spectrum ...\P2SE6\002001.1R [4.44] with the WIN-NMR. Enter the **Analysis | Interval** mode and press the **Regions...** button. **Load...** the predefined iteration region. Quit the dialog with **Ok** and press the **Export...** button to send interval data to **WIN-DAISY** (deny to send eventually present multiplet data: answer **No**). When WIN-DAISY is moved to the WINDOWS foreground accept the message box with **Yes** and press the **Connect** button. To change the experimental spectrum from ^{31}P to ^{77}Se open the **Spectrum Parameters** dialog box. Press the **Isotope** button and select **77Se** from the **Element-Data** table. Automatically the **Experimental** and **Calculated** file names change to the ^{77}Se data set. Run a **Simulation** and display the results in the **Dual Display** of WIN-NMR. Use the <Ctrl + Y> key to align the y-expansion of both spectra.

286   4 Spin Systems and the Periodic Table

Overall, the ^{77}Se spectrum is reproduced quite well, but both the chemical shift and linewidth are wrong. Iteration of the ^{77}Se spectrum is not feasible because the lineshape is not symmetrical, but the chemical shift and linewidth can be adjusted manually. The adaptation of the shifts can be done via an offset correction in WIN-DAISY.

Check it in WIN-NMR:

With the experimental and calculated ^{77}Se spectra displayed in the **Dual Display** mode of WIN-NMR switch the X-axis units to Hz, <Ctrl + X>. Click the **Move Trace** button. Drag the calculated spectrum across the experimental spectrum until they match. Make a note of the frequency shift, ca. "–267 Hz", displayed in the status line. Toggle back to **WIN-DAISY**.

Check it in WIN-DAISY:

Open the **Spectrum Parameters** dialog box and enter the frequency shift (with the opposite sign) into the **Offset [Hz]** edit field, e.g. <267> Hz. Press the button **Apply!** to apply the frequency correction to all the chemical shifts associated with the **Isotope** ^{77}**Se**. Check in the **Frequencies** dialog boxes that only the ^{77}Se shifts have been modified. Open the **Lineshape Parameters** dialog box and increase the ^{77}Se linewidth (nuclei with index 3) to <18> Hz. Run a spectrum **Simulation**. Compare the results using the **Dual Display** mode of WIN-NMR, using <Ctrl + Y> to adapt the Y-axis. For fine-tuning you may **Apply!** again a frequency correction for the ^{77}**Se** chemical shifts extracted via **Move Trace** option in WIN-NMR. To store the result with the experimental spectrum, use the **Export WIN-NMR exp.** option. It will then be possible to reconstruct the WIN-DAISY file via export from WIN-NMR **Multiplets** mode while the **Windaisy** button is depressed.

4.2.6 Spin Systems Containing ^{115}Sn / ^{117}Sn / ^{119}Sn

Tin has a number of naturally occurring isotopes of which three are NMR active and posses a nuclear spin quantum number I = ½; ^{115}Sn (0.4%), ^{117}Sn (7.7%) and ^{119}Sn (8.6%) [4.46], [4.1]. Due to its low natural abundance, ^{115}Sn is usually ignored and only the ^{117}Sn/^{119}Sn satellite signals in heteronuclear spin systems considered.

The structure of the compound used in this example is shown in Fig. 4.27 [4.47].

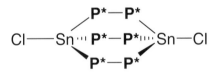

Fig. 4.27: Structure of hexa-t-butyl-1,4-dichloro-1,4-distanna-2,3,5,6,7,8,hexa-phospha-bicyclo[2.2.2]octane, **P*** = P-t-butyl.

Check it in WIN-NMR:

Examine the structure in Fig. 4.27 and determine the spin system in the absence of any spin active tin isotope (the main spin system). Load the experimental ^{31}P{^{1}H} spectrum ...\PROPELL\001001.1R into the spectrum window of WIN-NMR. Measure a **Linewidth** and then in the **Multiplets** mode define a singlet for the intense central peak. **Export...** the data to **WIN-DAISY**.

As expected, the main spin system containing only the phosphorus atoms (printed bold in Fig. 4.27) corresponds to a A_6 spin system and occurs as a singlet.

Table 4.8: Natural abundance of tin isotopes.

Isotope	a(^{x}Sn)
^{115}Sn	0.004
^{117}Sn	0.077
^{119}Sn	0.086
^{0}Sn = Σ	1– (0.004+0.077+0.086) = 0.833
spin inactive ^{x}Sn	

The probabilities of the various tin satellite spin systems may be calculated using the formula:

$$\binom{2}{2-n} \cdot a\left(^X Sn\right)^n \cdot a\left(^0 Sn\right)^{2-n} \cdot 100 \, [\%]$$

with n = 0, 1, 2.

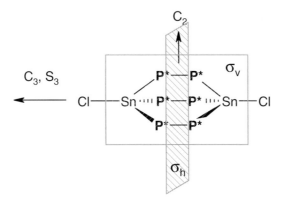

Fig. 4.28: Symmetry of the main spin system (not all symmetry elements mentioned, only one of three perpendicular C_2-axes and vertical mirror planes shown).

Check it in WIN-NMR and WIN-DAISY:

Using Fig. 4.29, determine the type of *spin system,* the *symmetry* and the *probability* (see formula and Table 4.8) for all the possible combinations of ^{0}Sn, ^{115}Sn, ^{117}Sn and ^{119}Sn.

It is necessary to build up a set of fragments to represent all the available spin systems Start by modifying the main A_6 spin system fragment; in the **Frequencies** dialog box increase the **Spins in group** to <6> and then switch to **Main Parameters** dialog box. Enter the probability (in percent) of the main spin system as the **Statistical weight** and then enter an appropriate **Title**, e.g. <Main spin system A6>. Run a **Simulation** and display the spectra in the WIN-NMR **Dual Display**.

Load the experimental ^{119}Sn{^{1}H} spectrum ...\PROPELL\002001.1R into the spectrum window of WIN-NMR, mark the central peak with the spectrum cursor in **maximum cursor** mode and make a note of the chemical shift value (about 21.19 ppm). Toggle back to **WIN-DAISY**.

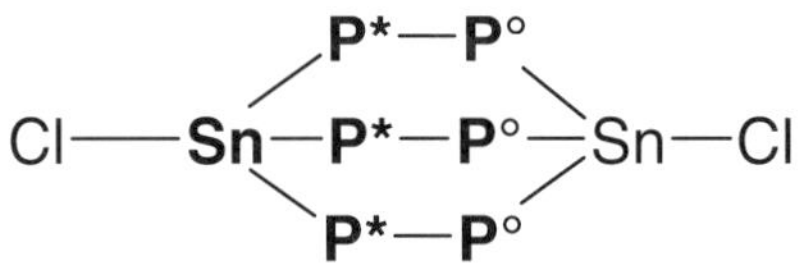

$$Cl-Sn\begin{matrix} P^*-P^° \\ -P^*-P^°- \\ P^*-P^° \end{matrix}Sn-Cl$$

Fig. 4.29: Satellite spin system containing one spin active **Sn** isotope, either ^{119}Sn, ^{117}Sn or ^{115}Sn, P*,P° = P-t-butyl.

Starting with the ^{0}Sn - ^{119}Sn spin system, create three new fragments to define the satellite spin systems which contain one spin active tin isotope.

Use the **Edit I Copy fragment** command (**Copy fragment** button) to duplicate the main spin system within the same WIN-DAISY input document and then modify the appropriate parameters. In the **Main Parameters** increase the **Number of spins or groups with magnetical equivalent nuclei to** <7>, enter the calculated probability as the **Statistical weight** and then enter a different **Title** string, e.g. <[AA']3X (C3v) mono-119Sn>. Select the appropriate **Symmetry Group** and enter the **Symmetry Description**. In the **Frequencies** dialog box decrease the **Spins in group** for Spin 1 from 6 to <1>. For Spin 4 select ^{31}P using the **PSE** button and copy the chemical shift from Spin 1. Select ^{119}Sn for Spin 7, check the **Disable** option and enter the ^{119}Sn chemical shift taken from the experimental spectrum. Finally, enter the **Scalar Coupling Constant** values taken from Table 4.9. **Change** the calculated spectrum name e.g. to ...\001998.1R, and run a spectrum **Simulation**. Display the experimental and calculated spectrum in the WIN-NMR **Dual Display**.

Introducing one spin active tin isotope removes the magnetic equivalence of the six phosphorus nuclei and leads to an [AB]$_3$X (C_{3v}) spin system, with P* = A, P° = B and Sn = X. This example clearly shows the difference between magnetic equivalence and isochronicity caused by spin system symmetry. The [AB] fragments represent the –P–P– bridges. Depending on whether the spin active tin isotope causes an isotope effect or not, the phosphorus atoms P* and P° can also be accidentally isochronous, so that the spin system can be described as [AA']$_3$X (C_{3v}). In this case the prime notation is necessary to distinguish between the isochonicity caused by the negligible isotope effect on the phosphorus chemical shifts and the isochronicity caused by the threefold spin system symmetry, symbolized by the brackets [...]$_3$. In the structure the phosphorus isotopes are labeled A, A'... up to A''''' with the even number of primes corresponding to P* and the odd number of primes to P°.

Table 4.9: Spin system parameter for the propeller compound, Sn = ^{119}Sn [4.47].

Parameter	Value [Hz]
$^1J_{PP}$	− 213
$^2J_{PP}$	− 5.6
$^3J_{PP}$	− 1.4
$^1J_{SnP}$	+ 1368
$^2J_{SnP}$	+ 48

Check it in WIN-DAISY:

Using the ^{0}Sn - ^{119}Sn spin system fragment, define the corresponding fragments for the ^{0}Sn - ^{117}Sn and ^{0}Sn - ^{115}Sn spin systems. In the main document window select the ^{0}Sn - ^{119}Sn fragment and execute the **Edit | Copy fragment** command (**Copy fragment** button). In the **Main Parameters** dialog box of the new fragment enter the probability for the ^{0}Sn - ^{117}Sn satellite spin system as the **Statistical weight** and give a suitable **Title** string for identification. Open the **Frequencies** dialog box, press the **PSE** button for Spin 7 and select **Nuclear Isotope 117** from the **Element-Data** table. Leave the **Frequencies** dialog box by pressing the **Next** button in the bottom line to enter the **Scalar Coupling Constants** dialog box. Before the dialog box is displayed WIN-DAISY asks whether to modify the current heteronuclear coupling constant values according to the ratio of the magnetogyric ratios of the selected isotope. Answer the question with **Yes**. **Change** the calculated spectrum name and run a spectrum **Simulation**. Display the experimental and calculated spectrum in the WIN-NMR **Dual Display**. **Zoom** in the satellite signals to see the increasing number of signals reproduced in the simulation.

Repeat the above procedure for the ^{0}Sn - ^{115}Sn spin system. Again, change the calculated spectrum name before running a **Simulation** and displaying the results in the **WIN-NMR Dual Display**.

In the next step the isotopomers with two different spin active Sn isotopes will be calculated.

Check it in WIN-DAISY:

Copy the ^{0}Sn - ^{119}Sn fragment. In the new ^{117}Sn - ^{119}Sn fragment **Main Parameters** dialog box enter an appropriate **Title**, e.g. <[AA']3XY (C3v) 119/117Sn>, increase the **Number of spins or groups with magnetical equivalent nuclei** to <8> and enter the calculated probability as the **Statistical weight**. The **Symmetry Group** and **Symmetry Description**

remain unchanged. In the **Frequencies** dialog box, select the ^{117}Sn isotope for *Spin 8* and check the **Disable** option. Open the **Scalar Coupling Constants** dialog box and enter the coupling constants J_{18} ($^2J_{SnP}$) and J_{48} ($^1J_{SnP}$), they can be copied from the ^{0}Sn - ^{117}Sn fragment. Change the calculated spectrum name e.g. to ...\001995.1R, and run a **Simulation**. Compare the experimental and calculated spectrum in the **WIN-NMR Dual Display** (Fig. 4.30).

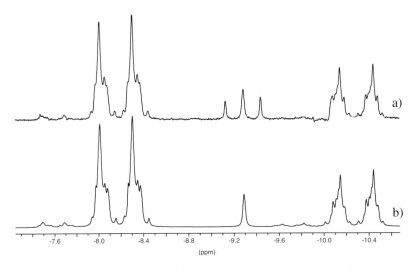

Fig. 4.30: a) Experimental 31P{1H} spectrum ...\PROPELL\001001.1R.
b) Calculated 31P{1H} spectrum ...\PROPELL\001995.1R.

The first calculation including a "mixed" ^{117}Sn/^{119}Sn isotopomer leads to the appearance of the middle signal of the pseudo-triplet as shown in Fig. 4.30b.

Check it in WIN-DAISY:

Create the fragments for the two remaining "mixed" spin systems as described in the previous "Check its". Again, **Change** the calculated spectrum name before running a **Simulation** to keep all the simulated spectra.

The last two "mixed" satellite spin systems are of very low probability, so that almost no changes are visible in the simulated spectra. The three remaining isotopomers to be included contain two identical tin isotopes and are of higher symmetry (D_{3h}).

Check it in WIN-DAISY:

Copy the ^{0}Sn - ^{119}Sn fragment. In the new ^{119}Sn - ^{119}Sn fragment **Main Parameters** dialog box enter an appropriate **Title**, e.g <[[A]2X]3 (D3h) bis-119Sn>, extend the **Number of spins or groups with magnetical equivalent nuclei** to <8> and enter the calculated probability as the **Statistical weight**. Open the **Scalar Coupling Constants** dialog and copy the value of J_{47} into J_{18}. (It simplifies the alterations to do this before changing the symmetry group.) Then open the **Symmetry Group** window and

4.2 Spin-½ Rare Isotopes

select the point group **D3h**. Use the **Next** or **Ok** button to display the **Symmetry Description** dialog box.

In comparison with C_{3v} symmetry, D_{3h} contains over twice the number of symmetry operators. The vertical reflection planes σ_v, $\sigma_{v'}$ and $\sigma_{v''}$ are now located in different rows and the original permutations need to be moved from lines 4, 5, 6 to lines 10, 11 and 12 respectively. The new permutations for C_2, C_2', C_2'' (the perpendicular twofold axes), σ_h (the horizontal mirror plane), S_3 and S_3^2 (the improper axis of rotation) now transfer the two ^{119}Sn isotopes into each other (index 7 to 8).

Check it in WIN-DAISY:

Move lines 4 to 6 into lines 10 to 12. Enter the new permutations and check your results with Fig. 4.31. Note: it is particularly important that the correct permutations are enter, otherwise a run time error may occur. The **Frequencies** dialog box then contains only one ^{31}P shift. Change the calculated spectrum name, run a spectrum **Simulation** and display the spectra in the WIN-NMR **Dual Display**.

Check it in WIN-NMR:

Zoom in the region of the pseudo-triplet, the outer line of the signal arises from the symmetrical ^{119}Sn isotopomer. Return to the experimental spectrum, switch to the **Analysis I Integration** mode and integrate the three peaks of the pseudo triplet. Mark the middle integral with the **perpendicular spectrum** cursor and in the **Options...** dialog box chose the **Manual Calibration** option and enter the value of <1.32>, the statistical weight of the ^{117}Sn/^{119}Sn isotopomer, for the **Selected Integral**. The normalized integral value of the right-hand side peak in Fig. 4.30a refers to the statistical weight of the symmetrical bis-^{119}Sn isotopomer and the left-hand side integral to the probability of the bis-^{117}Sn isotopomer. Toggle back to **WIN-DAISY**.

Symmetry Description								
Symmetry Group: D3H			No. of Descriptions:			12		
E	1	2	3	4	5	6	7	8
C3	2	3	1	5	6	4	7	8
C3^2	3	1	2	6	4	5	7	8
C2	5	4	6	2	1	3	8	7
C2'	4	6	5	1	3	2	8	7
C2"	6	5	4	3	2	1	8	7
SigH	4	5	6	1	2	3	8	7
S3	5	6	4	2	3	1	8	7
S3^2	6	4	5	3	1	2	8	7
SigV	2	1	3	5	4	6	7	8
SigV'	1	3	2	4	6	5	7	8
SigV"	3	2	1	6	5	4	7	8

Fig. 4.31: Symmetry Description for D_{3h} point group.

Check it in WIN-DAISY:

Copy the ^{119}Sn - ^{119}Sn **fragment** and in the **Main Parameters** modify the **Title** and the **Statistical weight**. Open the **Frequencies** dialog box, click the **PSE** button for Spin 7 and select **Nuclear Isotope 117** from the Element-Data table. Press the **Next** button in the bottom line to enter the **Scalar Coupling Constants** dialog box and to adapt the heteronuclear coupling constants to the new isotope (**Yes**). **Change** the calculated spectrum name, run a **Simulation** and display the spectra in WIN-NMR **Dual Display**.

Using the same procedure, create a fragment for the least probable bis-^{119}Sn isotopomer. Compare your final document with the model input document SNP.MGS which contains all possible 10 isotopomers.

For the present spin system a ^{119}Sn{^{1}H} spectrum is available, so that the spin system data can be checked against the experimental tin data.

Check it in WIN-NMR and WIN-DAISY:

Load the experimental ^{119}Sn spectrum ...\PROPELL\002001.1R into WIN-NMR. Enter the **Analysis | Interval** mode and click on the **Regions...** button to **Load...** the predefined region file. Export... the interval data to **WIN-DAISY**. In WIN-DAISY accept the message box relating the WIN-NMR data to an existing document with **Ok** and press the **Connect** button. Open the **Spectrum Parameters** dialog box and select **Nuclear Isotope 119** from the Element-Data table. Quit the Element-Data table with **Ok** to automatically change the **Spectrometer Frequency** from 250 to **400** MHz, the ^{119}Sn spectrum is measured at a different field strength. Open the **Frequencies** dialog box and remove the **Disable** flag in front of each ^{119}Sn spin. Open the **Control Parameters** and change the **Mode** to **Single Simulation**. In the **Spectrum Parameters** dialog box enter a **Lower frequency limit** of <5> ppm and an **Upper frequency limit** of <40> ppm. Run a spectrum **Simulation** and display the experimental and calculated spectrum in the WIN-NMR **Dual Display**. The linewidth in the experimental spectrum is broader than the calculated spectrum, so toggle back to **WIN-DAISY**, open the **Lineshape Parameters** dialog box and check the **Nuclei specific linewidth** option. Set the linewidth for all the ^{31}P nuclei to the value in the main A_6 spin system fragment and enter a value of about <8> Hz for all the tin nuclei (indices 7 and 8). Run another **Simulation** and compare the spectra using the WIN-NMR **Dual Display** adjusting the Y-axis if necessary using <Ctrl + Y>.

4.3 Spin Systems with Quadrupolar Nuclei

Nuclei with a nuclear spin quantum number I > ½ posses a quadrupolar moment. The quadrupolar moment provides a very effective relaxation process and the relaxation times of the quadrupolar nuclei are often very short (section 5.1.1.1). In many cases the short relaxation times gives rise to broad signals in the spectra of quadrupolar nuclei preventing the accurate measurement of chemical shifts and the resolution of scalar coupling splittings. However in a highly symmetrical environments the quadrupolar relaxation mechanism becomes less effective, relaxation times become longer and linewidths become smaller e.g. the ^{11}B spectrum of Na[BH$_4$] in NaOH/H$_2$O. In anistoropic solutions a large splitting due to the quadrupolar coupling constant can be detected but in isotropic solution quadrupolar coupling need not be considered.

Quadrupolar nuclei can also broaden the linewidths in the NMR spectra of neighboring atoms, e.g. the ^{1}H signal of NH (section 3.3.1.1), NH$_2$, OH groups. This line broadening should not be confused with chemical exchange processes which will be discussed in section 5.3.

Some quadrupolar nuclei are pure elements e.g. ^{93}Nb, while others can occur with spin-½ isotopes either with high natural abundance e.g. ^{14}N/^{15}N or very low natural abundance e.g. ^{2}H/^{1}H.

4.3.1 Basic Features of Quadrupolar Nuclei

The interaction between spin-½ nuclei has been discussed in detail in Section 2 and illustrated pictorially in Fig. 2.8. The interaction of a spin-½ nucleus with a spin-1 nucleus can be symbolized as follows: for the spin-1 nucleus (spin A) there is no difference in the resulting splitting pattern, it is still a doublet because it recognizes a single spin-½ nucleus (spin X).

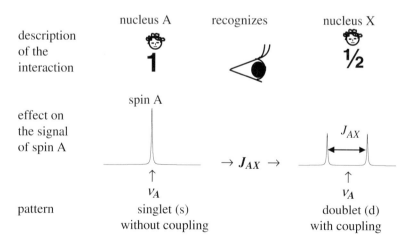

However, for the spin-½ nucleus the circumstances are different it recognizes a nucleus with spin quantum number I = 1. There are three possible spin states for a I = 1 nucleus, $m_z = -1, 0, +1$, and it splits the signal of the spin-½ nucleus into a 1:1:1 triplet. This type of spin system is called ^3AX.

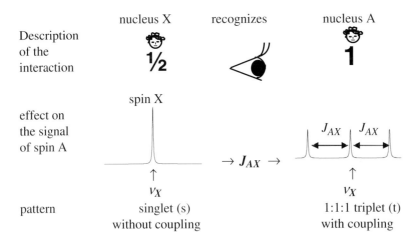

Check it in WIN-DAISY:

Load the input document AX_HETER.MGS. Open the **Frequencies** dialog box and change the isotope for the second spin, index 2, from **19 F** to **14 N**. Press the **PSE** button, select **N** in the **Element-Data** table and **14** from the **Nuclear Isotope** pull-down menu. Close the **Element-Data** table and **Frequencies** dialog box with **OK**. Open the **Spectrum Parameters** dialog box and press the **Auto Limits** button to update the spectrum limits. **Simulate** the spectrum and **Display** the **Calculated Spectrum** (red peak button) in WIN-NMR. The spectrum shows the expected 1:1:1 triplet pattern.

In the discussion on magnetic equivalence (Section 2.4.4) the X nucleus in a A_2X spin system (both A and X spin-½ nuclei) appears as a 1:2:1 triplet. Using the composite particle approach, the triplet in a A_2X spin systems arise from the superposition of the individual subspectra of the irreducible components $^3\{A_2\}X$ and $^1\{A_2\}X$ (refer to Fig. 2.56). The $^3\{A_2\}X$ subspectrum is *identical* with the ^3AX spin system and this fact is used in programs like WIN-DAISY for the calculation of ^3AX spin systems which essentially correspond to A_2X spin systems with the suppression of the $^1\{A_2\}X$ subspectrum.

The program can distinguished between the two different types of spin systems by the **Spin Value** of the individual spins I(A) and the number of **Spins in group**.

A_2X: $I(A) = ½$ Spins in group: $2 \rightarrow$ Calculation $^3\{A_2\}X + ^1\{A_2\}X$
^3AX: $I(A) = 1$ Spins in group: $1 \rightarrow$ Calculation $^3AX \equiv ^3\{A_2\}X$

Superscript numbers refer to the multiplicity g which is equal to $2 \cdot I + 1$. When g = 2, no superscript number is generally written and the absence of any superscript usually implies a multiplicity of 2. The superscript number belongs to the symbol written directly next to it on the right side which can be either a letter or a curved braces expression. In the first case the multiplicity relates to the spin quantum number of a nucleus and in the latter case it refers to the whole expression given in the curved braces. Expressions with curved braces refer to subspectra, not complete spin systems.

^{3}AX   means that spin A has a multiplicity 3 and therefore I(A) = 1.
^{3}{A$_2$}X   means the subspectrum in which group {A$_2$} exhibits a multiplicity of 3.
^{3}A$_2$X   means a spin system in which two magnetically equivalent nuclei A with I(A) = 1 are present.

In conclusion,

A single nucleus A, with a spin quantum number I(A), will split the signal of a coupled spin or group X into g = 2 · I(A) + 1 equidistant and equal intensity lines.

Check it in WIN-DAISY:

Load the input document AX_HETER.MGS and open the **Frequencies** dialog box. Change the **Frequency** of Spin 2, to <0> ppm. Press the **PSE** button, and select any isotope in the **Element-Data** table with a spin quantum number I = 1.5, e.g. ^{33}S, to form an A^{4}X spin system. Close the **Element-Data** table and **Frequencies** dialog box with **OK**. Open the **Spectrum Parameters** dialog box and press the **Auto Limits** button to update the spectrum limits. **Simulate** the spectrum and **Display** the **Calculated Spectrum** (red peak button) in WIN-NMR. Toggle back to **WIN-DAISY**.

Produce a series of spectra with increasing spin quantum number. Use the **PSE** button of the second spin in the **Frequencies** dialog box to access the **Element-Data** table. Some examples of isotopes with their spin quantum number are: ^{42}K (I = 2 → spin system A^{5}X), ^{17}O (I = 2.5 → spin system A^{6}X), ^{22}Na (I = 3 → spin system A^{7}X), ^{123}Sb (I = 3.5 → spin system A^{8}X) upto ^{93}Nb (I = 9 → spin system A^{19}X). After selecting a new isotope, use the **Auto Limits** button in the **Spectrum Parameters** dialog box to update the spectrum limits to include the additional multiplet lines in the spectral region **Simulate** the spectrum and **Display** the **Calculated Spectrum** (red peak button) in WIN-NMR.

The splitting caused by higher spin quantum numbers can also be reproduced using the subspectra of composite particles based on I = ½ spins.

Check it in WIN-DAISY:

Reload the AX_HETER.MGS input document containing a AX spin system. Open the **Frequencies** dialog box and for the second spin, increase the **Spins in group** to <3> to represent a AX$_3$ spin system. Open the **Main Parameters** dialog box and check the **Output Option Output of subspectra**. Update the spectrum limits. Run a **Simulation** and then **Generate** a

296 4 Spin Systems and the Periodic Table

sequence of **Subspectra**. **Display** the **Calculated Spectrum** (red peak button) in WIN-NMR.

In WIN-NMR choose the **Display | Multiple Display** option. Click the **Delete All** button and then select the two subspectra ...\999998.1R and ...\999997.1R. In the **Stack Plot** mode the 1:1:1:1 quartet of the first subspectrum is visible. Toggle back to **WIN-DAISY** and extend the spin system to AX_4. Update the spectrum limits. Run a **Simulation** and **Generate Subspectra**. Load the three subspectra into the **Multiple Display** mode of WIN-NMR. In the **Stack Plot** mode the 1:1:1:1:1 quintet of the first subspectrum can be clearly identified.

Continue to increase the number of **Spins in group** and generate a series of subspectra of composite particles of increasing size. Compare the composite particle spectra with the spectra generated in the previous "check it".

In Section 4.3.2.3 the concept of composite particles will be extended to magnetically equivalence quadrupolar nuclei.

4.3.2 Spin Systems Including ^{2}H

Deuterium (^{2}H or ^{2}D) is a very important isotope in NMR spectroscopy. If samples for ^{1}H NMR measurement were dissolved in protic solvents such as acetone, dimethylsulfoxide or chloroform the spectra would be dominated by the solvent signal causing dynamic range problems and obscuring signals from the sample. These problems can be overcome by using deuterated solvents which also have the additional advantage that they may be using as a lock signal for the NMR spectrometer. (The converse is also true, for ^{2}H NMR measurement samples must be dissolved in a proteo solvent!) Deuterated solvents are never 100% deuterated and humidity in the air can often cause proton - deuterium exchange. Thus a sample of acetone-d_6 will always contain a small amount of acetone-d_5 and this solvent signal is visible in the ^{1}H spectrum. Quite often NMR spectra are referenced with respect to the solvent signal, but it has to be remembered that chemical shifts depend on a number of factors including temperature, pH and concentration. Consequently, NMR spectra should always be referenced to a standard signal (refer to Table 2.1).

In the following section the different types of spin systems associated with chloroform-d_1, methylene chloride-d_2 and dimethylsulfoxide-d_6 in ^{1}H and ^{13}C spectra will be discussed. 100% deuterated solvents (pure ^{2}H spin systems) would give no signal in a ^{1}H NMR spectrum but practically all deuterated solvents contain a small amount of ^{1}H isotopomer. When the relative concentrations of the dissolved compound and deuterated solvent are considered, the amount of ^{1}H isotopomer even in a 99.6% deuterated solvent is high enough to give a significant signal in the ^{1}H spectrum.

4.3.2.1 Chloroform

The carbon isotopomers of $CHCl_3$ have already been discussed in detail in section 4.2.1.3; in the ^{1}H NMR spectrum the ^{12}C isotopomer occurs as a singlet and the ^{13}C isotopomer as a doublet (^{13}C satellites lines). Because chloroform contains only one hydrogen atom there will be no $^1H - {}^2H$ coupling interaction visible in the ^1H spectrum, even in partially deuterated chloroform.

The ^{13}C spectrum exhibits the typical pattern for the A part of an AX spin system with $I(X) = 1$. For quadrupolar nuclei $(I > \frac{1}{2})$ the multiplicity $(g = 2 \cdot I + 1)$ is indicated by a superscript numbers in front of the letter. In contrast to magnetic equivalence no curved brackets are used and it is possible to distinguish between a quadrupolar nucleus composite particle and a magnetic equivalence composite particle from the notation.

Check it in WIN-NMR and WIN-DAISY:

Load into WIN-NMR the $^{13}C\{^1H\}$ spectrum of $CDCl_3$, ...\CHCL3\002001.1R. There is an equidistant three-line pattern visible at 77 ppm. Measure a **Linewidth** and switch to the **Analysis I Multiplets** mode. Mark the middle line with the spectrum cursor and use the **Free Grid** with **3 distance lines** to define a multiplet. **Export...** the data to WIN-DAISY and *include* the non-assigned coupling constant.

chloroform-d₁

Open the **Main Parameters** dialog box and check the **Output Option Output of subspectra**. Run a spectrum **Simulation** and compare both spectra using the **Dual Display** mode of WIN-NMR.

The simulated pattern differs significantly from the experimental spectra. The simulated spectrum refers to the A part of an AX_2 spectrum as discussed in section 2.4.2.1, which leads to a 1:2:1 triplet. When data is exported to WIN-DAISY the multiplet table only contains information about the multiplicity of signals and there is no indication whether the coupling partner is a group of magnetically equivalent spins or a quadrupolar nucleus. When non-assigned coupling constants are included in an analysis, by default WIN-DAISY adds nuclei of the same experimental type (and spin quantum number) and the multiplicity is related to the number of magnetically equivalent nuclei according to the definition:

$$g = 2 \cdot n \cdot I + 1$$

If the experimental spectrum refers to a spin-½ nucleus, all non-assigned coupling partners will be defined with a spin-½ quantum number. So in the current example the multiplet analysis leads to a AX_2 spin system and not the required A^3X spin system. To illustrate the difference the subspectra need to be generated.

Check it in WIN-DAISY and WIN-NMR:

Toggle back to **WIN-DAISY** and check the entries in the **Frequencies** dialog box. **Generate Subspectra** and then **Display** the **Calculated Spectrum** (red peak button) in WIN-NMR. Load the two subspectra into the **Dual Display** mode of WIN-NMR (refer to Fig. 2.36 in section 2.4.2.1).

The $A^3\{X_2\}$ subspectrum leads to a 1:1:1 triplet pattern for the A nucleus and the $A^1\{X_2\}$ subspectrum a singlet at the chemical shift of A. The $A^3\{X_2\}$ subspectrum replicates exactly the pattern visible in the experimental spectrum. In fact the simulation of a A³X spin system with I(A) = ½ and I(X) = 1 is equivalent to the A³{X₂} subspectrum with the A¹{X₂} subspectrum suppressed. For clarity, the superscript multiplicity index for a doublet term is omitted for spin-½ nuclei.

It is now time to consider the difference between a AX₂ and a A³X spin system;

	CH₂	refers to	AX₂	with I(A) = ½ and I(X) = ½
and	CD	refers to	A³X	with I(A) = ½ and I(X) = 1

For both spin systems the main subspectra are identical: A³{X₂} = A³X, but the spin system AX₂ features a second subspectrum: A¹{X₂} which causes the intensity ratio of the pattern to be 1:2:1.

Check it in WIN-DAISY:

Open the **Frequencies** dialog box and click on the **PSE** button for the second spin. Select **H** in the **Element-Data** table and **2** from the **Nuclear Isotope** pull-down menu. Close the **Element-Data** table with **OK**. In addition to the atomic symbol, both the **ISO Value** and the **Spin Value** for the second spin have been modified. Change the **Spins in group** to <1>. Run a spectrum **Simulation** and display calculated and experimental spectra using the **Dual Display** mode of WIN-NMR. To align the Y-axis use <Ctrl + Y> and **ALL**. Using the **Move Trace** option, align the central peak of the CDCl₃ multiplet with the small additional peak at 77.18 ppm. This small signal corresponds to CHCl₃ and the shift difference of –0.20 ppm or –20.16 Hz is the isotope effect induced by the ²H.

There is a clear relationship between the coupling constants of isotopomeric spin systems which is of use in spectral analysis. Thus if the coupling constant ¹J$_{CD}$ of CDCl₃ is determined from the ¹³C spectrum, the corresponding ¹J$_{CH}$ coupling constant for the proton isotopomer CHCl₃ can be calculated using the equation:

$$J_{XH} = \frac{\gamma_H}{\gamma_D} \cdot J_{XD}$$

with $H = {}^1H$ and $D = {}^2H$.

Check it in WIN-DAISY:

Call the **Calculator** from the WIN-DAISY toolbar. Assuming that the magnetogyric ratio of ¹H and ²H is 26.7510 and 4.1064 respectively, calculate the ratio of the magnetogyric ratios. Open the **Scalar Coupling Constants** dialog box and copy the value of J₁₂ into the clipboard, <Ctrl + C>. Switch to the Calculator, enter <*> and paste in the value of J₁₂, <Ctrl + V>. Press the = button to obtain the result.

When the **Spin Value** of an isotope is changed in the **Frequencies** dialog box, the values of all the coupling constant associated with the isotope are automatically updated by WIN-DAISY. Open the **Frequencies** dialog box and make sure the units are **ppm**. Click on the **PSE** button for the second spin. and select **H** in the **Element-Data** table and **1** from the **Nuclear Isotope** pull-down menu. Close the **Element-Data** table with **OK**. Now use the **Next** button in the bottom line to open the **Scalar Coupling Constants** dialog box. Before the dialog box is displayed WIN-DAISY asks whether to modify the current coupling constant values according to the ratio of the magnetogyric ratios. Answer the question with **Yes**. Compare the value displayed in the dialog box with the calculated value.

Compare the result with the value of the coupling constant determined by the analysis of the ^{13}C satellite spectrum of CHCl$_3$ (section 4.2.1.3). Load the experimental spectrum ...\CHCL3\001001.1R into WIN-NMR and switch to the **Analysis | Multiplets** mode. Press the **Windaisy** button. Open the **Report...** box to display the spin system data.

Note that coupling constants will only be adapted when the spin value i.e. isotope of an element is altered and the Scalar Coupling constants dialog box is opened *immediately* afterwards using the **Next** button in the bottom line of the Frequencies dialog box.

4.3.2.2 Deuterium Substitution

The method just described is often used to determine "hidden" coupling constants that cannot be measured directly from the spectrum. For example, the ortho coupling constant in 1,2,3,4-tetrachlorobenzene could be determined by labeling the compound with one deuterium in the 5-position. In the ^{1}H spectrum of 1,2,3,4,-tetrachloro-5-deuterobenzene, the splitting pattern is caused by $^3J_{DH}$ from which the value of $^3J_{HH}$ can be easily calculated using the given equation.

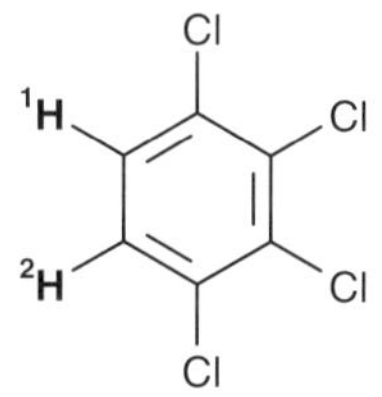

Fig. 4.32: Structure of *1,2,3,4-tetrachloro,5-deuterobenzene*.

Check it in the Calculator:

The ortho coupling $^3J_{HH}$ in 1,2,3,4-tetrachlorobenzene is 8.27 Hz. Calculate the value of $^3J_{DH}$ in *1,2,3,4-tetrachloro-5-deuterobenzene*.

Check it in WIN-DAISY:

Load the WIN-DAISY input document A.MGS. Open the **Main Parameters** dialog box and increase the **Number of spins or groups with magnetical equivalent nuclei** to <2>. Open the **Frequencies** dialog box and change the units to **ppm**. For both spins enter a **Frequency** of <7.26> ppm. Press the **PSE** button for one of them and select the deuterium isotope ^{2}H. Switch to the **Scalar Couplings** dialog box and enter the calculated value of $^3J_{DH}$. Call the **Spectrum Parameters** dialog box and press the **Auto Limits** to update the spectrum limits. **Simulate** the spectrum and **Display** the **Calculated Spectrum** (red peak button) in WIN-NMR.

The method of isotopic substitution is also useful to change the spin system type in order to extract formerly non-determineable coupling constants.

4.3.2.3 Methylenechloride

The deuterated solvent methylenechloride-d_2 (dichloromethane-d_2) always contains a small amount of methylenechloride-d_1 as shown in Fig. 4.33. The probability that there are two protons in the same molecule is very low compared with the probability that one deuterium is substituted by one proton.

$$\begin{array}{ccc}
\text{Cl} & \text{Cl} & \text{Cl} \\
| & | & | \\
{}^1\text{H}-{}^{12}\text{C}-{}^2\text{H} & {}^1\text{H}-{}^{13}\text{C}-{}^2\text{H} & {}^1\text{H}-{}^{13}\text{C}-{}^2\text{H} \\
| & | & | \\
\text{Cl} \quad a) & \text{Cl} \quad b) & \text{Cl} \quad c)
\end{array}$$

Fig. 4.33: Spin active nuclei in methylenechloride-d_1:
  a) ^{12}C isotopomer, spin system: A^3X ($A={}^1H$, $X={}^2H$),
  b) ^{13}C isotopomer, spin system A^3MX ($A={}^1H$, $M={}^2H$, $X={}^{13}C$),
  c) ^{13}C isotopomer with ^1H broadband decoupling: ^3AX ($A={}^2H$, $X={}^{13}C$).

In the ^1H NMR spectrum of methylene chloride-d_2 the main signal of the methylene chloride-d_1 isotopomer (Fig. 4.33a) refers to a A^3X spin system similar to the spin system just discussed for chloroform-d_1, but here spin A is ^1H, not ^{13}C.

Check it in WIN-DAISY:

Load the ^1H spectrum of CD_2Cl_2, ...\CH2CL2\001001.1R into WIN-NMR. Note the ^{13}C satellites and at 5.37 ppm a very small amount of the CH_2Cl_2 isotopomer. Measure a **Linewidth** and switch to the **Analysis | Multiplets** mode. Mark the middle line with the spectrum cursor and use the **Free Grid** with **3 distance lines** to define a multiplet. **Export...** the data to **WIN-DAISY** and *include* the non-assigned coupling constant. Open the **Frequencies** dialog box and click on the **PSE** button for the second spin. Select **H** in the **Element-Data** table and **2** from the **Nuclear Isotope** pull-down menu. Close the **Element-Data** table with **OK**. Change the **Spins in group** to <1>. Run a spectrum **Simulation** and display calculated and experimental spectra using the **Dual Display** mode of WIN-NMR. To align the Y-axis use <Ctrl + Y> and **ALL**.

The 1:1:1 splitting pattern typical of an A^3X spin system present in the experimental ^1H spectrum is reproduced exactly in the calculated spectrum. In comparison with the ^{13}C spectrum of $CDCl_3$ the coupling constant $^2J_{DH}$ is much smaller than $^1J_{CH}$. In the $^{13}C\{^1H\}$ spectrum of methylenechloride-d_2 the spin system CD_2Cl_2 is visible. The probability that a ^{13}C and a ^1H nucleus reside in the same molecule is very low.

4.3 Spin Systems with Quadrupolar Nuclei

Check it in WIN-DAISY:

Load the $^{13}C\{^1H\}$ spectrum, ...\CH2CL2\002001.1R into WIN-NMR. Measure a **Linewidth** and switch to the **Analysis | Multiplets** mode. Mark the middle line with the spectrum cursor and use the **Free Grid** with **5 distance lines** to define a multiplet. **Export...** the data to **WIN-DAISY** and *include* the non-assigned coupling constant. Open the **Main Parameters** dialog box and check the **Output** option **Output of subspectra**. Open the **Frequencies** dialog box and using the **PSE** button, change the second spin from **1 H** to **2 H**. Increase the **Spins in group** to <2>.

$^2H - {}^{13}C - {}^2H$ with Cl above and Cl below the carbon.

Methylenechloride-d_2

Run a spectrum **Simulation** and display calculated and experimental spectra using the **Dual Display** mode of WIN-NMR. If necessary, use <Ctrl + Y> and **ALL** to align the Y-axis. **Generate Subspectra**. Display the **Calculated Spectrum** (red peak button) in WIN-NMR. Switch to the **Display | Multiple Display** mode and load the *three* subspectra, select the **Stack Plot** option to display the individual subspectra and their relative intensities.

It is obvious that the multiplet pattern in the $^{13}C\{^1H\}$ spectrum of methylene chloride-d_2 is composed of three, equally weighted subspectra which contain a set of equidistant lines each with the same intensity:

$A^5\{^3X_2\}$	$1:1:1:1:1$
$A^3\{^3X_2\}$	$1:1:1$
$A^1\{^3X_2\}$	1
$A^3X_2 = $ sum:	$1:2:3:2:1$

The first subspectrum refers to the term $A^5\{^3X_2\}$ which describes an A^3X_2 spin system of nuclei with $I(A) = ½$ and $I(X) = 1$. The $^3X_2$ group can exist in three particle states $F_{max} = 2$ (quintet term), $F_{act} = 1$ (triplet term) and $F_{act} = 0$ (singlet term). The last state refers to a spin inactive nucleus.

Check it in WIN-DAISY:

Convert the input document for a A^3X_2 spin system into a AX_4 type system by changing the isotope for the second spin in the **Frequencies** dialog box from **2 H** to **1 H** and increasing the number of **Spins in group** to <4>. **Change** the calculated spectrum name to keep the simulations of the A^3X_2 spin system. Run a spectrum **Simulation** and **Generate Subspectra**. Display the **Calculated Spectrum** (red peak button) in WIN-NMR. Switch to the **Display | Multiple Display** mode, load the *three* subspectra and select the **Stack Plot** option.

There is no difference in the appearance of the individual subspectra, they are identical for both spin systems. The difference is found in the weighting of each subspectrum, which is clearly visible in the Stack Plot:

$A^5\{X_4\}$	(1:1:1:1:1)	• 1
$A^3\{X_4\}$	(1:1:1)	• 3
$A^1\{X_4\}$	(1)	• 2

AX_4 = sum: 1 : 4 : 6 : 4 : 1

Check it in WIN-NMR:

Display the simulated AX_4 spectrum in WIN-NMR. Switch to the **Display | Dual Display** mode and load the simulated A^3X_2 spectrum as the second trace. Compare the relative intensities of the multiplet lines.

4.3.2.4 Dimethylsulfoxide

Another common solvent for NMR measurements is dimethylsulfoxide-d_6. In the ^{1}H NMR spectrum the residual dimethylsulfoxide-d_5 leads to a significant line pattern.

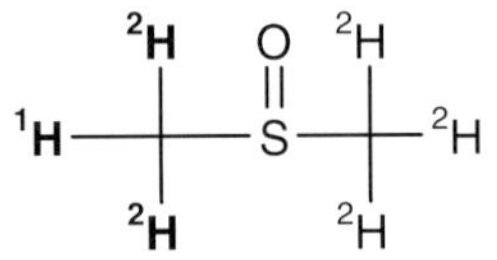

Fig. 4.34: Structure of dimethyl sulfoxide-d_5, spin system A^3X_2.

Inspection of the structure (Fig. 4.34) and definition of the spin system leads to the same spin system type as discussed in the previous section on the ^{13}C{^{1}H} spectrum of methylenechloride-d_2. However, here the splitting refers to $^2J_{DH}$ which is much smaller than $^1J_{CH}$ in methylenechloride-d_2.

Check it in WIN-DAISY:

Open... the experimental ^{1}H spectrum ...\DMSO\001001.1R with WIN-NMR. Focus on the signal at 2.5 ppm. (The small signal at 2.52 ppm corresponds to the dimethylsulfoxide-d_4 isotopomer.) Proceed in a similar manner to the previous "Check it", measure a suitable **Linewidth** and in the **Analysis | Multiplets** mode define a **5 distance lines** multiplet. Switch to the **Analysis | Interval** mode and define an iteration interval. **Export...** the interval and the multiplet data to WIN-DAISY and *include* the non-assigned coupling constant. Run a spectrum **Simulation** and display the spectra in the **Dual Display** mode of WIN-NMR.

As was to be expected the multiplicity of the experimental data is reproduced but not the intensity distribution. By default the data imported from the multiplet analysis

4.3 Spin Systems with Quadrupolar Nuclei

interprets the multiplicity based a nuclei I(X) = ½, so that the simulation input generates a AX_4 spin system. To reproduce the experimental spectrum the data has to be converted into the required A^3X_2 spin system.

Check it in WIN-DAISY:

Open the **Frequencies** dialog box. Change the isotope for the second spin in the **Frequencies** dialog box from **1 H** to **2 H** and decrease the number of **Spins in group** to <2>. Run a spectrum **Simulation** and display the spectra in the WIN-NMR **Dual Display**. Because of signal overlap it is not possible to determine the linewidth exactly. Run an **Iteration** and compare the experimental and iterated spectra using the **Dual Display** mode of WIN-NMR.

The value of the geminal coupling constant $^2J_{HH}$ in dimethyl sulfoxide can be determined based on the value of $^2J_{DH}$.

Check it in the WINDOWS Calculator:

Call the WINDOWS **Calculator** from the WIN-DAISY toolbar. Open the **Scalar Couplings** window and copy the value of J_{12} ($^2J_{DH}$) into the clipboard, <Ctrl + C>. Open the WINDOWS Calculator, paste in the value, <Ctrl + V> and multiply <*> by <6.5144>.

The ^{13}C spectrum of dimethylsulfoxide-d_6 exhibits the spin system shown (Fig. 4.35).

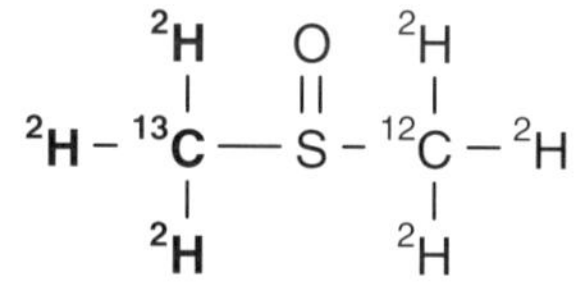

$$^2H-^{13}C(-^2H)(-^2H)-S(=O)-^{12}C(-^2H)(-^2H)-^2H$$

Fig. 4.35: Structure of the mono-^{13}C isotopomer of dimethylsulfoxide-d_6,
Spin System: A^3X_3. I(A) = ½, I(X) = 1.

Check it in WIN-NMR:

Open... the experimental spectrum ...\DMSO\002001.1R with WIN-NMR. Measure a **Linewidth** and switch to the **Analysis | Multiplets** mode. Define a **5 distance lines** multiplet. **Export...** the data to WIN-DAISY and *include* the non-assigned coupling constant.

Check it in WIN-DAISY:

Open the **Frequencies** dialog box. Change the isotope for the second spin in the **Frequencies** dialog box from **1 H** to **2 H** and decrease the number of **Spins per group** to <3>. Run a spectrum **Simulation** and display the spectra in the WIN-NMR **Dual Display**. Call the **Calculator** using the corresponding toolbar button.

Calculate the value of the $^1J_{CH}$, based on the $^1J_{CD}$ (J_{12}) constant determined from the ^{13}C spectrum.

4.3.2.5 Ammonia Hydrogen Isotopomers

In this section the ^{15}N spectrum of ammonia taking into consideration all possible hydrogen isotopomers will be calculated. Consult the literature for the experimental data [4.49], [4.50] and spectra [4.50]. The possible hydrogen isotopomers of NH_3 are given in Fig. 4.36.

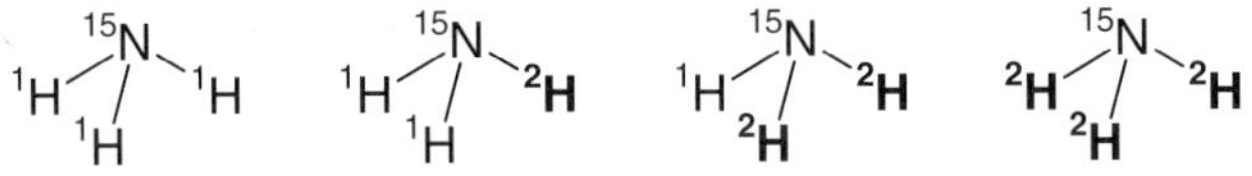

Fig. 4.36: Hydrogen isotopomers of $^{15}NH_3$, ^2H isotopes printed bold.

Check it in WIN-DAISY:

Load the WIN-DAISY input document NH3.MGS. This file contains the spin system information for the isotopomer $^{15}N^1H_3$. **Simulate** the ^{15}N spectrum and **Display** the **Calculated Spectrum** (red peak button) in WIN-NMR.

As expected a 1:3:3:1 quartet is obtained. Now the input will be modified to include the other isotopomers.

Check it in WIN-DAISY:

Copy the original fragment using either the **Edit | Copy fragment** command or the **Copy fragment** button. Open the **Main Parameters** and enter the **Title** string <ND3>. Change the isotope for the second spin in the **Frequencies** dialog box from **1 H** to **2 H**. Use the **Next** button in the bottom line to switch into the **Scalar Coupling Constants** dialog box and to modify the current coupling constant values according to the ratio of the magnetogyric ratios. Run a **Simulation** and **Display** the **Calculated Spectrum** (red peak button) in WIN-NMR.

There will be an isotope shift on the chemical shift of ^{15}N due to the substitution of three ^1H nuclei by three ^2H nuclei as illustrated in the ^1H spectrum of dimethyl sulfoxide-d_6. Isotope shifts are usually listed in ppb (parts per billion - 10^{-9}) and Table 4.10 gives the isotope shifts for the ^{15}N nucleus in the series of NH_3 isotopomers [2]. Any simulation must take this isotope shift into account.

Table 4.10: Isotope shifts of ^{15}N in NH_2D, NHD_2 and ND_3.

isotopomer	isotope shift in ppb
$^{15}ND_3$	−1868.7
$^{15}NHD_2$	−1249.1
$^{15}NH_2D$	− 626.4

Check it in WIN-DAISY:

Open the **Frequencies** dialog box for the ND3 fragment and enter the isotope shift for ^{15}N (first spin) in **ppm**. Run a **Simulation** and **Display** the **Calculated Spectrum** (red peak button) in WIN-NMR.

The isotope shift produces a significant change in the spectrum. In the next step the mixed ^{1}H/^{2}H isotopomer data will be defined.

Check it in WIN-DAISY:

Select the **NH3** fragment in the WIN-DAISY document main window and **Copy** the **fragment** again. Open the **Main Parameters** dialog box and enter the **Title** string <NH2D>. Change the **Number of spins or groups with magnetical equivalent nuclei** to <3>, because there are now three different isotopes. Open the **Scalar Coupling Constants** window and copy the coupling constant J_{12} into J_{13}. Enter a value of <–1.45> Hz for J_{23}, even though this value has no effect on the ^{15}N spectrum. Use the **Previous** button to open the **Frequencies** dialog box. For the third spin change the isotope from **1 H** to **2 H** and the **Frequency** from 0 ppm to <100> ppm. For the ^{15}N spin enter the isotope shift from Table 4.10, <-0.6264> ppm. Decrease the **Spins in group** for the ^{1}H isotope to <2>. Use the **Next** button to the open **Scalar Coupling Constants** dialog box and adapt the values according to the ratio of the magnetogyric ratios (**Yes**).

Select the third (new) fragment **NH2D** in the WIN-DAISY document main window and **Copy** the **fragment**. Open the **Main Parameters** and enter a **Title** line <NHD2>. Open the **Frequencies** dialog box and for the ^{15}N spin enter the isotope shift from Table 4.10, <–1.2491> ppm. Interchange the **Spins in group** entries for ^{1}H and ^{2}H. Run a **Simulation** and **Display** the **Calculated Spectrum** (red peak button) in WIN-NMR.

An experimental spectrum is given in reference [4.50] and [4.3]. The experimental data are not according to natural abundant isotopomer mixture, but obtained by a mixture of ^{15}NH$_3$ and ND$_3$. To get the same ratio of isotopomers as present in the experimental spectrum. use the Statistical weight parameter. For comparison you may use the WIN-DAISY input document NH3_ND3.MGS.*

Check it in WIN-DAISY:

Open the **Main Parameters** and enter a **Statistical weight** of <0.1> for the **ND3** fragment and <0.7> for the **NHD2** fragment. Run a **Simulation** and **Display** the **Calculated Spectrum** (red peak button) in WIN-NMR.

4.3.3 Spin Systems Containing ^{14}N

^{15}N is a rare spin ½ isotope and has been discussed in section 4.2.2. However, before polarization transfer experiments became popular, ^{14}N was the preferred nitrogen isotope for NMR measurements, its high natural abundance (99.63%) compensating to some extent for the quadrupole moment [4.36].

---

* The total intensity of a spin system is dependent on the spin quantum number due to the total intensity theorem. In cases where spin systems involving different spin quantum numbers are combined in one document the statistical weights of the individual spin systems need to be modified accordingly, refer to appendix, section 6.2.

306   4 Spin Systems and the Periodic Table

The only example to be discussed here is cis and trans difluoro diazine, the $^{15}N_2$ isotopomeric ^{19}F spectra have already been discussed in section 4.2.2 [4.40].

Check it in WIN-DAISY:

Load the input document N2F2_CIS.MGS and call the **Calculator** form the toolbar. The spin system data refers to the ^{15}N isotopomer of the cis isomer.

Open the **Frequencies** dialog box. Press the >>> option at the bottom and note the **Magnetogyric Ratio** of ^{15}N. Press the **PSE** button for the ^{15}N isotope. In the **Element-Data** table select ^{14}N and note the magnetogyric ratio. Use the **Cancel** button to leave the **Element-Data** table *without* changing the isotope. Open the **Scalar Coupling Constants** dialog box and use the **Calculator** to determine the ^{14}N isotopomeric coupling constant values. Take into account the sign of the new values and that $^1J_{NN}$ is a *homonuclear* coupling. Note your calculated values.

Open the **Frequencies** dialog box. Change the isotope for the first spin in the **Frequencies** dialog box from **15 N** to **14 N**. Use the **Next** button in the bottom line to switch into the **Scalar Coupling Constants** dialog box and to modify the current coupling constant values according to the ratio of the $^{14}N/^{15}N$ magnetogyric ratios (**Yes**). Compare the automatically and manually calculated values. Open the **Lineshape Parameters** dialog and change the **Global Linewidth** parameter to <60> Hz to represent the broadened quadrupolar lineshape. Open the **Spectrum Parameters** dialog box; change the **Isotope** from ^{19}F to ^{14}N and click the **Auto Limits** button to update the spectrum limits. Run a **Simulation** and **Display** the **Calculated Spectrum** (red peak button) in WIN-NMR.

Using the input file N2F2_TR.MGS, adapt the spin system data for the trans isomer of N_2F_2 from ^{15}N to ^{14}N. Again, compare the values of the automatically and manually calculated coupling constants. **Simulate** the spectrum and **Display** the **Calculated Spectrum** (red peak button) in WIN-NMR.

Experimental and calculated ^{19}F spectra are in the literature [4.51] and [4.40], corresponding ^{14}N spectra in [4.51].

4.3.4 Spin Systems Containing ^{10}B / ^{11}B

Boron has two naturally occurring isotopes, ^{10}B and ^{11}B. Both have a quadrupole moment, ^{10}B has a spin value of $I = 3$ and a natural abundance 19.6 % while ^{11}B has a spin value of $I = ^3/_2$ and a natural abundance 80.4 %. There is little interest in ^{10}B NMR due to the low natural abundance and low sensitivity. ^{10}B has a larger quadropolar moment than ^{11}B but when the higher spin value of ^{10}B is taken into account, the linewidth in ^{10}B and ^{11}B NMR spectra are comparable [4.52], [4.53].

The ^1H spectra of boronhydrides contains a mixture of ^{10}B and ^{11}B isotopomers, in the ratio of about 1 to 4, with very similar chemical shifts. Splitting patterns will depend on whether the ^1H is coupled to a ^{10}B (septet) or ^{11}B (quartet). The value of $^nJ(^{10}B, ^1H)$ and $^nJ(^{11}B, ^1H)$ are related by the ratio of the magnetogyric ratios.

$$J(^1H,^{10}B) = \frac{\gamma(^{10}B)}{\gamma(^{11}B)} \cdot J(^1H,^{11}B) = 0.335 \cdot J(^1H,^{11}B)$$

The ^{10}B multiplet (width: $6 \cdot {}^nJ(^{10}B, {}^1H)$ or $6 \cdot 0.335 \cdot {}^nJ(^{11}B, {}^1H)$) always falls inside the ^{11}B multiplet (width: $3 \cdot {}^nJ(^{11}B, {}^1H)$) and the spectrum of the ^{10}B isotopomer often appears as a hump under the ^{11}B isotopomer spectrum.

Check it in WIN-DAISY:

Open... the WIN-DAISY input document BH.MGS. The data refer to a system containing one ^1H and one ^{11}B isotope. The spectrum limits in the **Spectrum Parameters** refer to the ^1H spectrum. Run a **Simulation** and **Display** the **Calculated Spectrum** (red peak button) in WIN-NMR.

As expected from the multiplicity rule ($2nI + 1 = 4$ with $n = 1$ and $I = 3/2$) the simulation shows a 1:1:1:1 quartet. Normally the ^1H spectrum would also display the effect of the ^{10}B isotopomer.

Check it in WIN-DAISY:

Copy the original fragment using the **Copy fragment** command or button. Open the **Main Parameters**, modify the **Title** string <10B-1H> and enter a **Statistical weight** of ca. <11.4>* relating to the natural abundance and intensity norming of ^{10}B systems. Use the **Next** button in the header line to display the **Main Parameters** of the ^{11}B fragment. Change the **Statistical weight** to <80> to reflect the natural abundance of ^{11}B and its spin system norming. Close the dialog with **OK**. Select the second fragment in the fragment list in the main window. Open the **Frequencies** dialog box, make sure that the display units are in **ppm**, and change the boron isotope dialog box from **11 B** to **10 B**. Use the **Next** button in the bottom line to switch to the **Scalar Coupling Constants** dialog box and to modify the coupling constant values according to the ratio of the magnetogyric ratios. Run a **Simulation** and **Display** the **Calculated Spectrum** (red peak button) in WIN-NMR.

The 1:1:1:1:1:1:1 septet of the ^{10}B-^1H coupling pattern falls within the boundaries of the four-line ^{11}B-^1H signal. Both the ^{10}B and the ^{11}B NMR spectra show a doublet with a splitting of ${}^nJ(^{10}B, {}^1H)$ or ${}^nJ(^{11}B, {}^1H)$ respectively due to the coupling with a single ^1H.

Check it in WIN-DAISY:

Open the **Spectrum Parameters** dialog box and change the **Isotope** from ^1H to ^{11}B. Press the **Auto Limits** to update the spectrum limits and close the dialog with **OK**. Simulated the ^{11}B spectrum and **Display** the **Calculated Spectrum** (red peak button) in WIN-NMR. Open the **Spectrum Parameters**

---

* The correct isotopomeric ratio has to take into account not only the natural abundance but also the spin system normalization with respect to the spin quantum number, so that the ratio is not 80 / 20, but 80 / 11.4. See also footnote in section 4.3.2.5.

again, change the **Isotope** from ^{11}B to ^{10}B and update the spectrum limits. Run another **Simulation** and **Display** the **Calculated Spectrum** (red peak button) in WIN-NMR. The lineshapes are similar except for the magnitude of the splitting cause by the different boron-proton coupling constants.

Check it in WIN-DAISY:

Inspect the structure of $NaBH_4$ and decide what spin system is defined for the boron (either ^{10}B or ^{11}B) and the ^1H. Take into account the tetrahedral arrangement of ^1H around the boron. Modify the present input document BH.MGS to define the sodium borohydride spin system. Determine the splitting multiplicity using the 2nI+1 rule. Calculate the boron spectrum and **Display** the **Calculated Spectrum** (red peak button) in WIN-NMR.

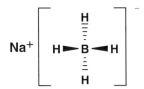

Sodium borohydride

The boron spectrum shows a 1:4:6:4:1 quintet as expected from the multiplicity rule (2nI + 1 = 5 with n = 4 and I = ½).

The multiplet pattern become much more complicated if more that one boron and one ^1H are involved. In most cases only directly bonded interactions need to be considered as neither ^{11}B-^{11}B, ^{11}B-^{10}B or ^1H-^xB-^1H coupling pathways lead to any visible splittings. In the following example the spectra of ^{11}B labeled diborane are considered and in this particular example long range couplings are visible.

There are two types of protons in the structure shown in Fig. 4.37. The terminal ^1H and the bridging ^1H involved in a two-electron-three-centre bond. The chemical shifts of these protons are significantly different. The spin system can be generated using the scheme in Fig. 4.37.

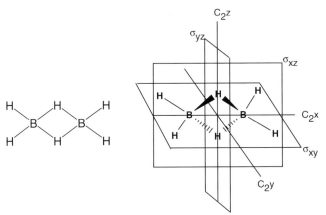

Fig. 4.37: Structure and Symmetry elements for diborane.

Taking into account the symmetry operations for the point group D_{2h} the spin system is $[[A]_2B^4X]_2\ D_{2h}$ where A and B are ^1H nuclei and X = ^{11}B with I = 3/2.

Check it in WIN-DAISY:

Load the input document B2H6_D2H.MGS. Open the **Symmetry Group** window and the **Symmetry Description** dialog box where the symmetry operations for D_{2H} are defined. In the **Spectrum Parameters** the spectral limits for the ^1H spectrum are given. **Simulate** the spectrum and **Display** the **Calculated Spectrum** (red peak button) in WIN-NMR. Open the **Simulation Protocol** and examine the sizes of the submatrices. Because of symmetry the problem is divided into eight independent quantum mechanical problems according to the eight irreducible components of D_{2h}. Note the calculation time given in the **Simulation Protocol** file is also displayed in the document main window. Because the **Output options Frequencies before degeneracy** has been selected it might not be possible to open the simulation protocol file using the default WIN-DAISY editor.

The corresponding ^{11}B spectrum can be quickly calculated. Open the **Main Parameters** dialog box, the **Reduce number of lines by** parameters relate to the extended quantum mechanical limits and are such that they cover all the transitions in both the ^{11}B and the ^1H spectra. The **Output options Frequencies before degeneracy** mean that all the transition frequencies relating to both the ^{11}B and ^1H spectra are stored in the file B2H6_D2H.FRE. Open the **Spectrum Parameters** and change the **Isotope** from ^1H to ^{11}B. Press the **Auto limits** button and quit the dialog with **OK**. Using either the **Run | Generate Lineshape** command or the **Generate Lineshape** button generate a new lineshape for the ^{11}B spectral limits based on the calculated transition list. In large calculations this method of using the transition table is considerably faster than repeating the quantum mechanical calculation, however the **Output options Frequencies before degeneracy** must have been selected.

Open the input document B2H6_C1.MGS. This input document contains the data for diborane *without* any symmetry considerations. Run a spectrum **Simulation**. After the calculation is finished compare the calculation time with that of the previous calculation using D_{2h} symmetry. Display the **Simulation Protocol** list and examine the sizes of submatrices.

Comparing the calculation times with and without symmetry considerations, the use of symmetry reduces the calculation time by about a factor of seven times. The main reason for this is the size of submatrices; without using symmetry the dimension of the largest submatrix is 196 but with symmetry this falls to 48 with a subsequent reduction in calculation time. In the next "Check it", the symmetry data for the D_{2h} spin system will be generated from the basic C1 input document.

Check it in WIN-DAISY:

Load the input document B2H6_C1.MGS. Open the **Symmetry Group** window and change the symmetry from **C1** to **D2H**, press the **OK** button and automatically the **Symmetry Description** dialog box is displayed. Using the numbering and axis system shown in the figure opposite, determine the effect of the various symmetry operations on spins 1 to 8 (refer to Fig. 4.38).

The original labels are displayed for the symmetry element **E**. Starting with row one, in the first edit column below index 1 enter the destination spin label into which spin 1 is transferred after execution of the **C2(z)** symmetry operation. Repeat for the other 7 labels. After completely all the symmetry operations compare your symmetry description with that in the document B2H6_D2H.MGS.

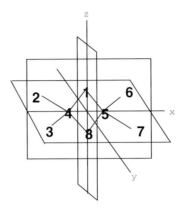

Fig. 4.38: Labeling of B2H6 for D2h symmetry description.

4.3.5 Spin Systems Containing ^{17}O

Oxygen has two isotopes, the most abundant (99.96%) is ^{16}O which does not have a nuclear magnetic moment and therefore is not NMR active. The other natural isotope ^{17}O, has an abundance of 0.04% and a spin quantum number of $I = \frac{5}{2}$. Despite its low natural abundance ^{17}O NMR is quite significance due to the biological important of oxygen. Linewidths in ^{17}O NMR spectra are often quite large and normally only the chemical shifts are reported, however coupled spin systems have been observed particularly for small molecules [4.3].

In the following section a mixture of water isotopomers will be discussed. In ^{17}O labeled water the following isotopomers are possible, (Fig. 4.39).

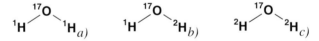

Fig. 4.39: Isotopomers of ^{17}O labeled water.
a) A_2^6X spin system, $A={}^1H$, $X={}^{17}O$.
b) A^3M^6X spin system, $A={}^1H$, $M={}^2H$, $X={}^{17}O$.
c) $3M_2^6X$ spin system, $A={}^2H$, $X={}^{17}O$.

Check it in WIN-DAISY:

Load the WIN-DAISY input document WATER.MGS. Open the **Frequencies** and **Scalar Coupling Constants** dialog boxes and inspect the values. The parameters correspond to the isotopomer A_2^6X (refer to Fig. 4.39a). The

4.3 Spin Systems with Quadrupolar Nuclei 311

Spectrum Parameters relate to the ^{1}H spectrum. Run a **Simulation** and **Display** the **Calculated Spectrum** (red peak button) in WIN-NMR.

The simulated ^{1}H lineshape relates to the A-part of a A_2^6X spin system and as expected is a 1:1:1:1:1:1 sextet, because the ^{17}O nucleus has $I = 5/2$ and a multiplicity of 6 (2·I+1). The corresponding X-part of the spin system is easily calculated. The X-nucleus couples to two magnetically equivalent A (spin-½) nuclei resulting in a standard 1:2:1 triplet.

Check it in WIN-DAISY:

Toggle back to **WIN-DAISY**. Open the **Spectrum Parameters** and change the **Isotope** from ^{1}H to ^{17}O. Press the **Auto limits** button to determine appropriate spectral limits for the ^{17}O spectrum and **Generate Lineshape**.

To calculate the all-deuterium isotopomer the isotope ^{1}H needs to be exchanged for ^{2}H. First of all think about the pattern expected for the ^{17}O spectrum. Setting up the isotopomer spin systems is very similar to the ammonia problem discussed in section 4.3.2.5. Table 4.11 gives the isotope shifts for the ^{17}O nucleus in the $H^{17}OD$ and $D_2^{17}O$ isotopomers [4.50]. Any simulation must take this isotope shift into account.

Table 4.11: Isotope shifts for ^{17}O in $H^{17}OD$ and $D_2^{17}O$ [4.50].

Isotopomer	Isotope shift in ppb
$H^{17}OD$	−1550
$D_2^{17}O$	−3090

Check it in WIN-DAISY:

The experimental data relates to the various isotopomers of water immersed in cyclohexane-d$_{12}$. Copy the original **H2O** fragment using the **Copy fragment** command or button. The **Main Parameters** open automatically, modify the **Title** string <D2O> and enter a **Statistical weight** of <0.2>. Change the isotope for the second spin in the **Frequencies** dialog box from 1 H to 2 H. For the ^{17}O spin insert the isotope shift from Table 4.11, <−3.09> ppm. Use the **Next** button in the bottom line to switch into the **Scalar Coupling Constants** dialog box and to modify the current coupling constant values according to the ratio of the magnetogyric ratios. Copy the original **H2O** fragment again. In the **Main Parameters** dialog box, modify the **Title** string <DOH> and increase the **Number of spins or groups with magnetical equivalent nuclei** to <3>. Open the window for **Scalar Coupling Constants** and copy the coupling constant J$_{12}$ into J$_{13}$. Use the **Previous** button to open the **Frequencies** dialog box. Decrease the **Spins in group** for the ^{1}H isotope to <1>. For the second or third spin change the isotope from 1 H to 2 H and the **Frequency** from 0 ppm to <3.3> ppm. For the ^{17}O spin give the isotope shift <−1.55> ppm. Use the **Next** button to the open **Scalar Coupling Constants** dialog box and adapt the values according to the ratio of the magnetogyric ratios. Open the **Spectrum Parameters** and determine **Auto Limits**. Enter a **Lower frequency limit** of <−5> ppm and **Upper** frequency limit of <2.5> ppm. Run a **Simulation** and **Display** the **Calculated Spectrum** (red peak button) in WIN-NMR. There is good agreement between theory and the literature [4.50], compare with the input document H2O_D2O.MGS.

4.3.6 Spin Systems Containing ^{9}Be

Beryllium is an example of a pure quadrupolar element. ^{9}Be has a spin quantum number $I = 3/2$. In highly symmetrical i.e. tetrahedral environments, the effect of the quadrupolar moment is minimized and the signals are often sharp enough to resolve ^{9}Be coupling [4.1]. With one exception, in compounds where ^{9}Be is coupled to ^{11}B no multiplet splitting is resolved. The only known example where the ^{9}Be - ^{11}B coupling is resolved is the compound (η^5-C$_5$H$_5$)BeBH$_4$ [4.57].

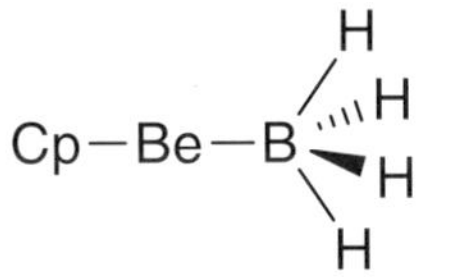

$$Cp-Be-B\begin{matrix}H\\H\\H\\H\end{matrix}$$

Fig. 4.40: Structure of (η^5-C$_5$H$_5$)BeBH$_4$

Check it in WIN-DAISY:

Open... the input document BEBH4.MGS in WIN-DAISY. This file refers to the ^{11}B isotopomer of (η^5-C$_5$H$_5$)BeBH$_4$. Open the **Frequencies** and **Scalar Coupling Constants** dialog boxes and inspect the spin system data. (The chemical shifts are just chosen to be apart from each other.)

$Cp - {}^9Be - {}^{11}B^1H_4$

^{11}B isotopomer:

spin system ^{4}AM$_4{}^4$X
(A=^{11}B, X=^{9}Be, M=^{1}H)

In the **Spectrum Parameters** dialog box the spectral limits are defined for the ^{11}B lineshape. Run a **Simulation** and **Display** the **Calculated Spectrum** (red peak button) in WIN-NMR. To visualize the splitting pattern and display the coupling tree above the calculated lineshape **Export WIN-NMR calc.** (red FID button). The experimental spectrum is shown in [4.57].

The ^{11}B spectrum displays a 1:4:6:4:1 quintet with a large coupling, ^{1}J(^{11}B, ^{1}H), due to the ^{11}B coupling to four magnetically equivalent ^{1}H. Each line of the multiplet is then split further into a 1:1:1:1 quartet by the ^{11}B-^{9}Be coupling. The pattern is well defined because ^{1}J(^{11}B, ^{1}H) is considerably larger than ^{1}J(^{11}B, ^{9}Be).

Check it in WIN-DAISY:

Open the **Spectrum Parameters** and change the **Isotope** from ^{11}B to ^{9}Be. Press the **Auto limits** button to determine the appropriate spectral limits for the ^{9}Be spectrum. Run a spectrum **Simulation**. Display the **Calculated Spectrum** (red peak button) in WIN-NMR.

The ^{9}Be spectrum looks quite different from the ^{11}B spectrum and it is not easy to identify the lines making up the multiplet. However, if the coupling tree is exported to the calculated spectrum the pattern becomes obvious. The four magnetically equivalent ^{1}H couple with the ^{9}Be to give a 1:4:6:4:1 quintet, each line is then split further into a 1:1:1:1 quartet by the ^{11}B, spin I=$\frac{3}{2}$. Complications arise because $^1J(^{11}B, ^9Be)$ is approximately a third of the value of $^2J(^9Be, ^1H)$, consequently the width of the ^{11}B-^9Be multiplet is comparable to the value of $^2J(^9Be, ^1H)$ giving rise to signal overlap.

The ^{9}Be spectrum will also contain signals from the ^{10}B isotopomer superimposed upon the spectrum of the ^{11}B isotopomer was an intensity ratio of 1:4.

Check it in WIN-DAISY:

With the input document BEBH4.MGS still loaded, **Change** the calculated spectrum name to keep the former ^{9}Be lineshape. Duplicate the spin system using the **Copy fragment** button.

$Cp - {}^9Be - {}^{10}B^1H_4$

^{10}B *isotopomer:*

spin system $^7AM_4{}^4X$
(A=^{10}B, X=^{9}Be, M=^{1}H)

Automatically the **Main Parameters** of the copied data set is opened, modify the **Title** string <10B Spin System> and enter a **Statistical weight** of ca. <11.4>* to reflect the ratio of natural abundance and spin value of the boron isotopes, the ^{11}B fragment is weighted with the natural abundance of the isotope 80. Open the **Frequencies** dialog box, make sure the display units are in **ppm**, and change the boron isotope dialog box from **11 B** to **10 B**. Use the **Next** button in the bottom line to switch into the **Scalar Coupling Constants** dialog box and to modify the values of J$_{12}$ and J$_{23}$ according to the ratio of the magnetogyric ratios. Make sure that the spectral limits for the ^{9}Be lineshape are still defined in the **Spectrum Parameters** and run a **Simulation**. **Display** the **Calculated Spectrum** (red peak button) in WIN-NMR.

The ^{9}Be spectrum, including ^{11}B and ^{10}B isotopomers, shows no significant difference compared to the original ^{11}B simulation.

Check it in WIN-NMR:

Load the simulated spectrum ^{9}Be of the ^{11}B isotopomer as the second trace into the **Dual Display** of WIN-NMR. Use <Ctrl + Y> and **ALL** to align the Y-axis. Think about the ^{11}B/^{10}B isotopomer simulations discussed in section 4.3.4 and the differences visible in the Dual Display.

As discussed in section 4.3.4, the magnitude of scalar coupling constants involving ^{10}B are about a third of the corresponding ^{11}B values, so that the ^{9}Be-^{10}B multiplet lies within the ^{9}Be-^{11}B pattern. In addition the multiplicity of couplings involving ^{10}B nucleus is 7 compared with a multiplicity of 4 for ^{11}B, so that in the ^{9}Be spectrum a 1:4:6:4:1 quintet of 1:1:1:1:1:1:1 septets is expected. The overall effect of the ^{10}B isotopomer on the ^{9}Be spectrum is to appear as a poorly resolved hump under the ^{11}B isotopomer multiplet.

---

* Refer to footnote in section 4.3.4.

Check it in WIN-DAISY:

Open the **Lineshape Parameters** of the ^{10}B isotopomer fragment. Enter a **Linewidth** of ca. <4.0> Hz for Spin 1 (^{9}Be) and Spin 2 (^{10}B) and <2.0> Hz for Spin 3 (^{1}H group of spins). Open the **Spectrum Parameters** and change the **Isotope** from ^{9}Be to ^{1}H. Press the **Auto limits** button. Close the dialog box with **Ok**. Run a spectrum **Simulation** and **Display** the **Calculated Spectrum** (red peak button) in WIN-NMR.

The ^{1}H spectrum shows an intense 1:1:1:1 quartet (^{11}B-^{1}H coupling) of 1:1:1:1 quartets (^{9}Be-^{1}H coupling) on top a 1:1:1:1:1:1:1 septet (^{10}B-^{1}H coupling) of 1:1:1:1 quartets (^{9}Be-^{1}H coupling). Because of the values of $^{2}J(^{9}$Be, ^{1}H) and $^{1}J(^{10}$B, ^{1}H) there is some overlap of signals.

4.4 References

[4.1] Harris, R.K., Mann, B.E. (Eds.), *NMR and the Periodic Table*, London: Academic Press, 1978.
[4.2] Lazlo, P. (Ed.), *NMR of Newly Accessible Nuclei*, London: Academic Press, *2*, 1984.
[4.3] Berger, S., Braun, S., Kalinowski, H.-O., *NMR Spectroscopy of the Non-Metallic Elements*, Chichester: Wiley, 1997.
[4.4] Cavalli, L., in *Nuclear Magnetic Resonance Spectroscopy of Nuclei Other than Protons*, Axenrod, T., Webb, G.A. (Eds.), New York: Wiley, 1974, 287–315.
[4.5] Seel, F., Budenz, R., Gombler, W., *Chem. Ber.*, 1970, *103*, 1701.
[4.6] Kutzelnigg, W., Fleischer, U., Schindler, M., in: *NMR Basic Principles and Progress*: Diehl, P., Fluck, E., Kosfeld, R. (Eds.). Berlin: Springer, 19xx, *23*, 165-262.
[4.7] Schindler, M., *J. Chem. Phys.*, 1988, *88*, 7638.
[4.8] Bauer, G., *PhD Thesis*, Düsseldorf, 1977.
[4.9] Olschner, R., Hägele, G., unpublished results.
[4.10] Mallory, F.B., *J. Amer. Chem. Soc.*, 1973, *95*, 7747.
[4.11] Hilton, J., Sutcliffe, L.H., *Progr. NMR Spec.*, 1975, *10*, 27.
[4.12] Abraham, R.J., Cooper, M.A., *J. Chem. Soc. B*, 1968, 1094.
[4.13] Suntioinen, S., Weber, U., Laatikainen, R., *Magn. Reson. Chem.*, 1993, *31*, 406.
[4.14] Weber, U., *PhD Thesis*, Düsseldorf, 1994.
[4.15] Laatikainen, R., Niemitz, M., Weber, U., Sundelin, J., Hassinen, T., Vepsäläinen J., *J. Magn. Reson.*, 1996, *120A*, 1.
[4.16] Kleemann, G., Seppelt, K., *Chem. Ber.*, 1983, *116*, 645.
[4.17] Pötter, B., Kleemann, G., Seppelt, K., *Chem. Ber.*, 1984, *117*, 3255.
[4.18] Hägele, G., *J. Fluor. Chem.*, 1995, *71*, 223.
[4.19] Jameson, C.J., *J. Chem. Phys.*, 1977, *66*, 4983;
Osten, H.-J., Jameson, C.J., *J. Chem. Phys.*, 1984, *81*, 4288;
Jameson, C.J., Osten, H.-J., *J. Chem. Phys.*, 1984, *81*, 4293;
Jameson, C.J., Osten, H.-J., *J. Chem. Phys.*, 1984, *81*, 4300;
Osten, H.-J., Jameson, C.J., *J. Chem. Phys.*, 1985, *82*, 4595;

Jameson, C.J., Osten, H.-J., *J. Am. Chem. Soc.*, 1985, *107*, 4158;
Jameson, C.J., Osten, H.-J., *Mol. Phys.*, 1985, *55*, 383.
[4.20] Hansen, P.E., *Progr. NMR Spectrosc.*, 1988, *20*, 207.
[4.21] Diemert, K., Kuchen, W., Mootz, D., Poll, W., Sandt, F., *Phosph., Sulf. and Silic.*, 1996, *111*, 100.
[4.22] Batz, M., *PhD Thesis*, Düsseldorf, 1989.
[4.23] Wilke, E., *PhD Thesis*, Düsseldorf, 1992.
[4.24] Grossmann, G., unpublished results.
[4.25] Bley, B., Bodenbinder, M., Balzer, G., Willner, H., Hägele, G., Mistry, F., Aubke, F., *Can. J. Chem.*, 1996, *74*, 2392.
[4.26] Ricci, J.S., Koetzle, T.F., Goodfellow, R.J., Espinet, P, Maitlis, P.M., *Inorg. Chem.*, 1984, *23*, 1828.
[4.27] Corio, P.L., *Structure of High-Resolution NMR Spectra*, New York: Academic Press, 1966.
[4.28] Karlinowski, H.-O., Berger, S., Braun, S., *Carbon-13 NMR Spectroscopy*, Chichester: Wiley, 1994.
[4.29] Wehrli, F.W., Wirthlin, T., *Interpretation of Carbon-13 NMR Spectra*, London: Heyden, 1976.
[4.30] Breitmaier, E., Voelter, W., *Carbon-13 NMR Spectroscopy*, 3. ed., Weinheim: VCH, 1989.
[4.31] Braun, S., Kalinowski, H.-O., Berger, S., *100 and More Basic NMR Experiments*, Weinheim: VCH, 1996.
[4.32] Dodrell, D.M., Pegg, D.T., Bendall, M.R., *J. Magn. Reson.*, 1982, *48*, 323.
[4.33] Bigler, P., *NMR Spectroscopy: Processing Strategies*, Weinheim: Wiley-VCH, 1997.
[4.34] Hägele, G., Kückelhaus, W., Quast, H., *Chem. Zeitg.*, 1985, *109*, 405.
[4.35] Peters, R., *PhD Thesis*, University of Düsseldorf, 1992.
[4.36] Witanowski, M., Webb, G.A., *Nitrogen NMR*, London: Plenum Press, 1973.
[4.37] Städeli, W., Bigler, P., von Philipsborn, W., *Organ. Magn. Reson.*, 1981, *16*, 170.
[4.38] Jakobsen, H.J., Yang, P.-I., Brey, W.S., *Organ. Magn. Reson.*, 1981, *17*, 290.
[4.39] Adams, C.M., von Philipsborn, W., *Magn. Reson. Chem.*, 1985, *23*, 130.
[4.40] Aubke, F., Hägele, G., Willner, H., *Magn. Reson. Chem.*, 1995, *33*, 817.
[4.41] Marsmann, H., in NMR Principles and Progress, Berlin: Springer, 1981, *17*, 65-235.
[4.42] Jeffery, J.C., Moore, I., Murray, M., Stone, F.G.A., *J. Chem. Soc. Dalton Trans.*, 1982, 1741.
[4.43] Klapötke, T.M., Broschag, M., *Compilation of Reported ^{77}Se NMR Shifts*, Chichester: Wiley, 1996.
[4.44] Karaghiosoff, K., Schuster, M., Philipp, M., *4. Workshop "NMR-Spektroskopie an Phosphorverbindungen"*, Chorin (FRG), 1997.
[4.45] Schuster, M., Eckstein, K., Karaghiosoff, K., *GIT*, 1996, *12*, 1249.
[4.46] Harris, R.K., Kennedy, J.D., FcFarlane, W., in *NMR and the Periodic Table*, Harris, R.K., Mann, B.E. (Eds.), London: Academic Press, 1978, 342-366.
[4.47] Bongert, D., Heckmann, G., Hausen, H.-D., Schwarz, W., Binder, H., *Z. Anorg. Allg. Chem.*, 1996, *622*, 1793.
[4.48] Emsley, J.W., Feeney, J., Sutcliffe, L.H., *High Resolution Nuclear Magnetic Resonance Spectroscopy*, *1*, Oxford: Pergamon Press, 1965.
[4.49] Litchman, W.M., *J Chem. Phys.*, 1969, *50*, 1897.
[4.50] Wasylishen, R.E., Friedrich, J.O., *Can. J. Chem.*, 1987, *65*, 2238.
[4.51] Noggle, J.H., Baldeschieler, J.D., Colburn, C.B., *J. Chem. Phys.*, 1962, *37*, 182.

[4.52] Eaton, G.R., Lipscomb, W.N., *NMR Studies of Boron Hydrides and Related Compounds*, New York: Benjamin, 1969.
[4.53] Nöth, H., Wrackmeyer, H. in *NMR Principles and Progress*, Berlin: Springer, 1979, *14*.
[4.54] Farrar, T.C., Johannesen, R.B., Coyle, T.D., *J. Chem. Phys.*, 1968, *49*, 281.
[4.55] Kintzinger, J.-P. in *NMR Basic Principles and Progress*, Berlin: Springer, 1981, *17*, 1–64.
[4.56] Mason, J., *Multinuclear NMR*, New York: Plenum Press, 1987.
[4.57] Gaines, D.F., Coleson, K.M., Hillenbrand, D.F., *J. Magn. Reson.*, 1981, *44*, 84.

5 Time Dependent Phenomena

The discussions in the previous chapters have not included the influence of time dependent effects on nuclei. In this chapter three time dependent phenomena will be considered; relaxation in section 5.1, the relaxation related Nuclear OVERHAUSER Effect in section 5.2 and finally exchange processes in section 5.3.

The analysis of NMR spectra with regard to the mentioned phenomena would take up an individual volume with its various aspects to discuss. Nevertheless it seems to be indispensable to discuss at least the basic methodical aspects in an interactive course dealing with the analysis of one-dimensional spectra. Thus theoretical aspects are only presented here as far as they are essential for the process of analysis. No claim is laid to completeness rather than the discussion concentrates on the practical aspects of the examples presented. Regarding relaxation times the focus is laid on the determination of spin-lattice relaxation times via the inversion recovery method, the spin-spin relaxation time is only mentioned. Especially the section about dynamic NMR in 5.3 merely hits the basics of the large field of applications conceivable and deals only with the line-shape analysis.

For the representation of detailed theoretical background and chemical applications the reader is referred to the corresponding chapters of standard NMR textbooks ([5.1] – [5.5]) or monographs (for section 5.2: [5.6] and [5.7], for section 5.3: [5.8]-[5.10]).

5.1 Relaxation Times

In a FOURIER transform NMR experiment, a pulse or series of pulses are applied to the sample before the free induction decay is acquired. This sequence of events, often referred to as a scan, is then repeated to build up an adequate signal-to-noise ratio before the data is finally transformed. At the start of the experiment the spin system population corresponds to the BOLTZMANN distribution, but the effect of the pulse(s) is to perturb this distribution. Consequently at the end of each scan there is a delay period to enable the spin system to return to the BOLTZMANN equilibrium. The spin system can return to equilibrium by a number of processes and the overall time constant is called the relaxation time. If the delay period between scans is too short, the spin system does not have time to fully return to equilibrium and the signal intensities in the spectrum will be distorted.

318   5 Time Dependent Phenomena

There are two different types of relaxation times in NMR spectroscopy: the time constant for the z magnetization (M_z) to return to its equilibrium magnetization in the z-direction (M_0) is called the spin-lattice (longitudinal) relaxation time T_1 while the time constant for the xy-magnetization to decay to zero i.e. dephase in the xy-plane, is called the spin-spin (transverse) relaxation time T_2. Although T_1 values can vary over several orders of magnitude, the following relationship is always true for spin ½ nuclei:

$$T_1 \geq T_2$$

For spin number I > ½ nuclei and medium molecular weights, the following equation is usually valid:

$$T_1 = T_2$$

As a rough guide, the delay period between scans should be at least $5 \cdot T_1$'s to allow the system to return to equilibrium, e.g. after $5 \cdot T_1$'s M_z is 99.3% M_0. For non-coupled spins the recovery of M_z is exponential but for spin-coupled systems the cross relaxation time has to be considered. Cross relaxation time will be discussed later in this chapter.

Relaxation is a temperature dependent phenomenon. Unlike chemical shifts and spin-spin coupling constants, relaxation times are not directly related to molecule structure but they do depend, amongst other things, on molecular mobility. An important application of relaxation times is the determination of the mobility of biomacromolecules and the examination of non-covalent interactions. Relaxation times are also used in the measurement and evaluation of the images obtained using nuclear Magnetic Resonance Imaging (MRI) methods.

5.1.1 Spin-Lattice Relaxation Time

5.1.1.1 Quadrupolar Nuclei

For quadrupolar nuclei, spin number I > ½, the interaction of the quadrupole moment and electric field gradients in the molecule produces a very powerful relaxation mechanism and $T_1 = T_{1\text{-}Q}$ The relaxation time depends upon the symmetry of the electron shell. For highly symmetrical electron shells the relaxation mechanism becomes ineffective, longer T_1 values result and smaller linewidths are obtained. For quadrupolar nuclei T_1 is equal to T_2 and the spin-lattice relaxation time can be determined directly from the half-height linewidth. For lower symmetries very short T_1's are measured and the linewidths can be so broad that it is difficult to detect a signal (refer to section 5.3 about spin systems involving quadrupolar nuclei). The quadrupolar relaxation time is also dependent on the value of the *spin quantum number I*, $T_{1\text{-}Q}$ becoming longer (and linewidths decreasing) as the value of *I* increases.

5.1.1.2 Spin-½ Nuclei

For spin number I = ½ nuclei the overall longitudinal relaxation time T_1 may have contributions from one or more of the following components:

- dipole-dipole relaxation T_{1-DD}
- spin-rotation relaxation T_{1-SR}
- relaxation by chemical shift anisotropy T_{1-SA}
- relaxation by scalar coupling T_{1-SC}
- relaxation by electron spins T_{1-E}

For spin-½ nuclei with directly attached protons and for protons themselves, the relaxation is usually dominated by the dipole-dipole mechanism. Spin-rotation is important in the gas phase and for small freely rotating units such as methyl groups. Chemical shift anisotropy contributes to the relaxation time of quaternary carbons and is the main mechanism for heavy metals, particularly at high magnetic field strengths. Scalar coupling relaxation arises when two nuclei are scalar coupled and one of the nuclei undergoes fast relaxation, e.g. due to quadrupolar relaxation. This relaxation mechanism is inversely proportional to the square of the frequency difference of the two coupled nuclei and may be neglected at high magnetic field strengths. Relaxation by electron spins is an extremely powerful relaxation mechanism and arises from the interaction of the nuclei with paramagnetic centers such as the unpaired electrons in dissolved oxygen and transition metal ions.

Longitudinal relaxation times are normally measured using the inversion recovery sequence. In this experiment the equilibrium magnetization is transferred from the +z to the −z direction, without producing any xy contribution, by a 180° pulse. After a variable evolution period τ, a 90° read pulse is applied to transfer the remaining z-magnetization into detectable transverse magnetization. As the evolution period τ increases, the detectable magnetization goes from being negative (short τ) through zero and to its equilibrium value (long τ). A series of inversion recovery experiments are performed using a range of τ-values to cover the expected T_1 values. The dependence of the z-magnetization M_z on the evolution time τ is given by the BLOCH equation (assuming that M_x and M_y are both zero):

$$\frac{dM_z}{d\tau} = -\frac{M_z - M_0}{T_1}$$

which represents a first order differential equation which can be solved by integration leading to the following exponential expression:

$$M_0 - M_z^\tau = A \cdot e^{-\frac{\tau}{T_1}}$$

The constant A depends on the initial conditions. M_z^τ refers to the actual M_z values at different evolution time τ, e.g. at τ = 0, $M_z = -M_0$ and $A = 2 \cdot M_0$.

The equation can be transformed into a semi-logarithmic form:

5 Time Dependent Phenomena

$$\ln\left[M_0 - M_z^{\tau}\right] = \ln 2 + \ln M_0 - \frac{\tau}{T_1} \qquad \ln\left[\frac{M_0 - M_z^{\tau}}{2M_0}\right] = -\frac{\tau}{T_1}$$

The longitudinal relaxation time T_1 can be easily determined from a linear regression analysis of a plot of $\ln\left[(M_0 - M_z)/(2 \cdot M_0)\right]$ against τ.

Basically the analysis of the data from a T_1 inversion recovery experiment is a comparative intensity evaluation. In NMR spectra, intensities are more susceptible to variation than chemical shifts and it is very important in a T_1 analysis to have good signal-to-noise ratio and adequate digital resolution. Consequently the processing of the raw inversion recovery data is just as important as the final T_1 analysis. In the following "Check its" both data processing and data analysis will be discussed using the ^{13}C data of the acetyl carbonyl groups in β-D-glucose pentaacetate and an oligosaccharide also discussed in the first volume of this series [5.11]. The ^1H spectra of these compounds have already been analyzed in section 3.3.3.1.

A T_1 analysis always starts with processing the data of the inversion recovery experiment acquired with the longest evolution time τ_∞. In this experiment the z-magnetization is closest to the equilibrium magnetization M_0 and all the peaks in the spectrum should be positive.

Check it in WIN-NMR:

Open... the τ_∞ inversion recovery FID, ...\GCT1\018001.FID, of the glucose data in WIN-NMR. Using the **Process | FT** command, perform a simple FOURIER Transformation without any manipulation of the FID. From the **Process** pull-down menu choose the **Phase Correction** command and click the **Automatic** button. Adjust the **Zero Order** phase using the phasing slider to get a properly phased spectrum. Click the **Return** button.

Zoom in on individual peaks; use the line-and-cross representation <Ctrl + O> to check the digital resolution and inspect the signal-to-noise ratio.

It is obvious that both the signal-to-noise ratio and the digital resolution are not adequate for a proper T_1 analysis of the data.

Check it in WIN-NMR:

Use the **File | Recall last** option to reload the τ_∞ file ...\GCT1\018001.FID. From the **Process** pull-down menu perform some manipulation of the FID, e.g. **Window Function, Exponential (LB>0)** and/or **Zero Filling**. Use the **Process | FT** command to obtain the spectrum. To phase the spectrum use the **Process | Phase Correction** command and click the **Automatic** button in the button panel. If necessary manually adjust the **Zero Order** phase. When the spectrum is correctly processed use the **File | Save as...** command and a different processing number to store the spectrum, e.g. ...\GCT1\018002.1R. (The original FID is not modified in any way.)

All the other inversion recovery data of this experiment must be processed using processing parameters *identical* to those used for the τ_∞ data.

Check it in WIN-NMR:

Open... the file ...\GCT1\001001.FID in WIN-NMR. To apply the identical processing parameters use the acceleration buttons in the button panel on the left hand side. With these buttons the parameters used for processing the *last* FID/spectrum are applied to the current data set. Press the buttons corresponding to the processing functions used for the τ_∞ data, e.g. **Window!**, **Zero Filling!** and **FT!**.

When the spectrum is displayed press the **Phase Corr.!** button to apply the same phase values as used for the τ_∞ data. Using the **File I Save as...** command, save the spectrum with the *same* processing number, but the correct experiment number as used for the τ_∞ data set, e.g. ...\001002.1R.

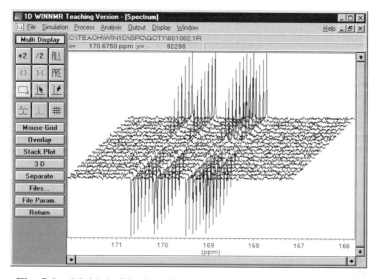

Fig. 5.1: Multiple Display of the 18 inversion recovery experiments.

Process the remaining inversion recovery experiments ...\GCT1\002001.FID up to ...\017001.FID in exactly the same way.

When all 18 experiments have been processed switch to the **Display I Multiple Display** mode and load all the inversion recovery spectra into the display window (Fig. 5.1).

The relaxation analysis is performed as follows.

Check it in WIN-NMR:

Load one spectrum, e.g. ...\GCT1\018002.1R into WIN-NMR and call the **Analysis I Relaxation I by Peak Picking** command. Press the **Execute** button in the **Peak Picking Options** dialog box. If another spectrum e.g. ...\GCT1\001002.1R containing negative peaks has been selected check the **Peaks sign both** option. The chemical shifts are reported at the top of the window and tick marks are placed on the tops of the peaks. Check that all five signals are picked.

322   5 Time Dependent Phenomena

Press the **Relax...** button. The **Relaxation Peaks List Parameter** dialog box shows the default file name for the variable delay list, VD.LST, that will be used in the analysis. The variable delay list is a simple ASCII text file, the first line contains the experiment number of the first spectrum of the series e.g. #1. The second line contains the VD value for the first experiment in µs and the following lines contain the VD parameters for the subsequent experiments. The list is terminated with a # symbol. If the correct **VD List File** is already selected click the **Execute** button otherwise click the **Select** button and chose the correct VD list. Automatically the **Relaxation** window is displayed beside the **Spectrum** window (refer to Fig. 5.2). Initially, the T_1 inversion recovery data that relates to the first picked peak starting from the left hand side of the spectrum is displayed. Every cross in the **Relaxation** window refers to the intensity of this particular line in the eighteen T_1 inversion recovery spectra. The *red colored cross* refers to the measured intensity in the spectrum currently displayed in the **Spectrum** window.

When the **Relaxation** window is highlighted (or has the focus) you will able to step through the relaxation curve for the five picked lines. Press the **Next Line** button and the **Relaxation** window display will change to show the intensity versus τ plot of the second line (the corresponding ppm value i.e. 170.2 ppm, is shown in the status line). Now put the focus on the **Spectrum** window. The button panel changes and you will now be able to step through the eighteen different T_1 inversion recovery spectra. As the **Prev. Spec. / Next Spec.** button is clicked the corresponding spectrum file name is displayed in the header status line and the red cross moves along the curve in the **Relaxation** window showing the measured intensity for the selected line in he currently displayed spectrum.

Once the T_1 inversion recovery data has been checked using the Spectrum/Relaxation windows the T_1 values can be calculated.

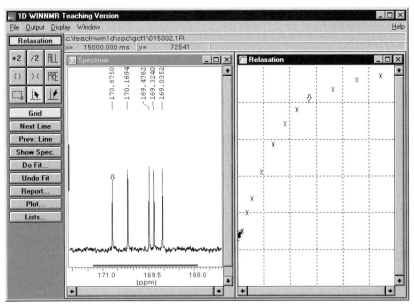

Fig. 5.2: Initial Relaxation Analysis Window of Glucose data.

Check it in WIN-NMR:

With the focus on the **Relaxation** window press the **Do Fit...** button. The **Relaxation Parameters** dialog box opens (Fig. 5.3). Make sure that the **T1 Inversion Recovery Experiment** option is selected and press the **Fit one Line** button. The relaxation window now includes a red curve corresponding to a regression analysis of the inversion recovery data. Open the **Report...** box where the chemical shift and the T_1 value (in milliseconds) are listed. Press the **Next Line** button and display the relaxation data for the next line. Click the **Do Fit...** button again but this time press the **Fit all Lines** button to analyze all the inversion recovery data sets. Open the **Report...** box, click the Options... button and set **Output Mode for Relaxation List** to **Reduced** and **Print...** the results. Change to **Full** mode and **Print...** the results again. Store the results using the **Store...** option.

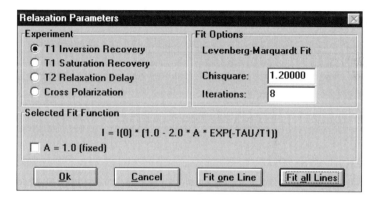

Fig. 5.3: Dialog box for Relaxation Analysis fitting procedure.

Because T_1 analysis is based upon intensity evaluation, the final result is strongly dependent on the processing of the raw inversion recovery data. Low digital resolution, poor signal-to-noise ratio's and experimental errors e.g. missetting the 90° and 180° pulse lengths, can all lead to biased results. Although *absolute* T_1 values can be determined, both the sample preparation and the experimental measurements are time consuming and often in practice T_1 values are compared *relative* to one another. Longer relaxation times indicating a greater mobility for the corresponding nucleus in the structure. The T_1 values for the C=O carbon atoms of the acetyl substituents in β-D-glucose pentaacetate leads to the conclusion, that the carbon at 170.7 ppm has a significant longer relaxation time than the others. Recalling the structure of β-D-glucose pentaacetate, the acetyl group at position 6 (attached to the exocyclic methylene group) is expected to be more flexible than all the other acetyl groups so that the assignment of the signal at 170.7 ppm is obvious. The chemical shift values used in the T_1 inversion recovery sequence are not yet referenced. For the correct chemical shifts use the already referenced full spectrum ...\GC\001001.1R.

324 5 Time Dependent Phenomena

Check it in WIN-NMR:

Experiment with different process options for the T_1 inversion recovery data and examine how the T_1 results vary. Use the **File I Serial Processing** option and select the eighteen FIDs ...\GCT1\001001.FID to ...\018001.1R. Select the **use common Job** option. Press the **Select common job** button and in the **Select Common Job** dialog box choose GCT1.JOB. Press the **Edit common Job...** button and examine the various processing options in the **Automatic Processing** dialog box. Click the **SPC→** button. The **Phase Correction Options...** will use the **Set Numerical Values (PHC0/1)** option and the set values of <–60.48> and <0.0> correspond to the **Zero Order** and **First Order** phase constants respectively used to phase the τ data set ...\GCT1\018001.FID. Click the **Save Spectrum Options...** button and check that the **Processing Number Increment** is set to <1> to keep the previously processed spectra. If you alter any of the processing parameters save the new job file under a different name, e.g. GCT1A.JOB using the **Save as...** option.

Execute the serial processing and perform a **Analysis I Relaxation I by Peak Picking** in a similar way as done before. Inspect the resulting T_1 values in the **Report...** box. Compare the results with the previous relaxation analysis.

In a similar way the T_1 inversion recovery data for the oligosaccharide can be analyzed.

Check it in WIN-NMR:

Use the **File I Serial Processing** command and select all eighteen FIDs ...\OCT1\001001.FID to ...\018001.FID corresponding to the T_1 inversion recovery data for the oligosaccharide. Select the **use common Job** option, press the **Select common job** button and load the job file OCT1.JOB. Press the **Edit common Job...** button and examine the various processing options. Close the dialog box with **OK** and then **Execute** the automatic processing procedure. When the automatic processing is completed the last spectrum ...\OCT1\018002.1R is displayed in the **Spectrum** window.

Call the **Analysis I Relaxation I by Peak Picking** mode. In the **Execute Peak Picking Options** dialog box press the **Execute** button. Check in the **Report...** box whether all peaks are picked and make sure that no noise spikes are present. Remove any unwanted peaks using the **Edit Cursor** mode; select the unwanted peaks with the left mouse button and delete the peak by clicking the right mouse button. When the required number of signals have been picked press the **Relax...** button. Click on the **Execute** button in the **Relaxation Peaks List Parameter** dialog box

Calculate the T_1 values in the same manner as for the β-D-glucose pentaacetate data set. Check the T_1 values in the **Report...** box and using β-D-glucose pentaacetate as a guide, interpret the values with respect to the proposed structure for the oligosaccharide.

Although not overlapping, the signals in the carbonyl region of the ^{13}C spectrum of the peracetylated oligosaccharide are rather close together and may be used to illustrate a second type of T_1 analysis using deconvolution. As discussed earlier, the intensity of an NMR signal refers not to its peak height but to its integrated area. Consequently with

5.1 Relaxation Times 325

signals that overlap or are close together, better results should be obtained using deconvolution of the signals instead of the peak height as used by the peak picking routine. For deconvolution to be successful, the spectrum must be adequately digitized so zero filling should be included in the serial processing of the oligosaccharide data.

Check it in WIN-NMR:

Call the **File | Serial Processing** command and select again all eighteen FIDs ...\OCT1\001001.FID to ...\018001.FID. Select the **use common Job** option but this time use the job file OCT1X.JOB. Examine the various processing options and notice that in the **Save Spectrum Options...** the **Processing Number Increment** has been set to <2> so that the spectra will be stored under a different processing number than the previous ones. Close the dialog box with **OK** and then **Execute** the automatic processing procedure. When the automatic processing is completed the last spectrum ...\OCT1\018003.1R is displayed in the **Spectrum** window.

Call the **Analysis | Relaxation | Deconvolution** mode. Press the **Options...** button and in the **Common Deconvolution Options** dialog box enter <0> in the **First Gaussian/Lorentzian Ratio** edit field. Leave the dialog box with **Ok**. Press the **/2** button in the button panel to decrease the y-scale. With the **Mouse PP** button highlighted, adjust the "rubber band" box so that all the peak tops are included. Press the **Report...** button and check whether all eleven peaks are picked, only eight peaks are shown on each page, use the **Next Page** button to look at the rest of the data. Press the **Pos. [ppm]** and **Intensity** buttons (all boxes checked) in order to select all positions and intensities for adjustment. Leave the dialog box with **Ok**. Press the **Do Fit...** button. In the **Fit Options** dialog box check that the **No. of Line Shape Parameters to fit:** is <22> and that the **Amplitude Estimation before fit** option is selected. Press the **Do Fit** button to start the analysis. When the deconvolution is completed click the **Relax...** button to start the relaxation analysis. In the **Relaxation List Parameters** dialog box check that the correct **VD List File** and the **use actual Deconvolution Table for fit** option are selected. Click the **Process** button.

Starting with the first spectrum a deconvolution with optimizing of chemical shift and intensities is performed. When the process is completed, the Relaxation window opens to display a plot of intensity versus τ.

Check it in WIN-NMR:

With the focus in the **Relaxation** window, press the **Do Fit** button and then the **Fit all lines**. The intensities for line 5 differ from a typical T_1 inversion recovery function and an error message **Line 5: No Inversion Recovery Data, abort?** will appear. Click **No** to continue with the analysis. Press the **Report...** button and compare the T_1 values obtained using deconvolution with those obtained using the peak picking method.

5.1.2 Spin-Spin Relaxation Time

The natural peak width at half-height of a LORENTZIAN line is given by the equation:

$$v_{1/2} = \frac{1}{\pi \cdot T_2}$$

where T_2 is the natural spin-spin or transverse relaxation time. Normally the external magnetic field is never completely homogeneous and nuclei in different parts of the sample experience slightly different magnet fields strengths and so resonate at slightly different frequencies. Consequently the experimental linewidth contains contributions from both the natural linewidth and the field inhomogeneity. The spin-spin relaxation time calculated from the experimental linewidths is called T_2^*.

$$v(obs)_{1/2} = \frac{1}{\pi \cdot T_2^*}$$

The spin-spin relaxation time T_2 is measured using a spin echo sequence. The initial 90° pulse tips the equilibrium magnetization into the y direction and then a series of 180° pulses separated by twice the evolution time 2τ are used to refocus the signal. The refocusing pulse is used to eliminate inhomogeneities in the B_0 field so that the true T_2 and not T_2^* is measured. The spectra are then analyzed in a similar manner to the T_1 analysis by plotting signal intensity against τ and then using regression analysis to fit a curve to the experiment data.

5.2 Nuclear Overhauser Effect

The *Nuclear OVERHAUSER Effect* (NOE) [5.6],[5.7] has already been mentioned in connection with the double resonance methods and in the WIN-DR program a Nuclear OVERHAUSER Effect correction can be applied (refer to section 3.2.1). The Nuclear OVERHAUSER Effect is inextricably associated with the dipole-dipole relaxation mechanism described in section 5.1 and does *not* depend on scalar coupling. If two spin-½ nuclei relax by dipole-dipole relaxation, then irradiating one nucleus will alter the intensity of the second nucleus. The NOE experiment is not to be confused with spin-decoupling experiment which requires considerably higher irradiation levels and simplifies the spectrum by the removal of J-couplings. In the NOE experiment irradiation of a selected proton or a group of selected protons is applied prior to data acquisition, while in the spin-decoupling experiment selected protons are irradiated during data acquisition. NOE enhancements build up and decay away at a finite rate, the exact time depending upon the value of T_1. Thus the NOE is a consequence of the steady state that builds up when perturbing the system by irradiation and its relaxation.

For a single spin-½ nucleus there is a population difference between the upper and lower energy levels given by the BOLTZMANN law. The transition probability and hence the signal intensity depends on the population difference. In a two spin-½ heteronuclear spin system there are now four energy levels, but again the population differences are given by the BOLTZMANN law. If nuclei A and X are mutually relaxed by the dipole-dipole mechanism and if nucleus A is preirradiated, then there will be a signal enhancement for nucleus X due to the Nuclear OVERHAUSER Effect and vice versa. For small molecules in non viscous solvents for which the extreme narrowing limit is valid, the maximum Nuclear OVERHAUSER Effect η_{max} is given by:

$$\eta_{max} = \frac{1}{2} \cdot \frac{\gamma_A}{\gamma_X}$$

with γ_A = magnetogyric ratio of the irradiated nucleus A
and γ_X = magnetogyric ratio of the observed nucleus X

and the NOE enhanced signal intensity for nucleus X is then given by:

$$I_X = I_X^0 \cdot (1 + \eta_{max})$$

with I_X^v = signal intensity without enhancement

> NOE data may be analyzed on a qualitative, a semi quantitative or a fully quantitative basis.

Check it in the WINDOWS Calculator:

Calculate the maximum OVERHAUSER enhancement η_{max} for the following proton decoupled nuclei: $^{13}C\{^1H\}$, $^{15}N\{^1H\}$, $^{29}Si\{^1H\}$ and $^{31}P\{^1H\}$. The magnetogyric ratios can be found in the **Element-Data** table (load the file A.MGS into WIN-DAISY and press the **PSE** button in the **Frequencies** dialog box) and Table 5.1.

Table 5.1: Magnetogyric ratios of selected isotopes.

isotope	magnetogyric ratio [10^7 rad T^{-1} s^{-1}]
^{1}H	26.7510
^{13}C	6.7263
^{15}N	−2.7116
^{29}Si	−5.3146
^{31}P	10.8289

For $^{13}C\{^1H\}$, η_{max} is 1.99, so that the ^{13}C signal intensity in an NOE enhanced ^{13}C spectrum is ca. three times (η_{max} + 1) larger than in the corresponding spectrum with no NOE enhancement. In contrast to $^{15}N\{^1H\}$ and due to the negative magnetogyric ratio of ^{15}N, negative peaks for the ^{15}N signals result for NOE enhanced spectra, with intensities which are ca. four times larger (η_{max} is −4.94) with respect to the intensities of the (positive) signals measured in the spectra with no NOE. The values calculated in the previous "Check it" are the *maximum* possible enhancement and in practice if no other nuclei or relaxation mechanisms contribute to the overall relaxation time the value of the OVERHAUSER enhancement will be reduced accordingly. This is particularly important for nuclei with a negative magnetogyric ratio e.g. ^{15}N and ^{29}Si, where it is possible to obtain an enhancement of "−1" which cancels out the signal completely. For homonuclear spin systems with $\gamma_A = \gamma_X$, e.g. in proton spectra, the maximum NOE is:

$$\eta_{max} = 1/2$$

Usually in homonuclear experiments, the NOE is reported as a percentage enhancement so that for proton spectra η_{max} is 50%. Because the NOE depends upon the distance between nuclei, it is possible to obtain information about the spatial arrangement of protons in a molecule. In the following "Check its", the NOE data for β-D-glucose pentaacetate will be considered and related to the protons on the pyranose ring. The discussion will then be extended to the peracetylated oligosaccharide which exhibits NOE responses between protons both on the same and adjacent sugar monomers.

With modern BRUKER NMR spectrometers 1D NOE spectra are normally recorded in pseudo-2D mode and corresponding data is stored as a serial file (extension *.SER). Before the pseudo-2D serial file can be processed using WIN-NMR the file has to be split up into a set of 1D files using the **Filecopy & Convert** command [5.11]. There is always one experiment, the reference experiment, where the irradiation frequency is set *off-resonance* while in the remaining experiments the irradiation frequency is set *on-resonance* and corresponds to the chemical shift of one of the target spins in the investigated molecule. Variants to the basic experiment which help to minimize *Selective Polarization Transfer* (SPT) effects (see later), and in which each signal multiplet is irradiated at a number of frequencies rather than at the center frequency exist. NOE enhancements can be quite small and consequently NOE spectra are usually displayed as a set of *difference spectra* whereby the *off-resonance* reference data is subtracted from the *on-resonance* NOE data. This data manipulation and data presentation makes it easier to recognize the enhancements. Subtraction of the off-resonance data from the on-resonance data is usually done using the free induction decays rather than the corresponding spectra and data manipulation is performed taking advantage of the serial processing option in WIN-NMR.

Check it in WIN-NMR:

Load the NOE reference FID, ...\GHNO\008001.FID, into WIN-NMR. Process the FID and phase the spectrum. Using the **File | Save** command save the spectrum using the default spectrum name ...\GHNO\008001.1R. Open the **Output | History...** window and examine the processing parameters.

Switch to the **File | Serial Processing** mode and select the seven NOE FIDs ...\GHNO\001001.FID up to ...\007001.FID. With the use **common Job option** selected, press the **Select common job...** button and choose the job file GHNO.JOB. Click the **Edit common Job...** button and inspect the processing parameters for both the **FID** and **Spectrum**. Leave the dialog box with **Ok** and **Execute** the serial processing.

To inspect the results switch to the **Display | Multiple Display** mode and select the seven automatically processed spectra.

The seven spectra represent the NOE differences as each of the proton attached to the pyranose ring is irradiated. The first experiment, ...\GHNO\001001.FID, corresponds to the irradiation of the anomeric proton at 5.97 ppm and the last experiment, ...\007001.FID, to the irradiation of C5-H at 4.09 ppm. The irradiated proton appears as a negative signal in the difference spectrum. Often in discussion about proton-proton distances and following a semi-quantitative analysis, it is convenient to categorize NOE enhancements as strong, medium or weak. Generally the larger the enhancement is the closer the protons are in distance.

The quantitative interpretation of NOE data is fraught with danger but in most cases the presence of an NOE enhancement is often all that is required to establish the stereochemistry at a particular center. Unfortunately the converse is not true, the *absence* of an NOE enhancement does *not* necessarily disprove a stereochemistry. If the total irradiation time is short compared with T_1 or if the populations are simply inverted by a 180° pulse, the relative *transient* NOE intensities approximate to the relative inter-proton distances. However, if the total irradiation time is long compared with T_1, the quantitative analysis of the *steady-state* NOE intensities that are measured are much more problematical. For the data used in the subsequent "Check it" exercises it is the *steady-state* NOEs that have been measured. In an AX two spin system, when there is no scalar coupling between A and X, the two A and the two X transitions are degenerate and the interpretation of the NOE data is relatively straight forward. Scalar coupling between A and X removes this degeneracy and both A and X appear as a doublet. Provided both of the lines of one of the doublets are saturated equally, a similar enhancement for both lines of the other doublet should be observed. However, if one of the doublet signals is irradiated preferentially the anti-phase Selective Polarization Transfer effects (SPT) may be superimposed upon the in-phase NOE enhancement. In such cases the phase of one of the doublet lines of the NOE enhanced signal can be partly inverted and the overall integral should be used to quantify the NOE enhancement. Isolated two spin systems are rare and most NMR spectra exhibit quite complicated multiplets often displaying second order effects.

In addition to the presence of scalar coupling in a larger spin system, the close proximity of more than one nucleus can also cause complications. The populations of the energy levels of the spins may be influenced by dipolar cross relaxation with neighboring nuclei. Thus in a homonuclear AMX spin system, if spin A is irradiated spin X may experience a positive *direct NOE* which arises from cross-relaxation between A and X and a negative *indirect NOE* caused by the cross-relaxation between A and M and then M and X (spin diffusion). It is possible that the *indirect NOE* could decrease or even erase any *direct NOE* enhancement. Different NOE build-up rates may be obtained if the inter-proton distances differ significantly e.g. if r_{AM} is twice the distance of r_{MX}, then when M is irradiated the rate of NOE build up for X will be 64 times faster than the rate of NOE build up for A. Similarly, if A is relaxed exclusively by its interaction with M, but M interacts with both A and X, then for the same inter-proton distance, different enhancements will be obtained when irradiating A and observing M and vice-versa.

In addition small dispersive signals can result in the subtraction of FIDs; if there is slight variation in line positions between the reference spectrum and the NOE enhanced spectrum, the difference spectrum will show a significant dispersive lineshape. To minimize this effect, an exponential window function is normally applied to broaden the lines. Consequently only signals with the correct phase can be properly evaluated.

With larger molecules or small molecules in viscous solvents the dependence of the NOE on correlation times should be taken into account. Whereas for small highly mobile molecules a maximum enhancement of 50% is obtained, with large molecules (biomolecules) a maximum value of –100% is approximated. As a consequence an unwanted zero-passing is observed for molecules of intermediate size. In such cases the ROE experiment, described later, is more suitable.

Check it in WIN-NMR:

Open... the reference spectrum ...\GHNO\008001.1R into WIN-NMR. Call the **Analysis I Integration** mode and define six integration regions for all of the signals. Define a single region for the two overlapping signals at 5.38 ppm. Activate the integral for the anomeric proton at 5.7 ppm, enter the Integral **Options...** dialog box and calibrate it as <100>. Use the **Regions... I Save...** option to store the defined regions for use in further "Check its". Save this spectrum under the same name with the **Save** option.

5.2 Nuclear Overhauser Effect

These integral regions shall form the basis of the quantitative evaluation of the NOE difference spectra. To help in this evaluation, construct a table similar to that shown in Table 5.2 and fill in the integrated values.

First the intensity enhancement for every signal i in each spectrum k is calculated and referenced to the intensity of the signal in the unperturbed spectrum. This first normalization is indispensable in cases where NOE enhancements shall be compared arising from different numbers of protons, which is not the case in the present example.

Enhancement of signal i when k is irradiatied [%]: $\quad NOE_i(k) = \dfrac{I_i^k - I_i^0}{I_i^0} \cdot 100$

with I_i^0 = intensity of signal i in the reference spectrum
and I_i^k = intensity of signal i in the NOE enhanced spectrum k.

Secondly and in order to compare this NOE with the other NOEs calculated for the same molecule this NOE is corrected ("normalized") by the saturation degree of the presaturated nucleus, so that the value is independent on the individual saturation degree S_k in spectrum k.

Saturation degree for NOE experiment k: $\quad S_k = \dfrac{I_k^0 - I_k^k}{I_k^0}$

with I_k^0 = intensity of signal k in the reference spectrum
and I_k^k = residual intensity of the preirradiated signal k in the NOE spectrum k

Normalized enhancement of signal i [%]: $\quad NOE_i^{norm} = \dfrac{1}{S_k} \cdot NOE_i(k)$

The normalization is indispensable in cases where NOE enhancements shall be compared arising from different numbers of protons. This is not the case in the present example, but the evaluation shall be done under consideration of the saturation degree.

Table 5.2: Nuclear OVERHAUSER Enhancement interpretation table. In the gray fields, you may fill in the corresponding saturation degree S_k.

Exp k	Irrad. nucleus	Number i Shift[ppm]	1	2	3	4	5	6	7
1	C?-H								
2	C?-H								
3	C?-H								
4	C?-H								
5	C?-H								
6	C?-H								
7	C?-H								

Check it in WIN-NMR:

Load now the first NOE difference spectrum ...\GHNO\001001.1R and enter the **Analysis I Integration** mode. Press the **Regions...** button and **Load...** the region file, ...\GHNO\008001.IR, generated from the reference spectrum. Press **Ok**. Push in the **Integral Options** dialog box the **Select Calibration File** button, select the reference spectrum ...\GHNO\008001.1R and activate the **File Compare Mode**. The WIN-NMR spectrum window shows the defined regions as red rectangles under the spectrum with the integral values written above the rectangle. In this case these values correspond directly to the $NOE_i(k)$ [%] normalized with respect to the intensity of the unperturbed signal I_i^0! Note however that for signals arising from more than only one proton these values must be divided by the number of protons, e.g. for methyl signals by 3.

To normalize the measured NOEs with respect to the saturation degree, set the mouse cursor in **perpendicular mode**, activate the integral for the irradiated (negative) signal, note its value (saturation degree in %) and its deviation from −100%. Press the **Options...** button and use the **Manual Calibration** option to set the **Selected Integral** to <−100>. All integral values of the NOE enhanced signals will now be normalized with respect to the saturation degree. Leave the options dialog with **Ok**. Open the **Report...** and note the normalized NOE values NOE_i^{norm} in Table 5.2.

Taking into account the various problems associated with the measurement and the evaluation of NOEs their accuracy should not be overestimated. For this reason a rigorous quantitative analysis is probably not always the best approach and in many cases a semi-quantitative analysis with a classification of the enhancements into three categories as shown in Table 5.3 is fully sufficient to solve structural problems.

Table 5.3: Nuclear OVERHAUSER Enhancement categories

Category	abbreviation	NOE^{norm} [%]
strong	s	10 – 20
medium	m	5 – 10
weak	w	< 5

Table 5.4: Nuclear OVERHAUSER Enhancement table for categories.

Exp k	Irrad. nucleus	Number i Shift[ppm]	1	2	3	4	5	6	7
1	C?-H		■						
2	C?-H			■					
3	C?-H				■				
4	C?-H					■			
5	C?-H						■		
6	C?-H							■	
7	C?-H								■

Check it in WIN-NMR:

Be careful when assigning NOEs especially in the case of partially overlapping signals. Examine closely the two overlapping signals at 5.38 ppm in each of the NOE difference spectra and try to evaluate in each case which proton shows the enhancement. Exploit the **Display I Dual Display** mode for this purpose and load the reference spectrum ...\GHNO\008001.1R as the second trace to overlay both spectra in order to simplify these assignments.

Using the notation in Table 5.3, fill in the entries in a table like Table 5.4 to evaluate the NOE results in categories. Try to establish the structure for the compound based on the proton-proton connectivities. Take into account that the NOE spectra are not correctly calibrated. The correct chemical shifts have to be taken from the spectrum ...\GLUCOSE\001001.1R.

In β-D-glucose pentaacetate the NOEs are observed only for protons in the pyranose ring. The data give information about the relative inter-proton distances in six-membered ring structure. In Table 5.5 the coupling pathways and the corresponding NOE enhancements for the chair conformation of the pyranose ring are given:

Table 5.5: Expected NOE interactions in a glucopyranose structure

coupling path	NOE
$^2J_{gem}$	s
$^4J_{axax}$	m
$^3J_{axax}$	w

Note that besides dispersive artifacts introduced into NOE difference spectra by algebraic inaccuracies, antiphase multiplet signals may appear due to selective population inversion SPT. These SPT effects are found for spins which are scalar coupled to the preirradiated spin(s) and are caused by a non-uniform perturbation of the individual multiplet lines of the preirradiated spin(s).

Check it in WIN-NMR:

Zoom in on the individual signals in each NOE difference spectrum (...\GHNO\001001.1R to ...\007001.1R) and look for any SPT effects. Using serial processing, process the data now with a suitable exponential window function and try to minimize these unwanted SPT effects.

Open... the reference free induction decay ...\GHNO\008001.FID and apply an **Exponential (LB>0)** window function with **LB** <3.0> Hz. Use the **File I Save as ...** command to save the modified FID under a different file name, e.g. ...\GHNO\008002.FID. Switch to the **File I Serial Processing** mode and select the seven NOE FIDs ...\GHNO\001001.FID up to ...\007001.FID. With the use **common Job option** selected, choose the job file GHNOW.JOB. Click the **Edit common Job...** button. **Select** the modified reference spectrum FID as the second file in the **File Algebra Options...**. Check the other processing options. **Execute** the job. Repeat the **Integration** of the difference spectra in the File Compare mode and then the normalized NOE enhancements and compare the values with the data in your table.

Check it in WIN-NMR:

When the analysis of the NOE data is completed, use the results to assign the protons in the reference spectrum starting with the anomeric proton. Load the original β-D-glucose pentaacetate spectrum, ...\GLUCOSE\001001.1R, into WIN-NMR. Switch to the **Analysis I Multiplets** mode and click the **Windaisy** button to load the previous WIN-DAISY analysis. Compare the Multiplet and WIN-DAISY assignment with the results of the NOE analysis.

Check it in WIN-DAISY:

To clarify the assignment of the overlapping signals at 5.38 ppm,. press the **Export...** button and transfer the data to **WIN-DAISY**. In the **Frequencies** dialog box **Disable** Spin 4. Run a spectrum **Simulation** and display both experimental and calculated spectra in the **Dual Display** mode of WIN-NMR. Use the **Separate** option to show which signal belongs to C2-H and to the suppressed multiplet C4-H as shown in Fig. 5.4.

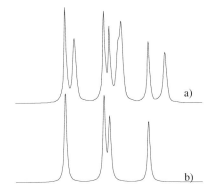

Fig. 5.4: a) Zoomed experimental spectrum ...\GLUCOSE\001001.1R.
b) Calculated spectrum with suppression of spin 4.

Check it in WIN-DAISY and WIN-NMR:

Toggle back to **WIN-DAISY** and **Display** only the **Calculated Spectrum** in WIN-NMR. Call the **Display I Dual Display** mode and load the reference NOE experiment ...\GHNO\008001.1R as the **second trace**. Use the **Move Trace** option to overlay both spectra and switch the X-axis into Hz <Ctrl + X> and note the x-shift in Hz. Toggle back to **WIN-DAISY** and open the **Spectrum Parameters**. Enter the **offset** value (Hz) and press the **Apply!** button. **Change** the calculated spectrum name, e.g. to ...\GHNO\999994.1R to store the calculated spectrum in the directory containing the NOE data. Run a spectrum simulation. Then change again the spectrum name, e.g. to ...\GHNO\999992.1R. Open the **Frequencies** dialog box and *remove* the disable option for spin 4 but check the **Disable** option for spin **2**. Run a spectrum **Simulation** and **Display** the **Calculated Spectrum** in WIN-NMR, now showing only the transitions of spin 4 under suppression of spin 2.

Check it in WIN-NMR:

Call the **Display | Dual Display** option and load one of the NOE difference spectra e.g. ...\GHNO\001001.1R as the second trace. **Zoom** in the region about 5.38 ppm and use the **Move Trace** button with depressed <Shift> key to align the Y-scale. Compare all the NOE difference spectra showing an effect at this distinct spectrum part with the simulations to decide if the enhancement can be assigned definitely to C4-H or C2-H respectively.

In addition to the difficulties already discussed, problems can also arise with larger molecules where the enhancements are either small or zero. In these cases the Nuclear OVERHAUSER experiment in the rotating frame (ROE) [5.11] is of particular value. Unlike the conventional NOE experiment, the enhancement in the ROE experiment normally evolve from elements of the *transverse* magnetization during the spin lock period and not the *longitudinal* magnetization. Consequently in a ROE experiment *direct* enhancements are always positive, *direct* and *indirect* enhancements can be distinguish by their sign and finally chemical exchange peaks have the opposite sign to *direct* enhancements. The ROE experiment is the rotating frame analogue of the *transient* NOE experiment, there is no *steady-state* NOE analogue. In a ROE experiment each FID represents the difference spectrum, and usually no reference spectrum is available. As a consequence ROE data is either interpreted on a qualitative level or series of ROE experiments with the spin-lock time incremented from experiment to experiment are performed from which ROE build-up rates may be calculated.

Check it in WIN-NMR:

Enter the **File | Serial Processing** mode and select the five ROE FIDs ...\GHRO\001001.FID up to ...\005001.FID. With the use **common Job option** selected, choose the job file GHROW.JOB. Click the **Edit common Job...** button and check the **Window Function Options...**, the **Phase Correction Options...** and the **Calibrate Options....** **Execute** the serial processing. Inspect one after another the difference spectra.

In the analysis of the oligosacchride data an important item has to be considered, the connection between the various sugar rings can be deduced from the NOE data. In β-D-glucose pentaacetate there was only one pyranose ring and consequently the NOE analysis is relatively straightforward. However in the peracetylated oligosaccharide there are three pyranose rings and NOE responses are expected between protons in the same and *adjacent* pyranose rings which will allow the connection between the rings to be determined. Using the β-D-glucose pentaacetate analysis as a guide, we have an idea what NOEs are to be expected within the individual glucose fragments of the oligosaccharide. To help in identifying the NOE responses between adjacent pyranose rings the difference spectra should be compared with the reference spectrum of the three 1D TOCSY experiments corresponding to the three individual glucose fragments.

Check it in WIN-NMR:

Load the NOE reference FID, ...\OHNO\014001.FID, into WIN-NMR. Process the FID, apply an **Exponential (LB>0)** window function with **LB** <3.0>. Use the **File | Save as ...** command to save the modified FID under a different file name, e.g. ...\OHNO\014002.FID. FOURIER transform the data. Switch to the **Process | Phase Correction** mode and phase the spectrum noting the numerical values of **PHC0** and **PHC1**. Enter the **Analysis | Calibration** mode and click **Calibrate** in the panel button. The spectrum does not contain a reference signal. Press the **Select Compare File** button and select the spectrum ...\TOCSY1D\001001.1R. Check the **File Compare Mode for X** option and quit the dialog box with **Ok**. Press the **Num. Comp.** button and make a note of the value of **X-Scaling (SR)**.

Switch to the **File | Serial Processing** mode. With the use **common Job option** selected, press the **Select common job...** button and choose the job file OHNOW.JOB. Click the **Edit common Job...** button and in the **File Algebra Options...** change **the Name of second NMR file** to the reference FID ...\OHNO\014002.FID which includes the window function. Check the **Phase Correction** and the **Calibrate Options....** Select the first thirteen FIDs and **Execute** the serial processing. When the processing is completed the last spectrum of the series ...\OHNO\013002.1R is displayed in the spectrum window.

The displayed spectrum refers to another reference spectrum recorded at the end of the series. This difference spectrum is an indication of the confidence level of the measurements and ideally should be a flat line.

Due to the complexity of the spectrum it is not feasible to quantify the enhancement using integration and consequently only a qualitative analysis can be performed. Initially the NOE difference spectra associated with a particular glucose fragment should be compared with the corresponding 1D TOCSY reference spectrum of the *same* fragment. The responses should be similar to those found in β-D-glucose pentaacetate. If any responses are found that do not correspond to the signals in the reference spectrum, compare the NOE difference spectrum with the TOCSY spectra of the other two glucose fragments. In this way NOE responses between adjacent pyranose rings can be identified and the bridgehead between glucose fragments determined.

Check it in WIN-NMR:

Call the **Display | Multiple Display** mode. Select the three 1D TOCSY spectra ...\TOCSY1D\002001.1R, ...\003001.1R and ...\004001.1R and one of the NOE difference spectra, e.g. ...\OHNO\001002.1R. Click the **Separate** button to be able to see the spectra clearly. Click on the **Equal X** button to equalize the ppm scale. Press the **Lock X** button to keep the ppm scale identical for all the spectra when zoomed. Select the bottom trace with a mouse click (spectrum will have a thick red frame). Using the ***2** button and the slider in the right hand side slider box, increase the intensity of the enhancement signals until they are comparable with the intensities in the TOCSY spectra.

For the NOE difference spectrum ...\OHNO\001002.1R it is obvious that the irradiated anomeric proton corresponds to the signal in the top TOCSY spectrum. Click the top spectrum window and the name of the spectrum, ...\TOCSY1D\002001.1R is displayed in the header line. If this spectrum corresponds to ring A, this difference spectrum only exhibits enhancements to protons in the *same* pyranose rings. Note the result.

Repeat this procedure for all the other NOE difference spectra, compare each difference spectrum with the three TOCSY reference spectra. Call the **Display | Multiple Display** option, click on the NOE spectrum name in the list box and press **Delete**. Select the next NOE spectrum and continue with the analysis. If signals are close together zoom in the spectral region.

There are three NOE spectra which display NOE responses between adjacent pyranose rings, one example is shown in Fig. 5.5.

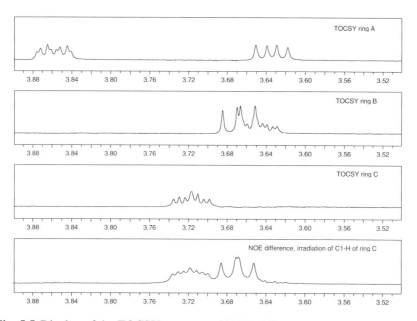

Fig. 5.5: Display of the TOCSY spectra and NOE difference spectrum 006001.1R

To identify the connections, the previous WIN-DAISY multiplet analysis of the oligosaccharide spectrum can be utilized.

Check it in WIN-NMR:

Open... the experimental spectrum ...\TOCSY1D\001001.1R in WIN-NMR and switch to the **Analysis | Multiplets** mode. Press the **Windaisy** button to display the iteration result for the three proton spin systems of the oligosaccharide and hence all three TOCSY spectra. Zoom in on the appropriate spectrum region, e.g. for the NOE difference spectrum, ...\OHNO\006001.1R with irradiation of C1-H in ring C the region is between 3.6 and 3.8 ppm (refer to Fig. 5.6). The WIN-DAISY multiplet labels clearly shows that the inter-ring NOE refers to C2-H in ring B, which is shifted to high-field due to that connection.

338   5 Time Dependent Phenomena

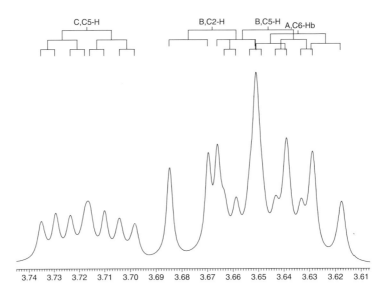

Fig. 5.6: Multiplet display of the WIN-DAISY result for the oligosaccharide.

The other inter-ring NOEs are found in spectrum ...\007001.1R (irradiation of C1-H in ring B, additional response to C6-H(B) and C6-H(A) of ring A) and in ...\010001.1R (irradiation of C6-H(A) in ring A, additional response to C1-H in ring B) which establish the connection between ring A and B. In conclusion, ring B is the central sugar ring connected at position C1 to C6 in ring A and at position C2 to C1 in ring C.

To confirm the results the 1D ROE data should be processed as well. They might be more suitable for larger molecules.

Check it in WIN-NMR:

Call the **File | Serial Processing** option and select all the FIDs in the directory ...\OHRO\. Press the **Select common job...** button and choose the job file OHROW.JOB. **Execute** the automatic processing procedure.

Switch to the **Display | Multiple Display** and select the three 1D TOCSY spectra ...\TOCSY1D\001001.1R to ...\004001.1R and one ROE spectrum. Again click the **Separate**, **Equal X** and **Lock X** buttons. Select the lowest trace and adjust the intensities in the ROE spectrum. Extract the inter-ring NOEs to establish the connections and compare with the results obtained from the normal NOE spectra.

As expected, the same results are obtained: in ...\OHRO\003002.1R perturbation of C1-H in ring C leads to an enhancement of C2-H in ring B, in spectrum ...\004002.1R perturbation of C1-H in ring B causes an enhancement at C6-H(B) and C6-H(A) in ring A and finally in spectrum ...\007002.1R perturbation of C6-H(A) in ring A results in a NOE enhancement in signal C1-H in ring B.

5.3 Magnetic Site Exchange

The analysis of spin systems exhibiting time-dependent phenomena requires the application of the time-dependent SCHROEDINGER equation. It is not the intention to discuss the density operator theory required to solve this equation, but to apply the results to spin systems discussed in earlier sections.

Strictly speaking, none of the molecules discussed so far is truly rigid. Every molecule dissolved in a solvent undergoes some sort of dynamic process such as BROWNian motion, vibrational effects and intra-molecular or even inter-molecular exchange processes. In those cases where the dynamic effects are fast in comparison with the NMR time scale (1s – 10^{-6}s) the effect on the NMR spectrum is small. Thus the protons of a methyl group undergo rapid rotation and appear to be magnetically equivalent (section 2.4) and consequently in the calculation of the spectrum the time-dependent factor can be neglected. However if the rotation is slowed down, e.g. due to a steric hindrance, this time-dependent effect has to be taken into account and is referred to as *chemical* or *magnetic site exchange*. If the exchange process requires the breaking of bonds it is a *inter*molecular exchange but if no breaking of bonds is required it is a *intra*molecular process. In this section we will focus on *intra*molecular processes and their effects on the lineshape.

There are many compounds which are in the fast-exchange region at room temperature, for these types of system as the temperature is lowered the appearance of the NMR spectrum changes and is dependent on the rate process until finally the slow-exchange limit is reached. The spectrum measured at this slow-exchange limit is often called the *limiting low temperature spectrum* and is a combination of the subspectra of the individual species. Cooling the sample below the slow-exchange limit has no further effect on the appearance of the NMR spectrum. Above the slow-exchange limit the NMR spectrum is again a combination of the subspectra but depending on the rate and temperature signals may have broadened or merged together. Finally beyond the fast-exchanging limit the signals in the NMR spectrum are again sharp but the observed NMR parameters are now the *weighted average* of the NMR parameters of the individual species.

In this first section we will focus on the basics principles of chemical exchange spectra. The first step is to define the limiting low temperature spectrum (slow-exchanging) spin system - and determine the so-called *static* NMR parameters. The next step is to define the type of exchange process. In a *mutual* exchange process the species involved in the dynamic process are identical, the spins simply change their positions and the spin systems are identical while in *non-mutual* exchange process the different species have spin systems with different static NMR parameters. The final step is to analyze the spectra recorded at different temperature and to determine the rate constants (dynamic parameter) based on the system defined in the previous steps. A major problem of this method is that the static NMR parameters (chemical shifts, coupling constants, linewidths and populations) are themselves temperature dependent. Consequently the temperature dependence of the static NMR parameters should be taken into account in

340   5 Time Dependent Phenomena

the chemical exchange analysis. Generally chemical shifts are more sensitive to temperature variation than coupling constants.

5.3.1 Non-Coupled Spin Systems

The simplest exchange process is the case where one spin A changes its magnetic environment by rotation around a single bond, e.g. in the following example of an N-dimethyl acetamide (R=CH$_3$) compound by the rotation around the C–N single bond.

$$\begin{array}{c} O \\ \parallel \\ R-C \\ \diagdown \\ N-CH_3(B) \\ | \\ CH_3(A) \end{array}$$

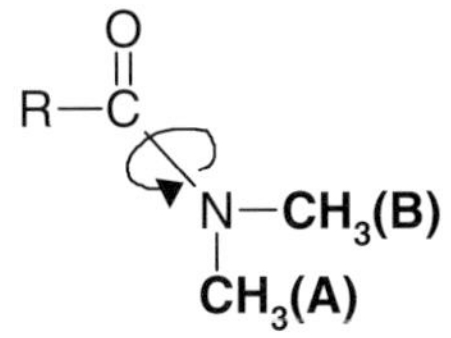

At low temperature the rotation around the C–N bond is slowed down and two separate signals are observed for the two methyl groups because of their different chemical environment. If we consider the $^{13}C\{^1H\}$ spectrum, the spin system would be labeled AB but because of the dilute nature of the ^{13}C nucleus there is no scalar coupling between A and B. Assuming fast rotation of the methyl groups, the corresponding ^1H spin system would be labeled A_3B_3. For the exchange process we will keep the same spin system notation as used in previous chapters. Now let us rotate the molecule around the C–N bond as shown in Fig. 5.7.

Because the molecule is identical before (I) and after rotation (II) around the C–N bond, the spin systems are identical, only the labeling is interchanged. The former carbon nucleus A in site (I) ends up as nucleus B in site (II) and vice versa. Regarding the protons, the A_3 methyl group in site (I) turns after rotation into B_3 in site (II).

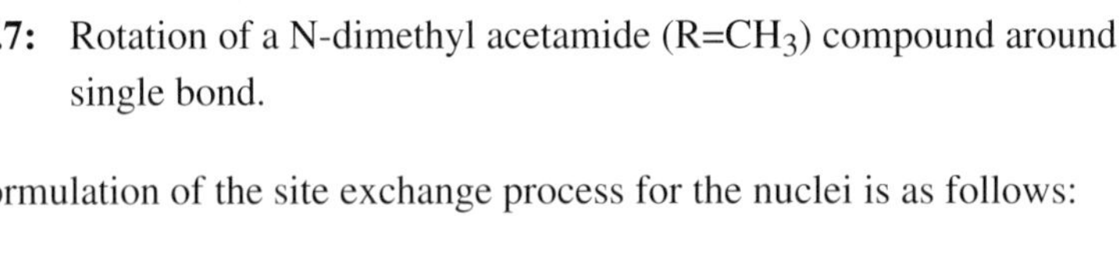

$$\begin{array}{c} O \\ \parallel \\ R-C \\ \diagdown \\ N-CH_3(B) \\ | \\ CH_3(A) \\ (I) \end{array} \quad \rightleftharpoons \quad \begin{array}{c} O \\ \parallel \\ R-C \\ \diagdown \\ N-CH_3(A) \\ | \\ CH_3(B) \\ (II) \end{array}$$

Fig. 5.7: Rotation of a N-dimethyl acetamide (R=CH$_3$) compound around the C–N single bond.

The formulation of the site exchange process for the nuclei is as follows:

$$^{13}C\{^1H\}: \quad AB \underset{k_{21}}{\overset{k_{12}}{\rightleftharpoons}} BA \quad \text{with } J_{AB} = 0.$$

The rate constant for the processes in the carbon spectrum k_{12} is identical with k_{21}. The system is mutual exchange because the compound on the left hand side (I) is identical with the one on the right hand side (II). Regarding the ^{1}H spectrum the spin system can also be reduced to an AB spin system (better AX), because there is no visible coupling between the two methyl groups:

$$^1H: \quad A_3B_3 \rightleftharpoons B_3A_3 \quad \text{or}$$
$$\text{because of } J_{AB} = 0: \quad AB \rightleftharpoons BA$$

The rate constant k_{12} is identical for the ^{1}H and the ^{13}C{^{1}H} spectrum. We will now define this exchange system starting from a WIN-DAISY spin system parameter set for the limiting low temperature spectrum.

Check it in WIN-DAISY:

Load the WIN-DAISY input document EXCH_1.MGS and inspect the values in the various parameter dialog boxes. Run a **Simulation** and **Display** the **Calculated Spectrum** (red peak button) in WIN-NMR.

The spectrum simply shows the limiting low-temperature spectrum with two singlets for the methyl groups at different chemical shifts.

Check it in WIN-DAISY:

Export to **WIN-DYNAMICS** the WIN-DAISY parameter set.

The program WIN-DYNAMICS will be started and the input document EXCH_1.DAT automatically loaded.

Check it in WIN-DYNAMICS:

Open the **Main Parameters** dialog box and inspect the basic spin system parameters. An *AB* spin system is defined and the **Number of exchanging sites:** refers to the number of AB spin systems that were defined in WIN-DAISY as different fragments. Because the spin system under consideration refers to a mutual exchange the **Number of exchanging sites:** is initially defined as "1". Use the **Next** button to open the **Static Parameters** dialog box. This dialog box contains the **Frequency [ppm]**, **Scalar Coupling [Hz]**, **Line width [Hz]** and **Population** of the spin system as defined in the original WIN-DAISY document. Run a spectrum **Simulation** and **Display** the **Calculated Spectrum** (red peak button) in WIN-NMR.

The displayed spectrum is identical to the spectrum simulated using WIN-DAISY. Now the exchange system will be defined.

Check it in WIN-DYNAMICS:

Toggle back to **WIN-DYNAMICS**. Open the **Main Parameters** dialog box and increase the **Number of exchanging sites:** to <2>. Press the **Next** button to switch to the **Static Parameters**. Check the **Mutual Exchange** option. The **Exchange Vectors** button will be enable. Press this button to open the **Exchange Vectors** dialog box. The **Exchange Vectors** dialog box contains the exchange permutation and in many ways is similar to the **Symmetry**

Description dialog box in WIN-DAISY. The first row contains the spin system letters A and B and the second row contains the spin labeling system for **Site 1**. In the last row, **Site 2**, the spin number that the original spin is changed into by rotation around the C–N bond is entered. Because spin A (label 1) is changed into spin B (label 2), enter number <2> in the first edit field under spin A and enter <1> in the second edit field under spin B. Leave the dialog boxes with **Ok**. Run a **Simulation** and **Display** the **Calculated Spectrum** (red peak button) in WIN-NMR.

Again, the displayed spectrum is identical with the previously calculated spectrum because no chemical exchange has been introduced yet. To study the effect of changes in the rate constant on the spectrum, a sequence simulation similar to that in WIN-DAISY can be used.

Check it in WIN-DYNAMICS:

Open the **Rate Constants** dialog box. The value of the only available rate constant between site 1 and 2 is currently zero. Open the **Simulation Parameters** dialog box. The only available **Rate constants between Positions 1-2** is selected. The **Sequence Simulation Control Parameters Step Width** is set to <1> s^{-1} and the **Number of Steps** to <10>.

The calculation sequence will generate eleven spectra; the first spectrum corresponds to the initial parameter set and is the reference spectrum, the following ten, stored with a decreasing processing number, correspond to an increase in the rate constant of 1 s^{-1} per spectrum. The value of the rate-constant used for the calculation is stored in the title file of each spectrum.

Check it in WIN-DYNAMICS and WIN-NMR:

Run a **Sequence** and **Display** the **Calculated Spectrum** (red peak button) in WIN-NMR. Switch to the **Display I Multiple Display** mode and select the eleven sequence spectra ...\999999.1R to ...\999989.1R. Click the **Stack Plot** button to display the spectra one upon the other.

The calculated sequence of spectra clearly shows how the lines broadening as the rate constant increases.

Check it in WIN-DYNAMICS and WIN-NMR:

Toggle back to **WIN-DYNAMICS** and open the **Simulation Parameters** dialog box. Enter a **Step Width** of <23.5> s^{-1} and increase the **Number of Steps** to <15>. Run a **Sequence** and **Display** the **Calculated Spectrum** (red peak button) in WIN-NMR. Switch to the **Display I Multiple Display** mode and select the sixteen sequence spectra ...\999999.1R to ...\999984.1R. Click the **Stack Plot** button to get an overview.

5.3 Magnetic Site Exchange

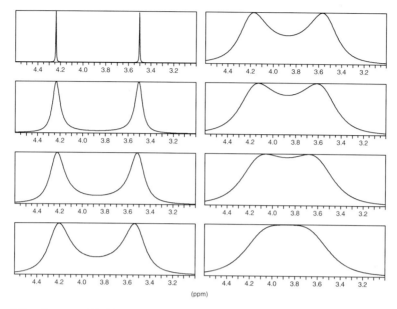

Fig. 5.8: Separate Multiple Display plot of the exchanging system.

The spectra show how the two singlets broaden until they merge into one peak. This process is called *coalescence*. At coalescence, the two singlets in the spectrum just merge to form one peak with a flat top. The temperature at which this occurs is called the coalescence temperature T_c.

As the rate increase the singlet gets sharper again. The final chemical shift of the singlet appearing in the high-temperature spectrum is the average of the static chemical shifts of A and B.

$$v_{average} = \frac{1}{n} \cdot \left(v_{A,I} + v_{A,II} \right)$$

with n = number of sites
$v_{A,I}$ = chemical shift of nucleus A in site I
$v_{A,II}$ = chemical shift of nucleus A in site II
(remember that the labeling A, B and so on is kept fixed due to the site I)

Because the exchange process is of the mutual type (no preferred site present) the population of the two sites is identical and is neglected in the formula above. The chemical shift at high-temperature is then simply the arithmetic mean of the individual shifts v_A ($v_{A,I}$) and v_B ($v_{A,II}$).

Check it in WIN-NMR:

Display the sequence calculations in the **Display | Multiple Display** mode. Press the **Overlay** button. The coalescence of the two signals to a singlet exactly in the middle can be clearly seen.

Check it in WIN-DYNAMICS and the Calculator:

Run the **Calculator** from WIN-DAISY. Open the **Static Parameters** dialog box in WIN-DYNAMICS and calculate the arithmetic mean of the two chemical shift values. Note the value.

Check it in WIN-NMR:

Open... the last calculated spectrum, ...\999984.1R, with WIN-NMR. Switch to the **Analysis | Peak Picking** mode and **Execute** a peak picking in **ppm**. Compare this value with the calculated value.

In the **Display | Multiple Display** mode load the sixteen calculated spectra. Press the **Separate** button to display the first eight spectra in separate windows (refer to Fig. 5.8).

At coalescence, the two singlets in the spectrum just merge to form one peak with a flat top to be seen in the last frame on the right side in Fig. 5.8.

Check it in WIN-NMR:

With the first eight spectra displayed in separate windows, click on the coalescence spectrum in the bottom right frame, the corresponding file name will be displayed in the header line. **Return** from the **Multiple Display** mode. Load the coalescence spectrum (...\999992.1R) into WIN-NMR and the open the **Output | Title...**, window. The entry displays the value of the rate constant used to simulate the (coalescence) spectrum. Using the formula below and the static NMR parameters, calculate the coalescence rate and compare it with the "experimental" value. Also calculate the life-time at coalescence.

$$k_c = \frac{\pi \cdot \Delta v}{\sqrt{2}} \quad \text{at temperature of coalescence } T_c$$

with k_c = rate constant at coalescence point
$\Delta v = v_A - v_B$ of the spin system in Hz

Note that the coalescence temperature depends on the separation of the lines in Hz. The life-times of the species at the given temperature is given by the reciprocal of the rate constant.

$$\tau_c = \frac{1}{k_c} \, [s]$$

τ_c = life time at coalescence point

The approximate value of the free energy of activation for the exchange process is given by the EYRING equation:

$$\Delta G^{\#} = 19.14 \cdot T_c \cdot \left(10.32 + \log \frac{T_c}{k_c}\right) [J \cdot mol^{-1}]$$

with T_c = absolute coalescence temperature [K]

Check it with the Calculator:

Assuming a coalescence temperature of 400 K, calculate the free energy of activation $\Delta G^{\#}$ using the EYRING equation given above.

The calculated free energy of activation will be approximately 82 kJmol^{-1} although it must be stressed that in practice it is quite difficult to measure the coalescence temperature accurately. To determine the values of the enthalpy $\Delta H^{\#}$ and entropy $\Delta S^{\#}$ for the exchange process it is necessary to measure the rate constant as a function of temperature. Using the substitution,

$$\Delta G = \Delta H - T \cdot \Delta S$$

the EYRING equation can be rewritten as,

$$\log \frac{k_i}{T_i} = 10.32 - \frac{\Delta H^{\#}}{19.14 \cdot T_i} + \frac{\Delta S^{\#}}{19.14} [J \cdot mol^{-1}]$$

From a plot of log (k/T) versus 1/T, the abscissa gives the value of $\Delta H^{\#}$ and the slope $\Delta S^{\#}$. Although individual values of $\Delta G^{\#}$ are usually reasonably accurate, the accurate measurement of $\Delta H^{\#}$ and $\Delta S^{\#}$ is very time consuming and any errors in temperature stability or systematic errors in temperature measurement will have a detrimental effect on the value of these thermodynamic parameters. To minimize the errors in $\Delta H^{\#}$ and $\Delta S^{\#}$, data from several sources such as different groups of exchanging nuclei or different magnetic field strengths should be combined and compared with complimentary techniques such as magnetization transfer.

The effect of the spectrometer frequency used to study of the exchange process must also be considered. The coalescence temperature depends upon Δv, the difference in Hz between the resonance frequencies of site A and site B. The resonance frequency in Hz depends upon the external magnetic field strength and the magnetogyric ratio of the observed nucleus. Consequently, the rate constant at which coalescence occurs will depend upon the spectrometer frequency and observed nucleus, e.g. in the current example the coalescence temperature will occur at different temperatures in the ^{1}H spectrum and the ^{13}C{^{1}H} spectrum.

Check it in WIN-DYNAMICS:

Open the **Spectrum Parameters** dialog box and change the **Units of Limits** to **ppm**. Double the **Spectrometer** frequency to <200> MHz and increase the **Size** of the spectrum to <2> K points (to provide proper resolution). Using the same **Simulation Parameters** run a simulation **Sequence** and **Display** the **Calculated Spectrum** (red peak button) in WIN-NMR.

Check it in WIN-NMR:

Load all sixteen simulated spectra into the **Display | Multiple Display** mode of WIN-NMR. Press the **Stack Plot** button. Identify the coalescence spectrum and load this spectrum ...\999985.1r into WIN-NMR. Open the **Output | Title...** window and note the value of the rate constant used to simulate the spectrum.

As expected from the direct relationship between Δv and k_c, doubling the spectrometer frequency doubles the rate at which coalescence occurs.

In the next example we will look at a related system; in this system only one methyl group is bonded to the nitrogen, the other has been substituted by another group.

```
      O                                    O
      ‖                                    ‖
   R—C                                   R—C
       \                                     \
        N—X              ⇌                    N—CH₃(B)
        |                                     |
        CH₃(A)                                X

       (I)                                  (II)
```

Again, assuming free rotation, the methyl group in species (I) is in a different chemical environment compared to the methyl group in species (II), but in contrast to the previous example there is only one spin (group) A in the basic spin system (I) which is transferred into spin (group) B at site (II). This is an example of non-mutual exchange because since (I) and (II) are different, more than one species is involved in the exchange process. The exchange can be represented as:

$$A \rightleftharpoons B$$

Because site (I) and site (II) have different NMR spin system parameters the free energy of both species may be different, so that both spin systems might not be equally populated. The first step is the determination of the populations of the two spin systems by integrating the signals in the limiting low-temperature spectrum.

Check it in WIN-DAISY:

Open... the input document EXCH_2.MGS in WIN-DAISY and examine the values in the parameters dialog boxes. Note that there are two fragments (site (I) and site (II)) and the **Statistical weight** in the **Main Parameters** dialog box relates to the different populations of the two different sites according to the equilibrium constant. Run a **Simulation** and **Display** the **Calculated Spectrum** (red peak button) in WIN-NMR.

The equilibrium constant is the ratio of the concentration at the two different sites:

$$K = \frac{[I]}{[II]}$$

with [I] = concentration at site (I)
[II] = concentration at site (II)

Check it in WIN-NMR and WIN-DAISY:

Enter the **Analysis | Integration** mode and determine the relative populations of the different sites (I) and (II). Compare the relative populations of the sites with the **Statistical weight** values in the **Main Parameters** in the WIN-DAISY document. **Export** to **WIN-DYNAMICS** the WIN-DAISY data.

Check it in WIN-DYNAMICS:

Open the **Main Parameters** dialog box, a **Non-coupled spin system** <1> spin(s) with **Number of exchanging sites:** <2> has been defined. Switch to the **Static parameters**. There are now two **Static Parameter** dialog boxes corresponding to **Exchange Site** 1 and **Exchange Site** 2. Again, each dialog box gives the **Frequency [ppm]**, **Population** and **Line width [Hz]** of the particular **Exchange Site**. Pressing the **Next Site** button will switch to the other site. Using the default **Sequence Simulation Control Parameters**, **Step Width** <1> s^{-1} and **Number of steps** <10> run a **Sequence** simulation. **Display** the Calculated Spectrum (red peak button) in WIN-NMR.

Check it in WIN-NMR:

Load the eleven calculated spectra into the **Display | Multiple Display** of WIN-NMR. Press the **Stack Plot** button and inspect the effect of the rate constant on the spectra.

It is obvious, that the less populated (low-field) site signal broadens faster than the other more populated (high-field) site and that the low-field signal tends to disappear at high temperature.

Check it in WIN-DYNAMICS:

Toggle back to WIN-DYNAMICS and open the **Simulation Parameters** dialog box. Increase the **Step Width** to <23.5> s^{-1} and the **Number of steps** to <15> and run another sequence simulation. **Display** the **Calculated Spectrum** (red peak button) in WIN-NMR.

Check it in WIN-NMR:

Using the **Display | Multiple Display** option, load the sixteen simulated spectra and press the **Stack Plot** button to get an overview. Press the

348 5 Time Dependent Phenomena

Overlay button. Using this perspective the way the low-field signal disappears and the position of the high-temperature chemical shift moves becomes obvious.

From the calculations it is clear that the less populated (low-field) site signal broadens earlier than the high-field site signal and that the averaged chemical shift is not simply the arithmetic mean of the two low-temperature chemical shifts. The reason for these differences is that the population in the two different sites are no longer equal.

The averaged shift, taking into account the population difference, is given by:

$$V_{average} = p_{A,I} \cdot V_{A,I} + p_{B,II} \cdot V_{B,II}$$

with the normalized overall population $\Sigma p_i = 1$

Check it in WIN-DYNAMICS and the Calculator:

Run the **Calculator** from the WIN-DAISY toolbar and determine the average chemical shift for fast exchange (high temperature) using the values given in the **Static Parameters** dialog box of WIN-DYNAMICS.

Check it in WIN-NMR:

Toggle back to **WIN-NMR** and **Open...** the last spectrum of the calculation sequence ...\999984.1R. Switch to the **Analysis | Peak Picking** mode and **Execute** a peak picking in **ppm**. Compare the chemical shift with the calculated value.

Although there appears to be only one signal present in last spectrum of the calculation sequence, the chemical shift has not yet reached the correct averaged value.

Check it in WIN-DYNAMICS and WIN-NMR:

Toggle back to **WIN-DYNAMICS**. Open the **Rate Constants** dialog box and enter <1e8> for the rate constant. Leave the dialog box with **Ok** and run a single **Simulation**. **Display** the **Calculated Spectrum** (red peak button) in WIN-NMR and then switch to the **Analysis | Peak Picking** mode and **Execute** a peak picking in **ppm**.

This example shows that for intermediate rates of exchange between two unequally populated sites it is not always obvious from the spectrum that exchange is actually taking place, the so-called *hidden exchange*. As shown in the simulation sequence, even at quite low exchange rates the signal of the least-populated site disappears completely and so it is not possible to observe the coalescence of the two signals. A typical example of this type of intermediate exchange occurs with acidic protons and the water impurities in the NMR solvent. The chemical shift of the acidic protons depends on the concentration (population) of the water and the sample temperature. The true chemical shift of the acidic proton is the static parameter measured at the slow-exchange limit, while the observed chemical shift is a weighted averaged chemical shift between the two exchanging sites.

5.3.2 Systems Involving Coupling

As discovered in the previous section, it is difficult to measure the exact coalescence point for unequally populated sites. The situation is even more complicated if there is scalar coupling in the spin system at the slow exchange limit.

As an example consider the aniline derivative shown in Fig. 5.9 [5.12]. In the ^{1}H NMR spectra measured at high temperature limit, the rotation around the C–N bond of the substituted amino group is fast, so that the two aromatic protons appear to be isochronous and a singlet is observed.

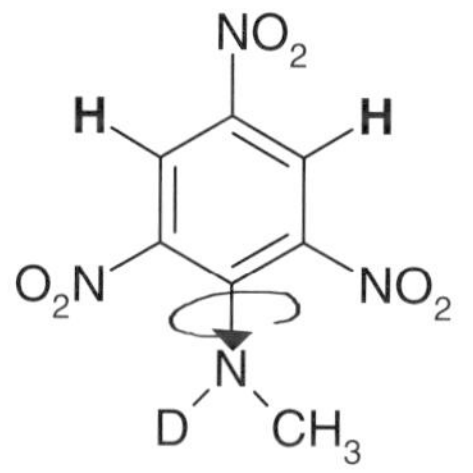

Fig. 5.9: *N-Methy,-N-deutero,2,4,6-trinitroaniline*

As the temperature is lowered, the rotation slows down until finally the rotation is frozen out and the two aromatic protons are no longer magnetically equivalent and not even isochronous. As the equivalence is canceled an ortho coupling, $^4J_{ortho}$, appears. The exchange process can be treated as a mutual exchange problem:

$$AB \underset{k^{-1}}{\overset{k}{\rightleftarrows}} BA \quad \text{with } J_{AB} \neq 0$$

site   (I)   (II)

Check it in WIN-DAISY:

Load the input document EXCH_3.MGS into WIN-DAISY and inspect the static **Parameters** defining the spin system in the various dialog boxes. **Export to WIN-DYNAMICS** the data set.

Check it in WIN-DYNAMICS:

Open the **Main Parameters** dialog box and set the **Number of exchanging sites:** to <2>. Switch to the **Static Parameters** dialog box using the **Next** button. Select the **Mutual Exchange** option. Press the **Exchange vectors** button and enter the exchange permutation row <2> and <1>, indication that A of site 1 turns into B of site 2. Quit the dialog with **Ok**. Run a **simulation** and **Display** the **Calculated Spectrum** (red peak button) in WIN-NMR.

The calculation shows the spectrum of an AB spin system as expected for a mutual two site exchange.

Check it in WIN-DYNAMICS:

Open the **Simulation Parameters** dialog box and enter a **Step Width** of <4> s^{-1} and a **Number of Steps** of <50>. Make sure that there is a minimum of 500 Kbytes free space on your hard disk. Run the simulation **Sequence** and **Display** the **Calculated Spectrum** (red peak button) in WIN-NMR.

Check it in WIN-NMR:

In the **Display | Multiple Display** mode, load the fifty one calculated spectra ...\999999.1R down to ...\999949.1R. Press the **Stack Plot** button and then use the **Mouse Grid** to obtain the display shown in Fig. 5.10.

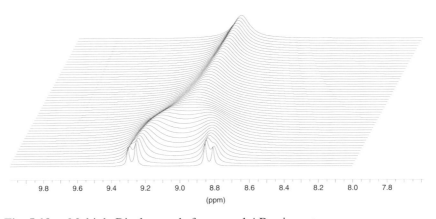

Fig. 5.10: Multiple Display mode for mutual AB spin system.

As the rate constant increases the AB four line pattern collapses to give a singlet at very high exchange rates. The coupling information is lost, because at the average resonance frequency the two protons are magnetically equivalent, so that the coupling constant has no effect on the spectrum. The simulation sequence covers the rate constant range from 0 up to 200 s^{-1} in 4 s^{-1} steps.

Check it in WIN-NMR:

Open... the last spectrum of the simulation sequence, ...\999949.1R, in WIN-NMR and check the entry in the **Output | Title...** window. Using this rate constant determine the life time of the individual species. Enter the **Analysis | Linewidth** mode and determine the peak width at half-height. Compare this measured value with the static **Line width** parameter in the **Static Parameters** dialog box of WIN-DYNAMICS.

The singlet obtained at a rate constant of 200 s^{-1} has a life time of $5 \cdot 10^{-3}$ s and the spectrum is obtained experimentally at $-26.3°C$ (246.9 K) and 60 MHz [5.12]. The fast exchange limit has not yet been reached because the linewidth of the calculated singlet is still much broader than the static linewidth parameter.

Check it in WIN-DYNAMICS and WIN-NMR:

Open the **Rate Constants** dialog box and enter value of <1e3>, run a **Simulation** and **Display** the **Calculated Spectrum** (red peak button) in WIN-NMR Enter the **Linewidth I Analysis** mode and measure the linewidth parameter. Toggle back to **WIN-DYNAMICS** open the **Rate Constants** dialog box again, enter a value 10 times bigger <1e4>. Run another **simulation**, **Display** the **Calculated Spectrum** (red peak button) in WIN-NMR and measure the **Linewidth**. Repeat the process, increasing the **Rate Constant** by a factor of 10 until the fast exchange limit is reached.

The linewidth remain constant for rate constants of 10^5 s^{-1} and above. The fast exchange limit is obtained experimentally at 36.2°C, so that at room temperature the linewidth of the two aromatic protons is still exchange broadened.

Check it in WIN-NMR and with the Calculator:

Calculate the coalescence rate constant based on Δv = 26.7 Hz. Note that the experimental ^{1}H spectrum was recorded at 60 MHz. Examine the sequence spectra and find out which spectrum refers to the coalescence temperature. Load the coalescence spectrum into WIN-NMR and inspect the rate constant in the **Output I Title...** window. Compare this value the calculated rate constant.

The calculated value of about 60 s^{-1} relates to the calculated spectrum ...\999984.1R where coalescence to a single peak is obvious.

The next example will examine the ^{1}H spectrum of *anti-1,6;8,13-bismethano-[14]-annulene* [5.13]. At room temperature a single AB spectrum is obtained for the methylene protons of the two bridge heads.

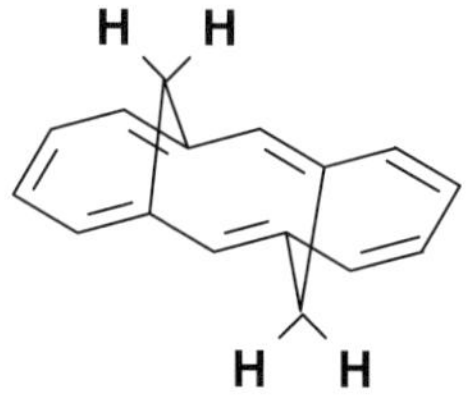

Fig. 5.11: Structure of *anti-1,6;8,13-bismethano-[14]-annulene.*.

The annulene structure in Fig. 5.11 shows, that the left CH$_2$ bridge head can be transferred into the right hand side CH$_2$ bridge head either by delocalization of the π–electrons or by valence tautomerism of the double bonds. For the last possibility two AB spin systems should be present in the slow exchange limit spectrum, so that at high exchange rates an averaged AB spin system is observed.

The 100 MHz ^{1}H spectrum measured in COS/CS$_2$ at –138°C [5.13] is shown in Fig. 5.12.

352   5 Time Dependent Phenomena

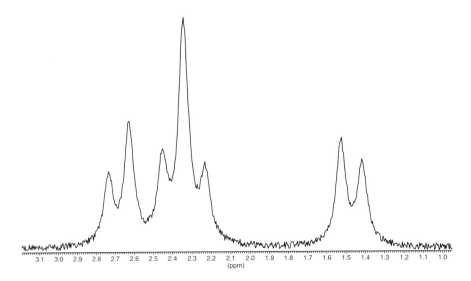

Fig. 5.12: Slow exchange limit spectrum of the annulene.

The exchange process can be analyzed in the following way. Two different AB type spin systems are present in the low temperature spectrum. The labeling AB and CD indicates two *different* AB spin systems not a single four spin system. According to the chemical shifts, the bridge head CH_2's should be labeled AC and BD, but this could led to confusion so consecutive letters will be used for the protons in the same CH_2 group.

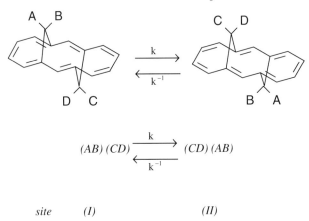

Fig. 5.13:   Exchange process definition for the annulene.

There are two concerted exchange processes taking place: AC ↔ BD and at the same time and with an identical rate constant BD ↔ AC (refer to Fig. 5.13).

5.3 Magnetic Site Exchange   353

Check it in WIN-DAISY:

Open... the input document EXCH_4.MGS. The WIN-DAISY document contains two fragments defining the two AB type spin systems. Because both spin systems reside in the same molecule they are equally weighted (**Statistical weight** in the **Main Parameters**). For both fragments, inspect the parameters in the various dialog boxes.

To define the exchange system the two individual fragments have to be combined into one fragment corresponding to the slow exchange spin system. The data can then be exported to WIN-DYNAMICS.

Check it in WIN-DAISY:

Use the **Edit | Merge fragment with...** command or button in the tool bar, to combine both fragments into one. Accept the **Fragment handling** dialog box with **Ok** and delete the merged fragment (**Ok**). In remaining fragment, the **Title** "AB + CD" or "CD + AB" is displayed depending on which of the original fragments was the origin and which the destination. **Export** to **WIN-DYNAMICS** the data set.

Check it in WIN-DYNAMICS:

Open the **Main Parameters** dialog box. The **Selected Spin system** is **ABCD** indicating a four spin system in the low-temperature spectrum. Increase the **Number of exchanging sites:** to <2>. Switch with the **Next** button into the **Static Parameters** and check the **Mutual Exchange** option. Note that the ABCD system (AB and CD) will be converted into CDAB by the exchange process. Press the **Exchange vectors** button and enter the appropriate permutations for the exchange process, e.g. <3>, <4>, <1> and <2>.

Open the **Rate Constants** dialog box and enter a large rate constant to calculate the fast-exchanging spectrum, e.g. <1e6>. Run a **Simulation** and **Display** the **Calculated Spectrum** (red peak button) in WIN-NMR. If the spectrum is not identical with Fig. 5.14 check the **Exchange vectors** in the **Static Parameters** dialog box.

Check it in the Calculator:

Calculate the average chemical shifts for the high temperature AB spin system. Using the formula given below, determine the value of the average coupling constant $J_{average}$.

$$v_{A,average} = p_I \cdot v_{A,I} + p_{II} \cdot v_{A,II}$$

$$v_{B,average} = p_I \cdot v_{B,I} + p_{II} \cdot v_{B,II}$$

with $v_{A,II} = v_{C,I}$
$v_{B,II} = v_{D,I}$
$p_I = p_{II} = 0.5$

$$J_{AB,average} = p_I \cdot J_{AB,I} + p_{II} \cdot J_{AB,II}$$
$$J_{CD,average} = p_I \cdot J_{CD,I} + p_{II} \cdot J_{CD,II}$$

In this case both the coupling constants are identical, so that the averaged coupling constants is the same as the static coupling constants.

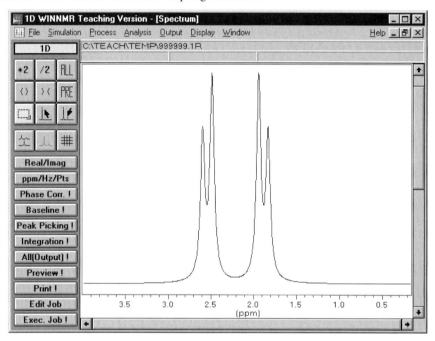

Fig. 5.14: WIN-NMR screen showing the fast exchanging annulene spin system.

Check it in WIN-DYNAMICS:

Open the **Rate Constants** dialog box again and change the rate to <0> s⁻¹ for a vanishing initial exchange rate. Open the **Simulation Parameters** dialog box and enter a **Step Width** of <50> s⁻¹ and a **Number of Steps** of <50>. Run a **Sequence** simulation and **Display** the **Calculated Spectrum** (red peak button) corresponding to the slow-exchange limit in WIN-NMR.

Note: It is possible that a "DYNNMR system error" may occur because many of the coupling constants in annulene are zero and the two non-zero coupling constants are the same, so that the matrix diagonalization routine fails. If this error does occur, open the **Static Parameters** dialog box and change the value of J_{12} to <11.1> Hz and run the sequence simulation again.

Check it in WIN-NMR:

Load all fifty one simulated spectra into the **Display | Multiple Display** to get an overview of the exchange process. Toggle **back to WIN-DYNAMICS**.

Check it in WIN-DYNAMICS:

Open the **Simulation Parameters** dialog box again and enter a smaller **Step Width** of <5> s^{-1} and a smaller **Number of Steps** e.g. <20> to study the beginning of the process in more detail. Run the sequence simulation again and **Display** the **Calculated Spectrum** (red peak button) in WIN-NMR. Load all twenty one simulated spectra into the **Display | Multiple Display**.

The next example to be discussed refers to *furan-2-aldehyde* (2-furfural) [5.14]. Here the rotation of the of the aldehyde group can be studied using variable temperature measurements Fig. 5.15.

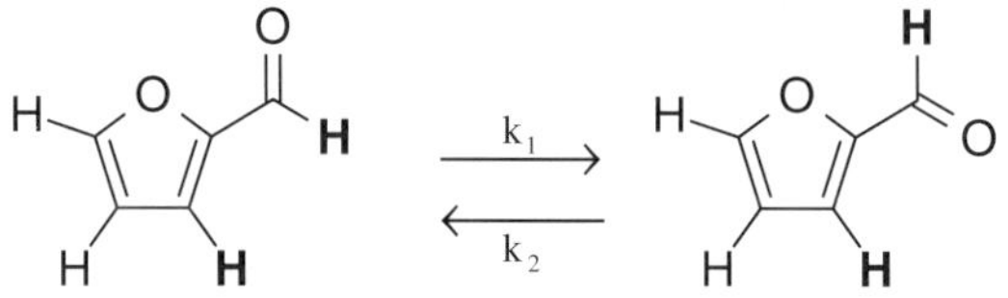

site   (I)   (II)

Fig. 5.15: Rotation around C–C bond in *2-furfural*.

Because the aldehyde proton has different chemical shifts in site (I) and (II) it is a non-mutual exchange process and the equilibrium constant must be used to define different populations for the two sites. In addition the aldehyde proton is also coupled to the proton at position 3 of the furan ring (printed in bold in Fig. 5.15), which itself is a member of the coupled spin system of the three furan protons. To correctly represent the exchange process a four spin system is needed. The exchange process can be formulated as shown in Fig. 5.16. The bold printed letters refer to the bold printed protons in Fig. 5.15.

$$AKMX \underset{k_2}{\overset{k_1}{\rightleftarrows}} BKNX$$

Fig. 5.16: Formalized exchange process of *2-furfural*.

Proton A and M of site (I) convert into B and N respectively in site (II). The protons printed in the regular type face coupled to them, but the shifts are not affected by the exchange process. In the low temperature spectrum the relative populations of the two sites have to be determined using signals A and B or M and N.

Check it in WIN-DAISY and WIN-NMR:

Load the input document EXCH_5.MGS. For both fragments defined as **site I** and **site II** inspect the values in the **Main Parameters**, **Frequencies** and **Scalar Coupling Constants** dialog boxes. Note that the shifts of nuclei 3 and 4 are identical in both fragments. Run a **Simulation** and **Display** the **Calculated Spectrum** (red peak button) in WIN-NMR.

Enter the **Analysis | Integration** mode and integrate the spectrum.

Check it WIN-DAISY and with the Calculator:

Toggle back to **WIN-DAISY** and run the **Calculator**. Using the integration data, calculate the relative populations of site I and II ensuring that $\Sigma p_i = 1$. Compare the calculated ratio with the values of the **Statistical weight** given in the **Main Parameters**. Open the **Frequencies** dialog box and using the normalized populations, calculate the averaged NMR parameters ν_A, ν_M, J_{AK}, J_{AX} for the fast exchange spectrum. (The notation refers to site I). **Export** to **WIN-DYNAMICS** the WIN-DAISY data.

Check it in WIN-DYNAMICS:

Open the **Rate Constants** dialog box and enter a large value, e.g. <1e6> to generate the averaged spin system. Run a **Simulation** and **Display** the **Calculated Spectrum** (red peak button) in WIN-NMR.

Check it in WIN-NMR:

In order to determine the averaged NMR parameters based on the high temperature spectrum enter the **Analysis I Multiplets** mode. Analyze the spectrum (define the multiplet trees and connect the coupling constants). Measure a suitable **Linewidth** and then switch to the **Analysis I Interval** mode and define a iteration region. Press the **Export...** button and transfer the interval data and the first order multiplet analysis to **WIN-DAISY**.

Check it in WIN-DAISY and WIN-NMR:

Run a spectrum **Iteration** and display the "experimental" and calculated spectrum using the **Dual Display** mode of WIN-NMR. Examine the statistical errors and the value of *R-Factor* in the **Iteration Protocol** file.

The iteration result shows slight differences between the fast exchanging spin system and the WIN-DAISY calculation due to the different linewidths of the four signals.

Check it in WIN-DAISY:

Open the **Lineshape Parameters** dialog box. Check the **Nuclei specific linewidths** option and press the **Iterate All** button. Run another **iteration**. Inspect the result in the WIN-NMR **Dual Display** and check the value of **R-Factor.**

Although the R-Factor in the iteration protocol list does not reach 0% (what would be the exact replication of the experimental of the spectrum) the value does fall below 0.05%. Note that the linewidths of the exchanging aldehyde proton and one of the furan proton are broader than the two remaining static furan protons. This might be due to either too low rate constant or long-range coupling from the aldehyde proton A to the furan proton X. It is quite possible that the value of J_{AX} is so small that no splitting will be visible in the spectrum and that the overall effect will be a contribution to the overall line broadening,

Check it in WIN-DAISY:

Open the **Scalar Coupling Constants** dialog box and enable J_{14} for iteration. In the **Lineshape Parameters** dialog box *deselect* the **Nuclei specific linewidths** and check the **Iterate** option for the **Global Linewidth**. Run an **Iteration**. Compare the optimized exchange averaged parameters in the **Iteration Protocol** file with the calculated parameters.

You may also select some intermediate values for the rate constant and examine how the parameters obtained using WIN-DAISY, assuming no chemical exchange, vary as a function of the rate constant.

Check it in WIN-DAISY, WIN-DYNAMICS and WIN-NMR:

Close the WIN-DAISY document and switch to WIN-DYNAMICS. Open the **Rate Constants** dialog box and enter the value <400> s⁻¹ for the rate constant. Run a **Simulation** and **Display** the **Calculated Spectrum** (red peak button) in WIN-NMR. As in the previous "Check its" analyze the **Multiplets**, measure a **Linewidth**, define an **Interval** and **Export...** all the data to WIN-DAISY. Open the **Scalar Coupling Constants** dialog box and select the **Iterate All** option to include the long-range couplings. In the **Lineshape Parameters** dialog box check the **Iterate** option for the **Global Linewidth**. Run an **Iteration** and check the values of the averaged NMR parameters and the *R-Factor*.

Enable the **Nuclei specific linewidth** option in the **Lineshape Parameters** and press the **Iterate All** button. Run an **Iteration** and inspect the **Iteration Protocol** file.

It is obvious from these results that for spectra still exhibiting exchange line broadening, the use of nuclei specific linewidths in the determination of the averaged NMR parameters is indispensable.

In the last part of this chapter we will examine symmetrical spin systems at either the low or the high temperature limit. The stereochemistry of dynamic process involving cyclohexane compounds is of particular interest. The full protonated compound would be too large a spin system to deal with, so the compounds are often partly deuterated. In the $^1H\{^2H\}$ spectrum the isotopomer 2,2,3,3,4,4,5,5-cyclohexane-d₈ displays a four spin system [5.15].

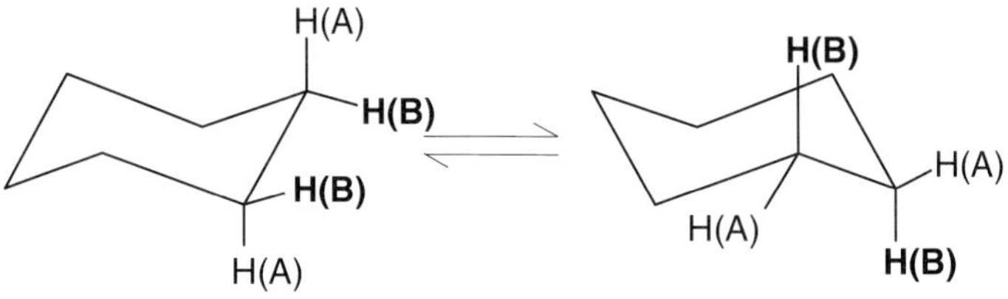

site   (I)   (II)

Fig. 5.17: Exchange between equatorial and axial protons in cyclohexane, for clarity, the deuterons are not shown in the structure.

The individual sites show C_2 symmetry and the exchange process is an example of mutual exchange:

$$[AB]_2 \underset{k_2}{\overset{k_1}{\rightleftarrows}} [BA]_2$$

Check it in WIN-DAISY:

Load the input document EXCH_6.MGS and inspect the parameters in the various dialog box. Look in particular at the **Symmetry Group** and **Symmetry Description**. **Export to WIN-DYNAMICS** the WIN-DAISY data.

Check it in WIN-DYNAMICS:

Open the **Main Parameters** dialog box and increase the **Number of exchanging sites:** to <2>. Switch to the **Static Parameters** dialog box using the **Next** button and check the **Mutual Exchange** option. Press the **Exchange vectors** button and enter the exchange row <3>, <4>, <1> and <2> because spin A turns into B and A' into B'. Quit the dialog box with **Ok**. Run a **Simulation** and **Display** the **Calculated Spectrum** (red peak button) in WIN-NMR.

Toggle back to **WIN-DYNAMICS** and open the **Simulation Parameters**. Enter a **Step Width** of <7.5> s^{-1} and a **Number of Steps** of <15>. Run a **Simulation** sequence and **Display** the **Calculated Spectrum** (red peak button) in WIN-NMR.

Check it in WIN-NMR:

Call the **Display | Multiple Display** mode and load the sixteen calculated spectra ...\999999.1R down to ...\999985.1R. Click the **Stack Plot** and/or **Separate** button to examine the dynamic process.

The process shows the broadening of the lines and the coalescence to a single peak, at the average frequency. The present case describes the equilibrium between two sites in a symmetrical spin system. A quite different type of exchange problem occurs when the mutual exchange is between two unsymmetrical spin systems and where fast exchange leads to a highly symmetrical averaged spin system. An example of this type of exchange is benzofuroxan (Fig. 5.18).

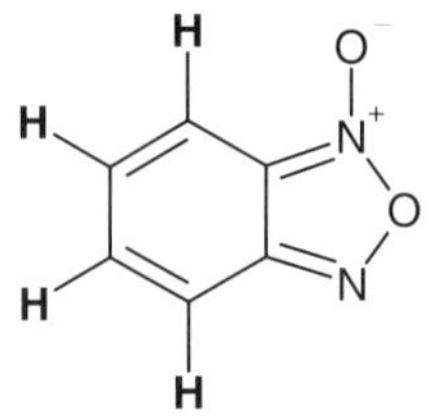

Fig. 5.18: Structure of *benzofuroxan*.

5.3 Magnetic Site Exchange

The ¹H NMR spectrum of *benzofuroxan* at room temperature exhibits the features of an [AB]₂ spin system type while the expected spin system based on the structure given in Fig. 5.18 would be an unsymmetrical ABCD spin system. Clearly there is some sort of dynamic process and one possible interconversion process is shown in Fig. 5.19.

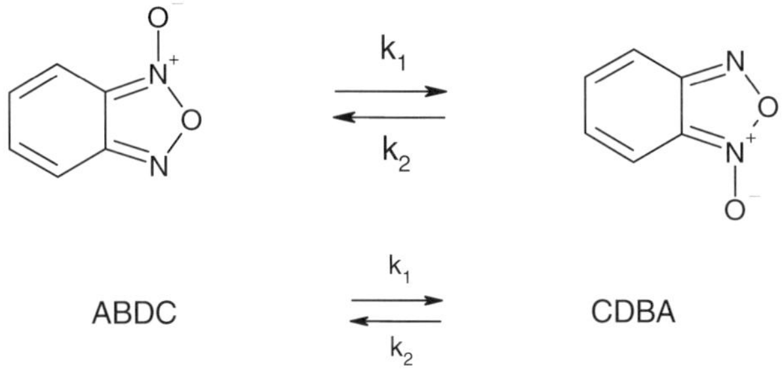

site   (I)   ABDC ⇌ CDBA   (II)

Fig. 5.19: Interconversion of *benzofuroxan*.

Check it in WIN-DAISY:

Load the input document EXCH_7.MGS, the parameters correspond to the unsymmetrical low-temperature ABCD spin system. **Export to WIN-DYNAMICS** the WIN-DAISY data.

Check it in WIN-DYNAMICS:

Open the **Main Parameters** dialog box and increase the **Number of exchanging sites:** to <2>. Use the **Next** button to change to the **Static Parameters** dialog box. Check the **Mutual Exchange** option, press the **Exchange vectors** button and enter the exchange row <4>, <3>, <2> and <1>. Open the **Simulation Parameters** and enter a **Step Width** of <15> s⁻¹ and a **Number of Steps** of <20>. Run a simulation **Sequence** and **Display** the **Calculated Spectrum** (red peak button) in WIN-NMR.

Check it in WIN-NMR:

Call the **Display | Multiple Display** and load the twenty one calculations ...\999999.1R down to ...\999979.1R and inspect the changing spectrum pattern.

The broadening of the spectrum and the formation of a broad rather unsymmetrical pattern can be clearly seen. The unsymmetrical pattern occurs because the difference in chemical shift between A and C and between B and D which are supposed to be averaged by the exchange process is not the same. Hence the fast exchange limit spin system can be labeled [EF]₂ if we define the averaged chemical shifts E and F as follows:

$$v_E = \frac{v_A + v_C}{2} \quad \text{and} \quad v_F = \frac{v_B + v_D}{2}$$

Check it in WIN-DYNAMICS:

Toggle back to **WIN-DYNAMICS** and open the **Simulation Parameters** dialog box. Enter a **Step Width** of <1000> s⁻¹ and **Number of Steps** <6> in order to reach the fast exchange limit.

Check it in WIN-NMR:

Inspect the seven spectra ...\999999.1R down to ...\999993.1R in the **Display | Multiple Display** mode. Load the last simulated spectrum into WIN-NMR and measure a suitable linewidth. Enter the **Analysis | Multiplets** mode, to estimate the chemical shifts define two singlets. Switch to the **Analysis | Interval** mode, define a iteration region and **Export...** the interval and the multiplet data *without* fragmentation of the spin system to WIN-DAISY.

Check it in WIN-DAISY:

Set up a symmetrical spin system; in the **Main Parameters** dialog box increase to <4> the **Number of spins or groups with magnetical equivalent nuclei**, in the **Symmetry Group** select **C2** or **Cs** symmetry and in the **Symmetry Description** enter <4>, <3>, <2> and <1>. Open the **Scalar Coupling Constants** dialog box and enter some small values for the coupling constants e.g. <1>, <2>, <3> and <4> Hz. In the **Control Parameters**, press the **Frequencies** button and enter a **Lower limit** e.g. <720> Hz and **Upper limit** <760> Hz for both spins. Close the dialog box with **Ok** and press the **Scalar Coupl.** button. **Select all** for iteration and set the **Lower limit** to <0> Hz and the **Upper limit** to <12> Hz. Close the dialog box with **Ok**. Select **Advanced Iteration Mode** with smoothing parameters, **medium Broadening** and **medium No. of cycles**. Check the option **Starting frequencies poor**. Run one or more **Iterations** and display the two spectra using the **Dual Display** mode of WIN-NMR. View the **Iteration Protocol** file and examine the statistics and averaged parameters.

The rate constant of 6000 s⁻¹ is still not large enough to reach the fast exchange limit and there are still some differences between the spectra. You may check the fast exchange limit by simulating the spectrum using an exchange rate of 10^8 s⁻¹ and then repeating the procedure in the previous "Check it" to extract the correct averaged values. In the light of the preceding discussion, the reason for the slight asymmetry in line position, intensity and broadening becomes clear, the asymmetry originates from an exchange process that has not quite reached the fast exchange limit.

In the experimental spectrum of *morpholine* ...\MORPHOL\001001.1R, discussed in section 3.2.3, the symmetry of the spin system arises from the same type of exchange process that occurs in *benzofuroxan* but with the additional complication that the quadrupole moment of the nitrogen broadens the A part of the [AB]₂ spectrum more than the B part but the fast exchange limit is already reached [5.16].

5.4 References

[5.1] Bovey F.A., *Nuclear Magnetic Resonance Spectroscopy*, 2nd ed., New York: Academic Press, 1988.
[5.2] Akitt, J.W., *NMR and Chemistry*, 3rd ed., London: Chapman & Hall, 1992.
[5.3] Günther, H., *NMR Spectroscopy*, 2nd ed., Chichester: Wiley, 1987.
[5.4] Friebolin, H., *Basic One- and Two-Dimensional NMR Spectroscopy*, VCH, Weinheim, 1991.
[5.5] Harris, R.K., *Nuclear magnetic Resonance Spectroscopy*, Harlow: Longman, 1986.
[5.6] Noggle, J.H., Schirmer, R.E., *The Nuclear Overhauser Effect*, Academic Press, London, 1971.
[5.7] Neuhaus, D., Williamson, M.P., *The Nuclear Overhauser Effect in Structural and Conformational Analysis*, Weinheim: VCH, 1989.
[5.8] Steigel, A., in: *Basic Principles and Progress*: Diehl, P., Fluck, E., Kosfeld, R. (Eds.) Berlin: Springer, 1971, *15*, 2-53.
[5.9] Jackman, L.M., Cotton, F.A. (Eds.), *Dynamic Nuclear Magnetic Resonance Spectroscopy*, London: Academic Press, 1975.
[5.10] Sandström, J., *Dynamic NMR Spectroscopy*, London: Academic Press, 1982.
[5.11] Bigler, P., *NMR Spectroscopy: Processing Strategies*, Weinheim: Wiley-VCH, 1997.
[5.12] Heidberg J., Weil J.A., Janusonis G.A., Anderson J.K., *J. Chem. Phys.*, 1964, *41*, 1033.
[5.13] Vogel E., Haberland U., Günther H., *Angew. Chem.,* 1970, *82*, 510.
[5.14] Dahlqvist K.-I., Forsén S., *J. Phys. Chem.*, 1965, 69, 1760 and 4062.
[5.15] Höfner, D., Lesko, S.A., Binsch , G., *Organ. Magn. Reson.*, 1978, *11*, 179.
[5.16] Harris, R.K., Spragg, R.A., *J. Chem. Soc., Chem. Comm.*, 1966, 314.

6 Appendix

6.1 Quantum Mechanical Theory

A good introduction into the quantum mechanical theory for the calculation of spin systems is given in the textbooks [6.1]-[6.3]. A very detailed description can be found in [6.4] and [6.5].

6.1.1 The AB Spin System and the General Case

When first order spin systems were explained in section 2.1 the HAMILTONian was truncated after the first perturbation term $\hat{H}_1$ by neglecting $\hat{H}_2$ in the so-called X approximation. Regarding the first order HAMILTONian $\hat{H}_0 + \hat{H}_1$ all m_T values are so-called *good quantum numbers*.

$$\hat{H}_0 = -\sum_{i=1}^{N} v_i \cdot \hat{I}_{z,i}$$

$$\hat{H}_1 = \sum_{i<k}^{N}\sum^{N} J_{ik} \cdot \hat{I}_{z,i} \cdot \hat{I}_{z,k}$$

Consequently all bpf's directly describe EIGENstates to the first order HAMILTONian. To introduce the second perturbation - given by $\hat{H}_2$ - of the nuclear energy levels the contributions of $\hat{H}_0 + \hat{H}_1$ must first written in a matrix representation of the following type:

$$\left(\hat{H}_0 + \hat{H}_1\right)|\phi_i\rangle = H_{ii}|\phi_i\rangle \quad \text{with } i = 1, \ldots 2N$$

Fig. 6.1 gives the matrix representation, to be read row by row, and illustrates that only *diagonal* contributions are present, which are identical with the EIGENvalues in the 1st order case.

$\hat{H}_0 + \hat{H}_1$	$\|\alpha\alpha\rangle$	$\|\alpha\beta\rangle$	$\|\beta\alpha\rangle$	$\|\beta\beta\rangle$	m_T
$\|\alpha\alpha\rangle$	$-½\nu_A-½\nu_B+¼J_{AB}$	0	0	0	1
$\|\alpha\beta\rangle$	0	$-½\nu_A+½\nu_B-¼J_{AB}$	0	0	0
$\|\beta\alpha\rangle$	0	0	$+½\nu_A-½\nu_B-¼J_{AB}$	0	0
$\|\beta\beta\rangle$	0	0	0	$+½\nu_A+½\nu_B+¼J_{AB}$	-1

Fig. 6.1: Hamilton matrix scheme containing contributions of $\hat{H}_0 + \hat{H}_1$.

Because the mixing term of the HAMILTONian $\hat{H}_2$ only consists of lowering and raising operators, only *off-diagonal elements* in the HAMILTONian matrix are generated when mixing different bpf's.

$$\hat{H}_2 = \frac{1}{2}\sum_{i<k}^{N}\sum^{N} J_{ik} \cdot \left(\hat{I}_{+,i} \cdot \hat{I}_{-,k} + \hat{I}_{-,i} \cdot \hat{I}_{+,k}\right)$$

For a two spin system $I = ½$ there are four basis product functions and a total of there would be $4\cdot 4=16$ matrix elements to fill in, following the equation given below:

$$\hat{H}|\phi_j\rangle = \sum_{i=1}^{2^N} H_{ij}|\phi_i\rangle$$

The matrix entries correspond to the integrals obtained by multiplying the equation above from the left-hand side by the conjugated complex function $|\phi_i\rangle$:

$$\langle\phi_i|\hat{H}|\phi_j\rangle = H_{ij}$$

The four diagonal elements remain unchanged by $\hat{H}_2$ and are already completely defined by the ZEEMAN and Coupling term, so that only 12 unknown contributions (with $i \neq j$) remain.

Based on the commutator probabilities between $\hat{H}$ and $\hat{F}_z$, the matrix entries between bpf's that are EIGENfunctions of $\hat{F}_z$ but belong to *different* EIGENvalues (different m_T values) vanish. Thus the functions $|\alpha\alpha\rangle$ and $|\beta\beta\rangle$ are also EIGENstates of the whole HAMILTONian, because there is no other function with $m_T = +1, -1$. Therefore no off-diagonal elements in the first and last row in Fig. 6.1 have to be considered. But the two functions $|\alpha\beta\rangle$ and $|\beta\alpha\rangle$ are no longer EIGENfunctions - they belong both to the same total spin value $m_T = 0$.

Consequently only two additional matrix elements H_{23} and H_{32} have to be considered for the two-spin case and only 6, not 16, elements have to be calculated as symbolized in Fig. 6.2:

6.1 Quantum Mechanical Theory

$\hat{H}$	αα	αβ	βα	ββ	m_T
αα	H_{11}	0	0	0	1
αβ	0	H_{22}	H_{23}	0	0
βα	0	H_{32}	H_{33}	0	0
ββ	0	0	0	H_{44}	-1

Fig. 6.2: HAMILTON matrix scheme [H] for the whole HAMILTONian $\hat{H}$.

To complete the HAMILTONian matrix for the whole Hamilton operator $\hat{H}$ the effect of the mixing term $\hat{H}_2$ on the bpf's with $m_T = 0$ need to be taken into account:

$$\hat{H}_2|\alpha\beta\rangle = \tfrac{1}{2} J_{AB}\left(\hat{I}_{+,A} \cdot \hat{I}_{-,B} + \hat{I}_{-,A} \cdot \hat{I}_{+,B}\right)|\alpha\beta\rangle$$
$$= \tfrac{1}{2} J_{AB}\left(\hat{I}_{+,A} \cdot \hat{I}_{-,B}\right)|\alpha\beta\rangle + \tfrac{1}{2} J_{AB}\left(\hat{I}_{-,A} \cdot \hat{I}_{+,B}\right)|\alpha\beta\rangle$$
$$= 0 + \tfrac{1}{2} J_{AB}|\beta\alpha\rangle$$

$$\hat{H}_2|\beta\alpha\rangle = \tfrac{1}{2} J_{AB}\left(\hat{I}_{+,A} \cdot \hat{I}_{-,B} + \hat{I}_{-,A} \cdot \hat{I}_{+,B}\right)|\beta\alpha\rangle$$
$$= \tfrac{1}{2} J_{AB}\left(\hat{I}_{+,A} \cdot \hat{I}_{-,B}\right)|\beta\alpha\rangle + \tfrac{1}{2} J_{AB}\left(\hat{I}_{-,A} \cdot \hat{I}_{+,B}\right)|\beta\alpha\rangle$$
$$= 0 + \tfrac{1}{2} J_{AB}|\alpha\beta\rangle$$

The full HAMILTONian matrix is as follows:

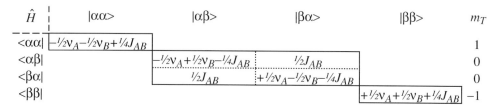

Fig. 6.3: Complete HAMILTON matrix [H] for an AB spin system.

The structure of the Hamilton matrix shows that the off-diagonal contribution is always $\tfrac{1}{2} J_{AB}$. The matrix is symmetrical with respect to the diagonal ($H_{23}=H_{32}$) so that in fact only 5 matrix elements need to be calculated.

This method where the complete HAMILTONian is subdivided into smaller parts according to the total spin values is named the *total spin (m_T) factorization*. The complete matrix of dimension 2^N is factored into smaller submatrices with dimensions resulting from the binomial coefficient (the so-called Pascal's triangle):

Table 6.1: Binomial coefficients to determine the sizes of submatrices, the row representing the two-spin system is printed bold.

total matrix dimensions	binomial coefficients dimensions of submatrices						no. of spins	no. of submatrices
$\Sigma=2^N$							N	N+1
1			1				0	1
2			1	1			1	2
4		1	2	1			2	3
8	1	3	3	1			3	4
16	1	4	6	4	1		4	5

The formula to calculate the binomial coefficients is as follows:

$$\text{dimension of submatrices} = \binom{N}{k} = \frac{N!}{k! \cdot (N-k)!}$$

with N = *number of spins*
and k = *1, ... N*

The next problem to solve is the way to extract the EIGENvalues from the factored Hamilton matrix. The EIGENvalues for the 'good quantum numbers' $m_T = +1, -1$ are already defined by the diagonal elements H_{11} and H_{44} of the matrix. They are identical with the first order values:

$$m_T = +1: \quad \hat{H}|\alpha\alpha\rangle = H_{11}|\alpha\alpha\rangle$$
$$m_T = -1: \quad \hat{H}|\beta\beta\rangle = H_{44}|\beta\beta\rangle$$

with $H_{11} = E_1 = -\frac{1}{2}\nu_A - \frac{1}{2}\nu_B + \frac{1}{4} J_{AB}$
with $H_{44} = E_4 = +\frac{1}{2}\nu_A + \frac{1}{2}\nu_B + \frac{1}{4} J_{AB}$

The case is different for the submatrix with the 'bad quantum number' $m_T = 0$:

$$\hat{H}|\alpha\beta\rangle = H_{22}|\alpha\beta\rangle + H_{23}|\beta\alpha\rangle$$
$$\hat{H}|\beta\alpha\rangle = H_{32}|\alpha\beta\rangle + H_{33}|\beta\alpha\rangle$$

The aim is to modify the wave functions |αβ> and |βα> in such a way, that they satisfy the SCHROEDINGER equation:

6.1 Quantum Mechanical Theory

$$\hat{H}|\Psi_2\rangle = E_2|\Psi_2\rangle$$
$$\hat{H}|\Psi_3\rangle = E_3|\Psi_3\rangle$$

In order to do that, we use the basic approach to develop the desired EIGENfunctions as a linear combination based on the bpf's:

$$|\Psi_2\rangle = c_{22}|\alpha\beta\rangle + c_{23}|\beta\alpha\rangle$$
$$|\Psi_3\rangle = c_{32}|\alpha\beta\rangle + c_{33}|\beta\alpha\rangle$$

Inserting, for example the first linear combinations for $|\Psi_2\rangle$ into the SCHROEDINGER equation leads to:

$$\hat{H}(c_{22}|\alpha\beta\rangle + c_{23}|\beta\alpha\rangle) = E(c_{22}|\alpha\beta\rangle + c_{23}|\beta\alpha\rangle)$$

Multiplication from the left with the conjugated complex bpf $\langle\alpha\beta|$ and integration yields in:

$$c_{22}\cdot\langle\alpha\beta|\hat{H}|\alpha\beta\rangle + c_{23}\cdot\langle\alpha\beta|\hat{H}|\beta\alpha\rangle = c_{22}\cdot E\langle\alpha\beta|\alpha\beta\rangle + c_{23}\cdot E\langle\alpha\beta|\beta\alpha\rangle$$

The integral values on the left-hand side of the equation correspond to the entries of the HAMILTONian matrix previously discussed, and the integrals on the right-hand side are given by the orthonormality definitions, so that the equation can be rewritten as:

1. $c_{22}\cdot H_{22} + c_{23}\cdot H_{23} = c_{22}\cdot E$

Repeating this method with $|\Psi_3\rangle$ as just done with $|\Psi_2\rangle$ with the other conjugated complex bpf $\langle\beta\alpha|$:

$$c_{22}\cdot H_{32} + c_{23}\cdot H_{33} = c_{23}\cdot E$$

The two types of equations developed above are called secular equations. To point out the structure, the right-hand side of the equations is equated to zero:

$$c_{22}\cdot(H_{22}-E) + c_{23}\cdot H_{23} = 0$$
$$c_{22}\cdot H_{32} + c_{23}\cdot(H_{33}-E) = 0$$

In matrix notation we obtain for the first secular equation:

$$\begin{pmatrix} H_{22}-E & H_{23} \\ H_{32} & H_{33}-E \end{pmatrix}\begin{pmatrix} c_{22} \\ c_{23} \end{pmatrix} = 0$$

To determine the EIGENvalue E by calculation of the secular determinant: When the determinant is zero, the energy level E is obtained:

$$\begin{vmatrix} H_{22} - E & H_{23} \\ H_{32} & H_{33} - E \end{vmatrix} = 0$$

In general the determinant has always the structure:

$$\begin{vmatrix} H_{ii} - E & H_{i,i+1} & \ldots & H_{i,n} \\ H_{i+1,i} & H_{i+1,i+1} - E & \ldots & H_{i+1,n} \\ \ldots & \ldots & \ldots & \ldots \\ H_{n,i} & \ldots & \ldots & H_{n,n} - E \end{vmatrix} = 0 \Leftrightarrow \|H - \delta \Lambda\| = 0$$

with [H] = HAMILTONian matrix
[Λ] = diagonal matrix containing the energy levels E on the main diagonal.
δ = KRONECKER symbol

Both energy levels (E_2 and E_3) can be calculated by solving the secular determinant.

In the actual problem of the 2x2 submatrix of the AB spin system the secular determinant may be solved using a quadratic equation that has two solutions, corresponding to the two desired energy levels:

$$(H_{22} - E)(H_{33} - E) - H_{23} \cdot H_{32} = 0$$

$$E_{2,3} = \frac{1}{2}\left(H_{22} + H_{33} \pm \sqrt{(H_{22} - H_{33})^2 + 4 \cdot H_{23} \cdot H_{32}} \right)$$

Introducing the quantities J_{AB}, Δ_{AB} and D:

$$H_{22} + H_{33} = -\tfrac{1}{2} J_{AB},$$
$$H_{22} - H_{33} = \Delta_{AB},$$
$$4 \cdot H_{23} H_{32} = J_{AB}^2 \quad \text{and} \quad D = \sqrt{\Delta_{AB}^2 + J_{AB}^2}$$

the expressions to determine both energy levels simplifies to:

$$E_{2,3} = -\tfrac{1}{4} J_{AB} \pm \tfrac{1}{2} D$$

The secular determinant of other spin systems with larger submatrices are more difficult to solve. Computer programs such as WIN-DAISY make use of matrix diagonalization routines. They take as input the basis product functions and the HAMILTON matrix and calculate the EIGENvalues and the EIGENfunctions. The EIGENfunctions correspond with the coefficients c_{ik} of the linear combinations already mentioned. As discussed for two-spin systems in the case of weakly coupled spins earlier in sections 2.1.2.1 and 2.1.2.2 the transition intensities correspond to the transition probabilities based on the EIGENfunctions. In order to determine the transition probabilities we need to calculate the EIGENfunctions for the present AB spin system.

6.1 Quantum Mechanical Theory

The knowledge available for the spin system is the functional basis consisting of basis product functions (bpf's) and according to that the HAMILTONian matrix [H]. We are searching for a transformation matrix [U] which turns the HAMILTONian matrix into a diagonal matrix [Λ] which contains the EIGENvalues on its main trace:

$$[H][U] = [\Lambda][U] \Rightarrow [U]^T[H][U] = [\Lambda]$$

with $[U]^T$ = transposed of matrix [U], rows and columns interchanged.

The unitary transformation matrix [U] for the AB spin system consists of six non-zero elements according to the HAMILTONian submatrices:

$$m_T = 1 \quad 0 \quad -1$$

$$\begin{pmatrix} c_{11} & & & \\ & c_{22} & c_{32} & \\ & c_{23} & c_{33} & \\ & & & c_{44} \end{pmatrix} = [U]$$

For the *good quantum numbers* $m_T = \pm 1$, the coefficients are 1: $c_{11} = c_{44} = 1$.

The coefficients for the *bad quantum number* $m_T = 0$ have to fulfill the normalization conditions:

$$c_{22}^2 + c_{23}^2 = 1 \quad \text{and} \quad c_{32}^2 + c_{33}^2 = 1$$

We may use the trigonometric functions sine and cosine to start with. The angle Θ is identical with the angle extracted in the graphical solution:

$$sin^2\Theta + cos^2\Theta = 1$$

To satisfy the orthogonality of the EIGENfunctions, the following conditions must also be fulfilled:

$$c_{22} = c_{33} \quad \text{and} \quad c_{23} = -c_{32}$$

Table 6.2: Coefficients c_{ik} of the bpf to form the EIGENfunctions.

C_{ik}	$\| \alpha\beta >$	$\| \beta\alpha >$
$\| \Psi_2 >$	cos Θ	sin Θ
$\| \Psi_3 >$	−sin Θ	cos Θ

The transition probabilities concerning the four possible transitions are as follows:

$$\langle \Psi_2|\hat{F}_-|\Psi_1\rangle = \langle \cos\Theta\cdot\alpha\beta + \sin\Theta\cdot\beta\alpha|\hat{F}_-|\alpha\alpha\rangle$$
$$= \langle \cos\Theta\cdot\alpha\beta + \sin\Theta\cdot\beta\alpha|\hat{I}_{-,A}|\alpha\alpha\rangle + \langle \cos\Theta\cdot\alpha\beta + \sin\Theta\cdot\beta\alpha|\hat{I}_{-,B}|\alpha\alpha\rangle$$
$$= \langle \cos\Theta\cdot\alpha\beta + \sin\Theta\cdot\beta\alpha|\beta\alpha\rangle \quad + \langle \cos\Theta\cdot\alpha\beta + \sin\Theta\cdot\beta\alpha|\alpha\beta\rangle$$
$$= \cos\Theta\langle\alpha\beta|\beta\alpha\rangle + \sin\Theta\langle\beta\alpha|\beta\alpha\rangle + \cos\Theta\langle\alpha\beta|\alpha\beta\rangle + \sin\Theta\langle\beta\alpha|\alpha\beta\rangle$$
$$= 0 + \sin\Theta + \cos\Theta + 0$$

The transition probability obtained for the four allowed transitions is listed below:

$$\langle\Psi_3|\hat{F}_-|\Psi_1\rangle = -\sin\Theta + \cos\Theta$$
$$\langle\Psi_2|\hat{F}_-|\Psi_1\rangle = \sin\Theta + \cos\Theta$$
$$\langle\Psi_4|\hat{F}_-|\Psi_2\rangle = \sin\Theta + \cos\Theta$$
$$\langle\Psi_4|\hat{F}_-|\Psi_3\rangle = -\sin\Theta + \cos\Theta$$

The transition intensity is connected with the square of the transition moments:

$$|\langle\Psi_3|\hat{F}_-|\Psi_1\rangle|^2 = (\sin\Theta + \cos\Theta)^2 = \sin^2\Theta - 2\cos\Theta\sin\Theta + \cos^2\Theta$$
$$|\langle\Psi_2|\hat{F}_-|\Psi_1\rangle|^2 = (\sin\Theta + \cos\Theta)^2 = \sin^2\Theta + 2\cos\Theta\sin\Theta + \cos^2\Theta$$
$$|\langle\Psi_4|\hat{F}_-|\Psi_2\rangle|^2 = (\sin\Theta + \cos\Theta)^2 = \sin^2\Theta + 2\cos\Theta\sin\Theta + \cos^2\Theta$$
$$|\langle\Psi_4|\hat{F}_-|\Psi_3\rangle|^2 = (\sin\Theta + \cos\Theta)^2 = \sin^2\Theta - 2\cos\Theta\sin\Theta + \cos^2\Theta$$

Under application of the following similarities,

$$2\cdot\cos\Theta\sin\Theta = \sin 2\Theta$$
$$\sin^2\Theta + \cos^2\Theta = 1$$

resulting in the transition intensities are already given in section 2.3.2.1, Table 2.10.

6.1.2 Composite Particle Theory

6.1.2.1 The A_2 Spin System

In case the A_2 spin system quantum mechanics will be treated with the second order two-spin HAMILTONian (according to the treatment of the AB spin system in section 6.1.1, Fig. 6.3) the HAMILTONian matrix will look like as follows:

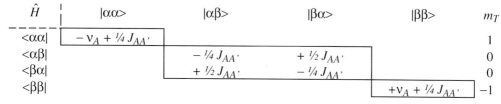

Fig. 6.4: Complete HAMILTON matrix [H] for an A_2 spin system.

The energy levels for the good quantum numbers $m_T = \pm 1$ are identical with the HAMILTONian entries, the intermediate energy levels for $m_T = 0$ can be obtained by using the solution for the corresponding modified secular equation:

To calculate the HAMILTONian matrix of composite particles (abbr. 'cp') we may define them separately for every irreducible component, because there is no transition possible between different irreducible components. Then it is possible to set up the HAMILTONian operator for every term j individually using the actual particle spin operators $\hat{I}(j)$ instead of the single spin operators $\hat{I}$. The accumulation of the HAMILTONian matrix entries is executed over the number of particles CP than over the number of spins N; no other changes are necessary. The complete isotropic time-independent HAMILTONian operator for the individual terms is given below:

$$\hat{H}(j) = -\sum_{i=1}^{CP} v_i \cdot \hat{I}_{z,i}(j)$$

$$+ \sum_{i<k}^{CP}\sum^{CP} J_{ik} \cdot \hat{I}_{z,i}(j) \cdot \hat{I}_{z,k}(j)$$

$$+ \frac{1}{2}\sum_{i<k}^{CP}\sum^{CP} J_{ik} \cdot \left(\hat{I}_{+,i}(j) \cdot \hat{I}_{-,k}(j) + \hat{I}_{-,i}(j) \cdot \hat{I}_{+,k}(j)\right)$$

The HAMILTONian necessary for the A_2 particle spin system is only the ZEEMAN contribution, because it has been already pointed out clearly that scalar inner-particle coupling constants are irrelevant for the transitions:

$$\hat{H}(j) = -v_A \cdot \hat{I}_{z,\{A_2\}}(j)$$

with $\hat{I}_{z,\{A_2\}}(j)$: z-component of the actual particle spin value in term j

The following schemes in Fig. 6.5 representing the complete HAMILTONian matrices for both terms describing the total A_2 spin system shows that the basis transformation does not change the total dimension (four basis functions) of the complete quantum mechanical system:

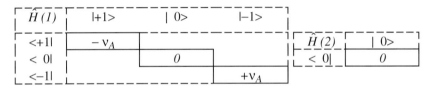

Fig. 6.5: left: HAMILTONian matrix [H] for $^3\{A_2\}$ term, right: for $^1\{A_2\}$ term.

The structure of the HAMILTONian matrices shows only the ZEEMAN contributions on the main trace, so that all total spin values are 'good quantum numbers' and the EIGENvalues are identical with the diagonal elements. The selection rule for calculation of the transition remains as usual, just applied to particles, not to single spins. No transitions between the different terms are allowed.

The possible transitions are determined using the total lowering operator $\hat{F}_-(j)$ composed of *particle lowering operators* $\hat{I}_{-,i}(j)$ for every particle i of the actual term j:

$$\hat{F}_-(j) = \sum_{cp=1}^{CP} \hat{I}_{-,cp}(j)$$

For the present A_2 particle the total lowering operator is as follows:

$$\hat{F}_-(j) = \hat{I}_{-,\{A_2\}}(j)$$

The spin lowering operator applied to spins or particles with any angular momentum I follows the operator equation given below:

$$\hat{I}_{-,cp}(j)|m_{cp}\rangle = \sqrt{F_{max} \cdot (F_{max} + 1) - m_{cp} \cdot (m_{cpt} - 1)}|(m_{cp} - 1)\rangle$$

with
$\hat{I}_{-,cp}(j)$: particle lowering operator,
$|m_{cp}\rangle$: particle basis function, symbolized by its angular momentum z-component,
F_{max} : angular momentum of the particle in the term $j \equiv$ maximum particle spin.

The intensity for each transition of 2.0 relative units arises from the normalization coefficient by calculating the transition moment using the total lowering operator, which consists only of a single particle lowering operator in the case of the A_2 system:

$$M_{2 \to 1} = \langle 0|\hat{I}_{-,cp}|+1\rangle = \langle 0|\sqrt{1 \cdot 2 - 1 \cdot 0}|0\rangle = \sqrt{2}$$
$$M_{3 \to 2} = \langle -1|\hat{I}_{-,cp}|0\rangle = \langle -1|\sqrt{1 \cdot 2 - 0 \cdot (-1)}|-1\rangle = \sqrt{2}$$

The intensity is calculated as the square of the transition probability.

$$G(F_{act}[i]) = B_1(i,n) - B_1(i-1,n)$$

with $i = 0,...t-1$ the term index
and $B_1(i,n)$: the ith element of the nth row of the *1*st order Pascal triangle

The first order Pascal triangle has been used earlier in Section 1.3.2.2 for determination of the sizes of submatrices in spin systems consisting of n nuclei (refer to Table 6.1):

$$B_1(i,n) = \binom{n}{i}$$

The definition of the spin multiplicities given above necessitates the definition of the element $B_1(-1,n)$:

$$B_1(-1,n) = \binom{n}{-1} \equiv 0$$

The introduced definition refers to zero's placed in front of each row. As a consequence, the particle multiplicity for the maximum particle spin F_{max} (refers to $i=0$) is always one:

$$G(F_{max}) = B_1(0,n) - B_1(-1,n) = 1 - 0 = 1$$

The statistical weight S of each subspectrum (irreducible component) is given by the product of the corresponding particle multiplicity of the involved particle states G:

$$S = \prod_{cp} G(F_{act})$$

In the present A_2X spin system both subspectra have equal intensity:

$$S(^3\{A_2\}^2X) = G(^3\{A_2\}) \cdot G(^2X) = 1 \cdot 1$$
$$S(^1\{A_2\}^2X) = G(^1\{A_2\}) \cdot G(^2X) = 1 \cdot 1$$

The total signal of spin X refers to a triplet. By default a triplet is connected with the intensity ratio 1:2:1. If a different intensity ratio is described, the intensity ratio has to be mentioned. For example, the subspectrum would be called a 1:1:1 triplet.

6.1.2.2 The A$_2$B Spin System

The inspection of the m_T diagram shown in section 2.4.6.1, Fig., 2.61, represents the following size of submatrices of the spin system, given in the following scheme in Fig. 6.6:

374   6 Appendix

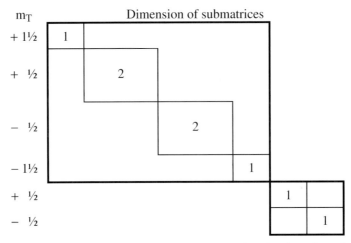

Fig. 6.6: HAMILTONian scheme for A_2B only representing the size of matrices.

According to the selection rule the maximum number of possible transitions in a HAMILTONian factored by subspectra and composite particle is given by the sum of the product of the dimensions of the sumbatrices being direct neighbors.

$$\sum_{i=1}^{g-1} dim_i \cdot dim_{i+1}$$

with $g = 2 \cdot F_{act} + 1$
and dim_i = dimension of submatrix i

In the present spin system the values are:

for the first subspectrum: $1 \cdot 2 + 2 \cdot 2 + 2 \cdot 1 = 8$ transitions
for the second subspectrum: $1 \cdot 1$ $= 1$ transition.

This circumstances can be worked out when the complete composite particle HAMILTONion is used to form the HAMILTONian matrix. The difference regarding the first order HAMILTONian of AX_2 spin systems is the presence of some off-diagonal elements within the 2 x 2 matrices. The off-diagonal elements refer to the third part of the composite particle HAMILTONian containing the flip-flop term. The interesting operator equations are the following taking into account the coefficients arising from the maximum particle spin F_{max}:

for $m_T = +\frac{1}{2}$

$$\frac{1}{2} J_{AB} \left(\hat{I}_{+,\{A_2\}} + \hat{I}_{-,B} \right) |0, \tfrac{1}{2}\rangle = \tfrac{1}{2} \sqrt{2} J_{AB} |1, -\tfrac{1}{2}\rangle = \tfrac{1}{\sqrt{2}} J_{AB} |1, -\tfrac{1}{2}\rangle$$

$$\frac{1}{2} J_{AB} \left(\hat{I}_{-,\{A_2\}} + \hat{I}_{+,B} \right) |1, -\tfrac{1}{2}\rangle = \tfrac{1}{2} \sqrt{2} J_{AB} |0, \tfrac{1}{2}\rangle = \tfrac{1}{\sqrt{2}} J_{AB} |0, \tfrac{1}{2}\rangle$$

for $m_T = -½$:

$$\tfrac{1}{2} J_{AB}(\hat{I}_{+,\{A_2\}} + \hat{I}_{-,B})|-1,\tfrac{1}{2}\rangle = \tfrac{1}{2}\sqrt{2} J_{AB}|0,-\tfrac{1}{2}\rangle = \tfrac{1}{\sqrt{2}} J_{AB}|0,-\tfrac{1}{2}\rangle$$

$$\tfrac{1}{2} J_{AB}(\hat{I}_{+,\{A_2\}} + \hat{I}_{-,B})|0,-\tfrac{1}{2}\rangle = \tfrac{1}{2}\sqrt{2} J_{AB}|1,\tfrac{1}{2}\rangle = \tfrac{1}{\sqrt{2}} J_{AB}|1,\tfrac{1}{2}\rangle$$

leading to the following sub matrix entries:

$\hat{H}$	$\|+1,-½\rangle$	$\|+0,+½\rangle$	$\|+0,-½\rangle$	$\|-1,+½\rangle$	m_T
$\langle+1,-½\|$	$-\nu_A+½\nu_B-½J_{AB}$	$2^{-½} J_{AB}$			$+½$
$\langle+0,+½\|$	$2^{-½} J_{AB}$	$-½\nu_B$			$+½$
$\langle+0,-½\|$			$+½\nu_B$	$2^{-½} J_{AB}$	$-½$
$\langle-1,+½\|$			$2^{-½} J_{AB}$	$+\nu_A-½\nu_B-½J_{AB}$	$-½$

Fig. 6.7: HAMILTONian submatrix scheme for A_2B only for $m_T \pm ½$.

Application of the pq-formula leads the energy levels, already given in section 2.4.6.1, Table, 2.28.

6.1.2.3 Total Intensity Theorem

It shall be only roughly mentioned that the quantum mechanical intensity sum of spin systems mentioned already in section 2.1.3.1 as

$$2^{N-1}$$

with N = number of spins,
is only valid for single spin systems with $I = ½$.

For general *spin quantum numbers I* (all nuclei in the spin system with the identical spin quantum number), the rule is given as

$$\tfrac{2}{3} \cdot N \cdot I(I+1) \cdot (2I+1)^N \ .$$

For the most general case with composite particles and different spin quantum numbers refer to the literature [6.5].

6.2 Spin System Probabilities

The relative probabilities of the possible isotopomers involving a spin-active dilute isotope is given by the binomial coefficient and the natural abundance's of the isotopes involved:

$$\binom{n}{k} \cdot a(^{X}\text{Element})^{n} \cdot a(^{0}\text{Element})^{k} \cdot 100 \, [\%]$$

with $\binom{n}{k} = \dfrac{n!}{k! \cdot (n-k)!}$ binomial coefficient

and n = number of spins
 k = n−r
 r = 0,...n
 a(^{X}Element) = natural abundance of isotope with mass number X of Element
 (100% = 1.0)
 a(^{0}Element) = natural abundance of all spin-inactive isotopes of Element

Applied to the practical example given in section 4.2.5 we obtain with formulae

$$\binom{n}{k} \cdot a(^{77}\text{Se})^{n} \cdot a(^{0}\text{Se})^{k} \cdot 100 \, [\%]$$

and a(^{77}Se) = 0.076
 a(^{0}Se) = 0.924
 n = 6
 k = 0,... 6

the seven probabilities listed in Table 4.1 for the case all places for selenium atoms are identical - not distinguishable. In the present structure, given in Fig. 4.1 this is not the case. For n=2,3 and 4 it makes a difference at which phosphorous position the spin-active isotopes are situated. Thus the example refers to a combinatorical problem with two types of distinguishable, e.g. "white and black balls to be taken from an urn".

In the present case there appear to be two types of place that are distinguishable, so that we can substitute

n = i + j with i = j = 3

6.2 Spin System Probabilities

what means three identical places at one phosphorous P1 and three at the other phosphorous atom P2. The following standard formula can be exploit to treat the problem:

$$\binom{i+j}{k} = \sum_{r=0}^{k} \binom{i}{r} \cdot \binom{j}{k-r}$$

For k = 2 (two ^{77}Se): the parameter r can be r = 0,1,2.
Thus there are three possible arrangements:
0 means no ^{77}Se at P1 but two at P2,
1 means one ^{77}Se at each P1 and P2,
2 means two ^{77}Se at P1 but none at P2

So for the spin system ABX$_2$ (Se = X$_2$) the coefficients for r=0 and r=2 have to be added:

$$\binom{3}{0}\cdot\binom{3}{2} + \binom{3}{2}\cdot\binom{3}{0} = 3 + 3 = 6$$

and for the spin system [AX]$_2$ r=1 refers to

$$\binom{3}{1}\cdot\binom{3}{1} = 9$$

The sum is 6+9=15 the "normal" probability with identical placed. The ratios $6/15$ for the one and $9/15$ for the other spin system in Table 4.7 are used to outline this.

In the same way the values for k=3: three ^{77}Se with r=0,1,2,3 are obtained:

r=0,3: 3 ^{77}Se at either P1 or P2, spin system ABX$_3$:

$$\binom{3}{3}\cdot\binom{3}{0} + \binom{3}{0}\cdot\binom{3}{3} = 1 + 1 = 2$$

and r=1,2: one ^{77}Se at P1 and two ^{77}Se at P2 and vice versa, spin system ABX$_2$Y

$$\binom{3}{2}\cdot\binom{3}{1} + \binom{3}{1}\cdot\binom{3}{2} = 9 + 9 = 18$$

The total sum refers to the binomial coefficient 2+18=20.

6.3 The Lineshape Fitting

The number of programs for optimizing NMR parameters is large as reported in [6.6]. Most of them differ in characteristics for simulating the spectra, in optimization functionality there are two basically different methods used: the line assignment algorithms and the line-shape fitting.

All methods make use of an iterative method: Based on the trial parameter set a theoretical spectrum is calculated, compared with the experimental data and a correction is applied to the parameters to be optimized to minimize the error. The differences of the methods are found on the nature of the experimental data. In the case of line assignment the experimental data are transition frequencies - in practice a peak-picking of the spectrum (the LAOCOON type programs, e.g. by Castellano and Bothner-By). This method is applicable for well-resolved spectral pattern if the trial parameter set is close enough to perform an assignment of theoretical calculated transitions to the experimental peak-picking or deconvoluted lines. The corrections for the parameters are usually determined by the NEWTON-RAPHSON method.

For line-shape fitting algorithms the digitized spectrum is directly taken as experimental data and compared with the calculated line-shape involving the line width parameter(s). The experimental data are the equidistant intensity data points of the spectrum. One important advantage of the line-shape fitting is the inclusion of intensity information. For example the LAOCOON iteration of an ABX spectrum will give identical error values for normal and conjugated solution, a line-shape fitting is able to distinguish between both solutions (refer to section 2.3.3.2). The problem with the line-shape fitting is that every time a theoretical and experimental signal overlap a local minimum arises [6.7]. This can be illustrated that e.g. the error hyper-space of a two-parameter fit looks like "eggboxes". The dimension of the error hyper-space is multidimensional n+1 with n = number of parameters to fit.

A special smoothing algorithm is applied to such problems that in the beginning of the line-shape fitting the surface of the error function is smooth featuring only a broad funnel. The smoothing is stepwise taken back to guide the line-shape analysis to the global minimum, because there is no mathematical algorithm resulting in the global minimum of a problem, the solutions found are always local minima.

There are two minimization algorithms used in WIN-DAISY. The **Standard Iteration Mode** uses the original spiral algorithm used in DAVINS [6.7], [6.8]. The smoothing is achieved my multiplying the target function (the GAUSS root mean square) with a mixing function, here of exponential type. The exponent used is controlled by the **Broadening** parameter and refers to the number given for the *Correlation Factor* in the iteration protocol. **None** = 1.1, **low** = 10^{-2}, **medium** = 10^{-4} and **high** = 10^{-6}. The back-taking of the smoothing till the original spectra are compared in the iteration progress is directed by the **No. of Cycles** parameter which refers to the *Multiplier* in the iteration protocol. **Low** = 10, **medium** = 6 and **high** = 4. The *Total number of cycles* calculated in the actual iteration is determined by how many times the *Multiplier* has to be multiplied with the *Correlation Factor* that the latter becomes bigger than "1.0", that's the end of the iteration. If **none Broadening** is selected, the *Correlation Factor* is already bigger than one, so that no smoothing is applied no matter which entry for **No. of Cycles** is selected. The actual *Exponential smoothing factor* for each cycle is reported in the iteration protocol. Because the multiplication's needed for strong smoothing are time-consuming. In the beginning of the iteration the *Number of correlated points* (non-zero elements in the mixing matrix) is equivalent to the *Number of points* in the experimental line-shape, which is decreased with back-taking of the smoothing to zero. To speed up the calculation not every digital point of the spectrum is used in the beginning of the iteration, if smoothing is applied. The intensity of a number of neighbored data points are added together (*Step Width*) to lead to the *Number of data groups*, which are used instead of the original digitization. The determination of the *Step Width* is done automatically based on the ratio of line width parameter and digital resolution of the spectrum. In the progress of iteration the *Step Width* is decreased to 1 (every digital point is used, *Number of data groups = number of points +1*). In case line width parameters are to be optimized the *Correlation factor lineshape* defines the *exponential smoothing factor* at which the line width parameter are included in the optimization procedure.

The method to determine the corrections is based on the idea that the solution of the minimization problem is found in the area defined by the negative gradient and the GAUSS-NEWTON correction. These **Correction vectors** can be reported using the corresponding **Output Option** in the **Control Parameters**. The Spiral algorithm is based on studies made by LEVENBERG [6.9] and MARQUARDT [6.10], according to an algorithm designed by JONES [6.11] searches for error reduction on test-points located on spirals in this area. The spiral parameters can be written into the iteration protocol using the **Output Option Search for error reduction**. The option **Parameter limit warning** prints a warning into the iteration protocol always when a correction reaches the upper or lower iteration limits. The exact formulae for determination of the iteration control parameters are described in [6.12].

The benefit of this method of iteration is that the trial parameters can be far away from the solution. As a rule of thumb the poorer the starting parameter set is, the stronger the smoothing should be. The digital resolution to obtain reliable results should be approximately at least 10 points per half-height linewidth, otherwise the peaks are not enough well defined. The base line of the experimental spectrum should be corrected properly - it should not neither positive nor negative, because in such cases an intensity contribution is misled. This can be checked in WIN-NMR by displaying the Y-axis <Ctrl + A> and zooming in the base line region or even by comparing the baselines of experimental with the simulated spectrum in WIN-NMR **Dual Display**.

The **iterate baseline** option in the **Control Parameters** only takes into account a base line tilt and shift of the experimental spectrum if only one spectral region is present. For two or more **Iter. Regions** are selected in the **Spectrum Parameters** no baseline tilt and shift is taken into account.

The **Advanced Iteration mode** [6.13] does not use the spiral method, but a successive execution of negative gradient correction followed by a quasi-NEWTON method. The first method is fast, but can overshoot the mark, so that the quasi-NEWTON method which is slower, but more accurate is used in the end. The selection of the option

Starting frequencies poor starts the iteration with a double-sum target function which is more sensitive to frequency shifts than the normal GAUSS target function which is used afterwards. The Advanced method is faster in cases where strong smoothing is necessary. It has one disadvantage if all the starting parameters are zero. Thus for Advanced iterations give random numbers for coupling constants. There cannot be given a rule which iteration method shall be used for which problem, but for complicated optimization tasks where strong smoothing is necessary the Advanced mode shall be favored.

In case of first order spin systems analyzed in the WIN-NMR **Multiplets Analysis** mode, the initial iteration **Control Parameters** in WIN-DAISY display the **Standards Iteration Mode** with **none Broadening** assuming that the experimental and initially calculated lines already overlap, so that the trial parameter set is close to the global minimum. If the spectrum could not be completely analyzed in means of first order rules the imported spin system data need to be modified in WIN-DAISY as outlined in the chapters 2, 3 and 4, e.g. to include non-determined couplings, setup of symmetry and application of smoothing parameters.

The iteration results can be examined visually in the WIN-NMR **Dual Display** for comparison of experimental and calculated line shape and the *statistical information* given in the **Iteration Protocol** file can be examined. The statistics include the mean square (*Final sum of squares*), the *Number of spectral points*, the *Degrees of Freedom* (number of spectral points - number of parameters to be iterated), the RMS value (*Standard Deviation of Measurements*) and the *R-factor (%)* which is the most important value. The latter one refers to the deviation between experimental and finally calculated line-shape in per-cent, which is more evidentiary in intensity iterations than the RMS value. It is not possible to give a general limit below which the result can be assumed to be reliable, because the deviation is dependent on the signal-to-noise level and presence of impurity signals in the experimental spectrum. But a value bigger than 3-4% implies that there might be an error in the result. Beside the *initial* and *best parameter vector* the *standard deviation* for every individual parameter is listed. In case one or more standard deviations are given as "-1.#IND0" or very large one or more parameters you want to iterate cannot determined from the experimental spectrum, e.g. a shift or nuclei specific line width of a hetero nucleus, or a coupling constant that has no effect on the spectrum or you try to optimize parameters independently which are dependent on each other, e.g. by symmetry. In such cases there appears singularity in the matrix of second derivatives which is reported for the Standard Iteration mode with the following string in the iteration protocol as a warning: *Matrix of second Derivatives is singular, pseudoinverse is calculated.*

In case the optimization result is not satisfying check the following points:
- In case smoothing parameters are selected you may run simply another iteration with the final parameters obtained by the previous iteration as the initial values for the next iteration with the same smoothing parameters. Check the iteration protocol file if the result is better. In case no changes are achieved other modifications are necessary.
- Change the smoothing parameters. E.g. if the correct result is almost achieved, change the broadening to none. Do not run a final adjustment with strong smoothing

6.3 The Lineshape Fitting

parameters. This may guide the iteration process out of the present minimum, even if it is the global minimum.
- Change one ore more parameter values to modify the initial calculated spectrum. Every change in parameter values.
- To check where the problem in obtaining a reliable iteration result you may turn on all available Output Options in the Control Parameters.
- Check the digital resolution of your experimental spectrum visually in WIN-NMR or the Hz/point value in the iteration output protocol. Check the spectral interval(s) for iteration in the protocol. At least 10 point per half-height line width - better 20 - ensure that a successful iteration is possible.
- If the parameter limit warning option is checked before the iteration is run check if one parameter reaches a limit during optimization. If this is the case enlarge the parameter limits. It might be as well necessary to allow sign changes, although the sign might be known. Even the parameter limits, not only the parameter values, influences the characteristics of the optimization problem strongly.
- For spectra with impurity signals or phase, baseline or line-shape problems it might be necessary to disable the iteration of the line width parameter(s) in the first steps of iteration or even for the whole optimization procedure.
- In case there are parts of the spin system with first order and some parts with strongly second order relation disable the second order coupling constants from iteration and adjust in the first steps the well-defined first order couplings and the resonance frequencies. But make sure that approximate values are given for the second order coupling constants, because they may influence the spectrum outlook significantly and also the values of all the other first-order coupling constants. Optimize the second order coupling without or with merely low smoothing parameters in the end together with all the other parameters.

6.4 Glossary

[...]	concentration
[...]	matrix
[...]	symmetry element
$[...]^T$	transposed matrix
$^x\{...\}$	term of multiplicity x
$\{yA\}$	term consisting of spins with spin multiplicity y
|...>	bra, DIRAC notation of a real function
<...|	ket, DIRAC notation of the conjugated complex function
α	spin basis function (for $m_z = +½$)
$a(^0Y)$	sum over the natural abundances of spin inactive isotopes of element Y (100% = 1.0)
$a(^xY)$	natural abundance of isotope x of element Y (100% = 1.0)
β	spin basis function (for $m_z = -½$)
$B_x(k,n)$	binomial coefficient of order x, "n over k"
bpf	basis product function
cp	composite particle
δ	chemical shift
δ	KRONECKER symbol
Δ_ν	chemical shift difference (in frequency units)
DEPT	Distorsionless Enhancement by Polarization Transfer
DMSO	dimethylsulfoxide
η_{max}	maximum Nuclear Overhauser Enhancement factor
E	energy, EIGENvalue
$\phi, \|\phi>$	basis function
FID	free induction decay
F_{act}	actual particle spin value
F_{max}	maximum particle spin value
FT	FOURIER Transformation
$\hat{F}_-$	total lowering operator
$\hat{F}_z$	total spin operator
γ	magnetogyric ratio
g	multiplicity
$G(F_i)$	particle multiplicity for particle spin F_i
h	PLANCK constant
$\hbar$	PLANCK constant divided by $(2\cdot\pi)$
$\hat{H}$	HAMILTONian operator
[H]	HAMILTONian matrix
HWB	half-height linewidth
Hz	Hertz, unit: s^{-1}
I	spin quantum number, angular momentum

INEPT	Insensitive Nuclei Enhanced by Polarization Transfer
$\hat{I}_+$	raising operator
$\hat{I}_-$	lowering operator
IUPAC	International Union of Pure and Applied Chemistry
$\hat{I}_z$	spin operator (z-contribution)
$^xJ_{ik}$	scalar coupling constant between nuclei i and k over x bonds
K	equilibrium constant, population
$[\Lambda]$	diagonal matrix containing EIGENvalues
ln	natural logarithm
μ	magnetic moment
M	transition probability
m_{cp}	composite particle actual spin value
$|m_{cp}>$	actual composite particle basis function
M_0	equilibrium magnetization
MRI	Magnetic Resonance Imaging
m_T	total spin value, EIGENvalue of total spin operator
m_z	actual spin value, EIGENvalue of spin operator (z-contrib.)
M_z	z magnetization
ν	frequency
$ν_0$	LARMOR frequency
$ν_{½}$	half-height linewidth
N	number of spins
NMR	Nuclear Magnetic Resonance
NOE	Nuclear OVERHAUSER Effect
OAc	acetyl
p	population
ppb	parts per billion
ppm	parts per million
R	organic substituent (in chemical formulae)
ROE	NOE in the rotating frame
σ	shielding constant
S	statistical weight of subspectra, product of particle multiplicities G
s	second
τ	evolution time
T	transition frequency
T_1	longitudinal relaxation time, spin-lattice relaxation time
T_2	transversal relaxation time, spin-spin relaxation time
T_2*	effective transversal relaxation time
TMS	tetramethylsilane
TOCSY	TOtal Correlation SpectroscopY
[U]	EIGENvector, unitary transformation matrix
$[U]^T$	transposed EIGENvector
Ψ, |Ψ>	EIGENfunction

6.5 References

[6.1] Harris, R.K., *Nuclear Magnetic Resonance Spectroscopy*, Harlow: Longman, 1986.
[6.2] Günther, H., *NMR Spectroscopy - Basic Principles, Concepts and Applications in Chemistry*, 2nd ed., Chichester: Wiley, 1995.
[6.3] Canet, D., *Nuclear Magnetic Resonance - Concepts and Methods*, Chichester: Wiley, 1996.
[6.4] Corio, P.L., *Chem. Rev.*, 1960, *60*, 363.
[6.5] Corio, P.L., *Structure of High-Resolution NMR Spectra*, London: Academic Press, 1966.
[6.6] Weber, U., Thiele, H., Spiske, R., Hägele, G. in *Software-Development in Chemistry*, Moll, R. (Ed.), Frankfurt: GDCh, 1995, *9*, 269-281.
[6.7] Stephenson, D.S., Binsch, G., *J. Magn. Reson.*, 1980, *37*, 395.
[6.8] Stephenson, D.S. in *Encyclopedia of Nuclear Magnetic Resonance,* Grant, D.M., Harris, R.K. (Eds.), Chichester: Wiley, 1996, *1*, 816-821.
[6.9] Levenberg, K., *Quart. Appl. Math.*, 1944, *2*, 164.
[6.10] Marquardt, D.W., *SIAM J. Appl. Math.*, 1963, *11*, 431.
[6.11] Jones, A., *The Computer Journal*, 1970, *13*, 301.
[6.12] Weber, U., *PhD Thesis*, University of Düsseldorf, 1994.
[6.13] Höffken, H.-W., *PhD Thesis*, University of Düsseldorf, 1994.

Index

A_2 75, 99, 370
A_2B 121, 244, 279, 373
$[[A]_2BX]_2$ 308
A_2X 108, 167, 301, 302, 310
A_2X_3 116, 201
$[[A]_2X]_3$ 290
A_3 106
A_3B_3 340
A_4 107
AA'X   see ABX
AB   64, 180, 200, 349, 351, 363
AB 'quartet'   181
ab subspectrum   79, 137, 195
AB_2 see A2B
$[AB]_2$ 149, 358
$[AB]_2$ at high temperature   359
$[AB]_2C$ 201
$[AB]_3X$ 289
$[AB]_3XY$ 289
ABCD   95, 359
ABM_3X 255
ABMX   93, 195, 197, 219
ABX   77, 183, 184, 185, 231, 233, 246, 265
– conjugated solution   82, 91
ABX_2 282
$[ABX]_2$ 239
ABX_2Y 282
ABX_3Y 282
ABX_3Y_2 282
acetamide
– N-dimethyl-   340
acetic acid
– 4-chloro-2-oxobenzothiazoline-3-yl   55
– methyl ester   34

acetophenone
– 2,3,6-trifluoro-   237
– 3-fluoro   235
$AKMRSX_3$ 237
AKMRX   178, 186, 235, 249
allyl group   169, 186, 198
allyl position   172
AM_2X 275
ammonia   304
$AMNR_2X$ 168
$AMNR_3X_3$ 200
AMNRX   275
AMNX see ABMX   93
$AMNX_2$ 168
$AMNX_3$ 176
AMNXY   177
AMRX   57, 233, 355
$AMRX_3$ 187
AMX   46, 62, 167, 183, 188, 219, 245, 310
AMX_2 169
$[AMX]_2$ 189, 192, 253
$A[MX]_2R$ 275
analysis
– graphical   69
– manual   69, 81, 125, 156
– multiplets as a starting guess   88
aniline
– 2-chloro-6-nitro-3-phenoxy-   64
– N-methyl,N-deutero,2,4,6-trinitro-   349
anisole
– 2,3,4-trichloro-   68
– 2,3,6-trichloro-   68, 73
– 2,6-dichloro-   112
– 3,5-dichloro-   112

annulene
– anti-1,6,8,13-bismethano-[14]- 351
anomeric proton 203, 329
ascorbid acid 197
Aspirin see salicylic acid
asymmetric center 174
Au(PF$_3$)$_2$SO$_3$F 257
AX 38, 64, 75, 297
AX$_2$ see A2X
[AX]$_2$ 134, 253, 276
[AX$_2$]$_2$ 282
AX$_3$ 113, 303
A[X$_3$]$_2$ see AX$_6$
[AX$_3$]$_2$ 282
AX$_6$ 118, 200
AXY see ABX

bad quantum number 71, 369
basis function see function
basis product function see function
benzamide
– 5-chloro-N-(2-chloro-4-nitrophenyl)-2-hydroxy- 59
benzene 34
– 1,2,3,4-tetrachloro- 265
– 1,2,3,4-tetrachloro,5-deutero- 299
– 1,2,3-trihydroxy- 154
– 1,2,4-trichloro 47
– 1,2-dichloro- 147, 155
– 1,2-difluoro- 239
– 1,2-dihydroxy- 152
– 1,2-dimethoxy- 154
– 1,2-diphosphonic acid 253
– 1,3,5-trimethyl- 34
– ethyl- 201
– hexafluoro- 234
benzofuroxan 358
benzoic acid
– 2,6-dichloro- 128
benzonitrile
– 2,6-dichloro- 126
beryllium-9 312
bicyclo[2.2.1]-hept-2-ene
– 1,2,3,4,7,7,hexachloro-5,6-bis(chloromethyl)- 192

– 5-bromomethyl-1,2,3,4,7,7-hexachloro- 178
binomial coefficient 119, 243, 366, 376
bipyridine, 2,2' 57
Bloch equation 319
Bohr 19
boron-10/11 306, 312
boson 19
bpf see function

C$_6$F$_6$ see benzene, hexafluoro-
calculation
– iteration 86
– sequence 65, 77, 151
– simulation 24
carbonic acid 2-sec.-butyl-4,6-dinitrophenyl-isopropyl ester 200
catechol see benzene, 1,2-dihydroxy-
CDCl$_3$ see chloroform
CFCl$_3$
– isotope effect 242
– standard 231
CH$_2$Cl$_2$ see methylenechloride
CHCl$_3$ see chloroform
chemical equivalence see isochronicity
chemical shift 20
– average 343, 348, 359
– calculation 233
– difference 68, 70, 124, 351
chemical shift range 231, 244, 260
chiral
– shift reagent 173
– solvent 173
chloroform
– signal 262
– solvent 185
– spin system 296
chrysanthemic acid
– 3-phenoxy-benzyl ester 171
cinnamic acid, trans 160
combination line 80, 122
composite particle 102, 295, 370. see also equivalence, magnetic

concentration 347
configuration 113
– absolute 173
conformation 113
correlation factor 378
coupling constant
– aromatic 49
– average 192, 354
– CD 298
– CH 261, 266
– cis/trans 161, 169, 171, 178, 198, 267, 276
– connection *see* multiplet,connect
– DH 299
– first order 36
– geminal 113, 160, 171, 176, 195, 197
– heteronuclear 236
– hidden 114, 299
– long range 118, 178, 180, 191, 202, 207, 210, 221, 225
– PP 269
– scalar 36
– second order 48
– sign 40, 86, 161, 183, 191, 210, 220
– spin-spin *see* scalar
– through space 237
– vicinal 115, 160, 176, 189, 190, 203
– weak *see* first order
– WIN-DAISY dialog box 37
cp *see* composite particle
cyclobutane
– 1,1,2-trifluoro-2,3,3,4,4,5,5-heptachloro- 233
cyclohexane
– 1,2-diphosphonic acid 271
– 2,2,3,3,4,4,5,5-octadeutero- 357
cyclopropyl group 171

d *see* doublet
dd *see* doublet of doublets
ddq *see* doublet
deceptively simple 185, 197
deconvolution 73, 325
degeneracy 330

– of energy levels 19, 30, 100
– of transitions 30, 123
DEPT 261
deuterium *see* hydrogen-2
deuterium substitution 299
diastereomer 173
diastereotopy 174, 180, 255
diazine
– cis/trans-difluoro- 276
diborane 308
difference spectrum
– NOE analysis 329
dimethylsulfoxide
– solvent 181, 186
– spin system 302
disabled spins *see* suppression
DMSO *see* dimethylsulfoxide
double resonance
– decoupling 167, 205, 260, 274, 327
– off resonance 329
– spin tickling 163, 183
doublet 37
– of doublets 46, 56, 77, 177
– of doublets of quartets 177
– of quartets 121
– of quartets of quartets 172
– signal 108
– term 104
Dual Display 44

eigenstate *see* function, eigen
eigenvalue 26, 100, 123
electronegativity 21, 56, 176
enantiomer 173
enantiotopic
– coupling path 132
enantiotopy 180
α-endosulfan 189
energy
– free activation 345
– level 71, 100, 164
– level diagram 25, 30, 32, 39, 102, 104, 107
– Zeeman 26, 38, 71
– zero order *see* Zeeman

enthalpy   345
entropy   345
equilibrium constant   347
equivalence
– chemical   see isochronicity
– magnetic   98, 132
ethane
– 1,2,2-tris-(methyl phosphinic acid isopropylester)   245
– 1,2-bis(phosphinic acid diisopropylester)-P,P'-dimethyl-   268
– 1,2-dimethoxy   34
– 1-phenyl-1,2-diol   185
– 1-t.-butyl,1,2,2,-tris-(phosphonic acid aniline salt)   248
–1,1,2,2-tetrakis-(methyl phosphinic acid Na)   247
ethanol   120
ether
– allyl-2,3-epoxypropyl-   186
– t.-butyl-methyl   34
evolution time   320
exchange
– hidden   348
exchange vector   341
Eyring equation   345

F_3SSCl 231
F_3SSF 232
fermion   19
field dependency
– of coalescence   346
file algebra   138, 261
first order
– splitting rule   see n+1 rule
fluorenone-9   93
fluorine-19   230, 257, 276, 277, 306
– heteronuclear spin system   235
– homonuclear spin system   231
fragment
– copy   285
– copy to...   214, 246
– delete   171
fragmentation   58, 117, 198, 212

frequency
– effective Larmor   39, 80
– Larmor   26
– resonance   24
– spectrometer   41, 65, 180, 230
– transition   24, 39, 55, 72, 100
– WIN-DAISY dialog box   24
fumaric acid   268
function
– basis   26
– basis product   29, 52
– eigen   26, 101, 368
– orthogonal   369
– spin   25
– target   380
furan
– 2-aldehyde-   355
furfural-2   see furan

g   see multiplicity
glucose   202
– relaxation time   320
glucose pentaacetate   see glucose
glycosidic bond   211
good quantum number   26, 363
grid
– coupled   47
– free   43

H_2O
– solvent   247, 253
– spin system   310
half spectrum   137
Hamiltonian
– complete   71
– flip-flop   see mixing
– mixing term   365
– off-diagonal element   364
– Zeeman term   26
Heisenberg uncertainly principle   27
homotopic
– nuclei   132, 175
hydrogen-1   21
hydrogen-2   296
hydroxyl group

– coupling constant 183, 186
– linewidth 168, 181, 187

I *see* spin quantum number
IGLO 233
imidazole
– 2-(allyloxy-2,4-di-
    chlorophenylethyl)- 198
increment system 237
INEPT 273
installation 8
integral 27, 169
interconversion 359
inversion recovery experiment 320
irreducible component 103
ISO value *see* X approximation
isochronicity 30, 98, 116, 132
isomer
– cis/trans 160
– configuration 160
– conformation 175
– mixture 169, 172
isopropyl group 118, 200
isotope
– Multiplets hetero list 235
– Spectrum Parameters 236
isotope effect 242, 265, 270, 298, 306
isotopic labeling 274
isotopomer 243, 263, 275, 280, 313

J *see* coupling constant

Karplus 176

Larmor *see* frequency, Larmor
life time at coalescence 345
linear combination 25, 75, 368
lineshape 27, 87
– dispersive 330
linewidth
– analysis in WIN-NMR 28, 43
– exchange broadened 351
– half-height 27, 326
– in tickling experiments 165
– nuclei specific 235, 357

– WIN-DAISY Lineshape Parameters
    28

magnetic field strength 19, 95
magnetic moment 19
magnetic site exchange 339
magnetogyric ratio 19, 23, 26, 41, 229,
    328
maleic acid 267
maleic acid dimethyl ester 34
matrix
– Hamilton 363
– parameter correlation 143
– singular 143
mesitylene *see* benzene, 1,3,5-
    trimethyl-
methane
– dichloro- see methylenechloride
methanol 34
– solvent 253
methine proton 176
methyl group 113, 172, 175
methyl-2-nitrobenzoate
– 5-(2,4-dichlorophenoxy)- 92
methylene group 198
– diastereotopic protons 174, 176,
    177, 200
– magnetic equivalent protons 168,
    202
methylenechloride 300
molecule size 330
morpholine 192, 360
m_T *see* total spin
multiplet
– connect 43
– definition 43
– designate 43
– export to WIN-DAISY 44
– export to WIN-NMR 44
– heteronuclear 235
multiplicity
– distance lines 176
– of nuclei 22, 294
– of particles 103, 106, 111, 294, 373
– of signals 36, 119, 297

mutual exchange 339

N lines 83, 137, 147, 194, 269
n+1 rule 37, 109, 114
natural abundance 243, 376
Newman projection 115, 175, 193, 203
NH 187, 188
nicotine 222
nitrogen-14 305
nitrogen-15 273, 304
NMR time scale 339
NOE
– analysis 327
– direct/indirect 330
– enhancement factor 327
NOE in the rotation frame *see* ROE
notation of spin systems *see* notation, spin system
nuclear overhauser effect *see* NOE

oligosaccharide *see* trisaccharide
operator
– permutation 135
– spin I_z 25
– spin lowering 372
– total spin F_z 29
options 232
oxazolidine-2,4-dion
– 3-(3,5-dichlorophenyl)-5-methyl-5-vinyl- 167
oxygene-17 310

parameter
– auxiliary 139
– averaged 339
– correlation 143
– dynamic 339
– perturbation *see* perturbation, parameter
– static 339
– WIN-DAISY 60
particle spin
– actual 103
– maximum 103

Pascal triangle *see* binomial coefficient
peak picking 73, 158, 321
perturbation
– first order 38, 71
– of spin systems 164
– parameter 36, 42, 63, 99, 134, 181, 197
– second order 71
phenol
– 2,4-dichloro- 50
– 2-tert.-butyl-4,6-dinitro- 43
phosphonic acid anhydride
– tris(1,1-dimethylethyl)- 244
phosphorothioate
– O,O-diethyl-2-chloro(α-cyanobenzyliden amino)- 255
phosphorus-31 244, 257, 269, 281, 287
piperine 219
Planck constant 25
platinum-195 279
population 347, 348
ppm *see* chemical shift
p-q formula 72, 123, 368
probability
– of isotopomers 243
– of isotopomers 376
– transition *see* transition, intensity
processing
– of inversion recovery experiments 320
prochiral 180
propane
– 1,2-dichloro- 173, 176
– 1,3-diol,2-bromo,2-nitro- 180
– 2-amino-1-ol- 187
propene
– 1,3-dichloro- 169
propene-3-ol 168
propionic acid
– 2-(4-chloro-2-methylphenoxy)- 115
– 2,3-dibromo- 183, 184
propionic acid ethyl ester

– 2-[4-(6-chloro-2-benzoxazolyloxy)-
  phenoxy]- 116, 200
propionic acid methyl ester
– 2-(2,4-dichloropehnoxy)- 50
proton-proton distance 329
pseudo 2D spectra 328
Pt$_3${C(OMe)C$_6$H$_4$Me-4}$_3$(CO)$_3$ 279
pulse angle 261
pure element 229
pyranose ring 202
pyridine
– 2-acetyl- 273
pyrimidine 274
pyrogallol see benzene,1,2,3-
  trihydroxy-
pyrrole carbonitrile
– 4-(2,3-dichlorophenyl)-1H- 89, 188
Pythagoras theorem 70

qdd see quartet
quadruplet see quartet
quadrupole moment 243, 293, 318
quartet 181
– of doublets 121
– of doublets of doublets 177
– signal 114, 296, 307
question mark see coupling constant,
  second order
quinone
– 2,3-dichloro-1,4-naphto- 147
quintet
– signal 296, 301, 312
– term 301

racemate 173
rate constant 339
reference standard 21, 231
relaxation analysis 321
relaxation time 317
– effective spin-spin 326
– spin-lattice 318
– spin-spin 27, 318, 326
resolution 233
R-Factor 50
Rh$_4$(C$_5$Me$_5$)$_4$H$_4$(BF$_4$)$_2$ 258

rhodium-103 258, 272
ring inversion 357, 360
ROE 335
roof effect 42, 43, 51, 65
rotamer 175
– mixture 249
rotation 340, 346
rule of repeated spacing see n+1 rule

salicylic acid
– ortho-acetyl- 95
satellite spectrum 263
saturation degree 331
Schroedinger equation
– time dependent 339
– time independent 366
screening constant see shielding
  constant
sec. butyl group 200
secular determinant 368
selection rule 31
selective polarization transfer see SPT
selenium-77 281
septet
– signal 118, 307, 313
serial processing 262, 324
sextet
– signal 311
shielding constant 20
silicon-29 277
singlet
– define 96, 126, 153, 201, 239
– signal 67, 132
– term 103
smoothing 378
solvent 180
– chiral 173
– deuterated 296
spin angular momentum see spin
  quantum number
spin quantum number 23, 229
spin system
– coupled 349
– field dependency 182, 197, 208, 227
– first order 36, 108

– heteronuclear 37, 41
– homonuclear 38, 42, 63, 231
– labeling 140
– large 198, 222
– non-coupled 20, 340
– notation 33, 63, 99, 133, 275
– second order 63, 132, 195, 219, 239
– symmetrical 134
spiral algorithm 378
SPT and NOE 329
statistical weight
– of fragments 170
– of subspectra 115, 373
strychnine 195
subspectrum 295
– (a_2) of $[AX]_2$ 137
– (ab) of $[AX]_2$ 137
– (ab) of ABX 79
– antisymmetrical 138
– composite particle 103
– symmetrical 138
– weighting 111, 114
substitution criterion 133, 175
suppression 74, 93, 151, 197
symmetry
– by molecular mobility 192
– C_2, C_S 134
– C_{2v} 141
– C_{3v} 288
– D_{2h} 308
– D_{3h} 290
– description 136
– molecular 134, 142
– operation 113, 135
– point group 133
– relaxation time of quadrupolar nuclei 318
– spin system 113, 131, 189
– T_d 258
– twofold see C2, Cs

T_1 see relaxation time, spin-lattice
T_2 see relaxation time, spin-spin
temperature 339
– coalescence 343

term see irreducible component
tetramethylsilane 35, 277
tetrazine
– 3,6-bis(2-chlorophenyl)-1,2,4,5- 58
tickling see double resonance
time scale 115
tin-115/117/119 286
TMS see tetramethylsilane
TOCSY 211, 336
total intensity theorem 305, 375
total spin
– diagram 53, 81, 108, 114, 117, 165
– eigenvalue 53
– factorization 365
transition
– double quantum 31
– forbidden 102, 110
– frequency see frequency, transition
– intensity 27, 31, 33, 85, 101, 327, 370
– maximum number 374
– moment see transition intensity
– progressive connected 164
– regressive connected 164
– single quantum 23
– suppression see suppression
– zero quantum 31
triplet
– of doublets of doublets 168
– signal 55, 93, 109, 294
– term 103
trisaccharide
– relaxation time 324
– TOCSY 1D 211

unitary transformation 369

valence tautomerism 351
variable delay list 322
vinyl group 160
vitamin C 197

water see H_2O
weak outer line 67, 80, 122, 194, 195, 196, 201, 234, 241

WIN-DR   163, 183, 206
WIN-DYNAMICS   341

X approximation   38, 63, 110, 363

xylene-1,4   34

z magnetization   23, 319